Mixture Toxicity

Linking Approaches from
Ecological and Human Toxicology

Other Titles from the Society of Environmental Toxicology and Chemistry (SETAC)

For information about SETAC publications, including SETAC's international journals, *Environmental Toxicology and Chemistry* and *Integrated Environmental Assessment and Management*, contact the SETAC office nearest you:

SETAC
1010 North 12th Avenue
Pensacola, FL 32501-3367 USA
T 850 469 1500 F 850 469 9778
E setac@setac.org

SETAC Office
Avenue de la Toison d'Or 67
B-1060 Brussells, Belguim
T 32 2 772 72 81 F 32 2 770 53 86
E setac@setaceu.org

www.setac.org
Environmental Quality Through Science®

Mixture Toxicity

Linking Approaches from Ecological and Human Toxicology

Edited by

Cornelis A.M. van Gestel

Martijs J. Jonker

Jan E. Kammenga

Ryszard Laskowski

Claus Svendsen

Coordinating Editor of SETAC Books
Joseph W. Gorsuch
Copper Development Association, Inc.
New York, NY, USA

CRC Press
Taylor & Francis Group
Boca Raton London New York

CRC Press is an imprint of the
Taylor & Francis Group, an informa business

Published in collaboration with the Society of Environmental Toxicology and Chemistry (SETAC)
1010 North 12th Avenue, Pensacola, Florida 32501
Telephone: (850) 469-1500 ; Fax: (850) 469-9778; Email: setac@setac.org
Web site: www.setac.org

International Standard Book Number: 978-1-4398-3008-6 (Hardback)

Library of Congress Cataloging-in-Publication Data

Mixture toxicity : linking approaches from ecological and human toxicology / editors, Cornelis A. M. van Gestel ... [et al.].
    p. cm.
Includes bibliographical references and index.
ISBN 978-1-4398-3008-6 (hardcover : alk. paper)
1. Environmental toxicology. 2. Mixtures--Toxicology. 3. Environmental risk assessment. I. Gestel, Cornelis A. M. van.

RA1226.M59 2011
615.9'02--dc22   2010032498

Visit the Taylor & Francis Web site at
http://www.taylorandfrancis.com

and the CRC Press Web site at
http://www.crcpress.com

and the SETAC Web site at
www.setac.org

SETAC Publications

Books published by the Society of Environmental Toxicology and Chemistry (SETAC) provide in-depth reviews and critical appraisals on scientific subjects relevant to understanding the impacts of chemicals and technology on the environment. The books explore topics reviewed and recommended by the Publications Advisory Council and approved by the SETAC North America, Latin America, or Asia/Pacific Board of Directors; the SETAC Europe Council; or the SETAC World Council for their importance, timeliness, and contribution to multidisciplinary approaches to solving environmental problems. The diversity and breadth of subjects covered in the series reflect the wide range of disciplines encompassed by environmental toxicology, environmental chemistry, hazard and risk assessment, and life-cycle assessment. SETAC books attempt to present the reader with authoritative coverage of the literature, as well as paradigms, methodologies, and controversies; research needs; and new developments specific to the featured topics. The books are generally peer reviewed for SETAC by acknowledged experts.

SETAC publications, which include Technical Issue Papers (TIPs), workshop summaries, newsletter (*SETAC Globe*), and journals (*Environmental Toxicology and Chemistry* and *Integrated Environmental Assessment and Management*), are useful to environmental scientists in research, research management, chemical manufacturing and regulation, risk assessment, and education, as well as to students considering or preparing for careers in these areas. The publications provide information for keeping abreast of recent developments in familiar subject areas and for rapid introduction to principles and approaches in new subject areas.

SETAC recognizes and thanks the past coordinating editors of SETAC books:

A.S. Green, International Zinc Association
Durham, North Carolina, USA

C.G. Ingersoll, Columbia Environmental Research Center
US Geological Survey, Columbia, Missouri, USA

T.W. La Point, Institute of Applied Sciences
University of North Texas, Denton, Texas, USA

B.T. Walton, US Environmental Protection Agency
Research Triangle Park, North Carolina, USA

C.H. Ward, Department of Environmental Sciences and Engineering
Rice University, Houston, Texas, USA

Contents

List of Figures

List of Tables

Preface

Mixture toxicity is a major challenge for scientists and regulators. The area of combined and complex exposure is a main topic, often defined as "cumulative stress." The integration of human and environmental risk assessment is another important issue. The project NoMiracle (Novel Methods for Integrated Risk Assessment of Cumulative Stressors in Europe; 2004–2009), financially supported by the European Commission within the 6th Framework Program, addressed these issues, which also are receiving continued and increasing interest from the scientific community organized within the Society of Environmental Toxicology and Chemistry (SETAC). For these reasons, NoMiracle and SETAC joined forces in autumn 2005 to organize a workshop addressing these issues.

As a result, on 2–6 April, 2006, in Krakow, Poland, SETAC Europe and NoMiracle organized an international workshop on mixture toxicity, focusing on the state of the art of mixture toxicity research and its use in environmental and human health risk assessment. The workshop was attended by 31 invited experts from the United States, Canada, and Europe, representing academia, business, and governmental agencies, and covering the fields of ecotoxicological and human health effects and risk assessment. The aim of the workshop was to discuss concepts and models for mixture toxicity assessment being used in human and environmental toxicology, and to develop a mutual understanding and check that the terms used by scientists in one area are meaningful for those in the other. This was in fact the first attempt to bring experts from human and environmental toxicology together in a workshop aiming at the elaboration of common concepts. Experts exchanged views on the current state of the art, across Europe and America, academia, regulators, and industry. Workshop participants also represented different EU projects related to the toxicity and risk assessment of chemicals, single and in mixtures.

During the workshop, separate groups worked in parallel on 5 topics important for grasping mixture toxicity whether for humans or for other organisms in the environment. These topics are

- exposure (how to measure the amounts of chemicals that may enter the living organism);
- kinetics and metabolism (how the chemicals travel within the organism and how they are metabolized and reach the target site);
- toxicity (what are their detrimental effects on the organism);
- test design and complex mixture characterization (how to measure effects of mixtures and identify responsible chemicals); and
- risk assessment to man and the environment.

In the evening sessions, the hard work of the day was presented to the plenum and exposed to harsh criticism, which might have intimidated the speakers if their presentation had not been based on solid scientific ground. Very little time was left

Figure 1   Venue of the SETAC-NoMiracle workshop in Krakow, 2–6 April 2006. (Photo by Ryszard Laskowski.)

for enjoying the excellent site, a Polish manor house (see Figure 1), well equipped for housing an international workshop. The Polish colleagues organized this very well, including a workshop dinner in old Krakow.

The workshop yielded this book, which aims at providing an overview of the state of the art of the different scientific aspects of ecotoxicological and human health risk assessment of mixtures. This book is useful for advanced-level students who want to become familiar with issues of mixture toxicity and for scientists who want a quick update of the field. The book may also prove valuable for regulators who are faced with questions related to the risk assessment of cumulative exposures.

We acknowledge the effort of the reviewers, who did a great job in reading and commenting on the manuscript and helping the editors and authors to meet a high quality standard.

Dr. Fred Heimbach
Past President of SETAC Europe

Dr. Hans Løkke
Coordinator of NoMiracle

General Introduction

Chemical exposure, both for humans and for organisms living in the environment, rarely consists of single chemicals but in many cases concerns mixtures of chemicals, often of fluctuating composition and concentrations. In some cases, exposure is to simple, well-defined mixtures of a few known compounds. In other cases, organisms are exposed to complex mixtures consisting of large numbers of chemicals of unknown composition. Both in human toxicology and in ecotoxicology there is a long-term history in mixture toxicity research. Nevertheless, it seems both areas have developed somewhat independently, resulting in similar but also differing approaches and priorities.

This book has its basis in discussions started at an international workshop on mixture toxicity, held 2–6 April, 2006, in Krakow, Poland. The aim of the workshop was to produce an updated review of the state of the art of mixture toxicity research and the potential for integrating its use in environmental and human health risk assessment. Since the previous key book on mixture toxicity (Yang 1994) there has been great progress in the development of concepts and models for mixture toxicity, in both human and environmental toxicology. However, due to the different protection goals of the 2 fields, developments have often progressed in parallel and with little integration. The workshop was therefore aimed mainly at developing mutual understanding, generating awareness across the fields, and investigating options for cross-fertilization and integration. In the time since the workshop, exchange of views and ideas has continued, resulting finally in this volume.

This book presents an overview of developments in both fields, comparing and contrasting their current state of the art to identify where one field can learn from the other. In terms of subject matter, the book progresses from exposure through to risk assessment, at each step identifying the special complications that are typically raised by mixtures (compared to single chemicals). Five chapters are included, each addressing a specific step from exposure to risk assessment for mixtures:

1) exposure (how to quantify the amounts of chemicals that may enter the living organism);
2) kinetics, dynamics, and metabolism (how the chemicals enter and travel within the organism, how they are metabolized and reach the target site, and finally, the development of toxicity with time);
3) toxicity (its detrimental effects on the organism);
4) test design and complex mixture characterization (how chemicals interact, how to measure effects of mixtures, and how to identify responsible chemicals); and
5) risk assessment to man and the environment.

This book refers to concepts generally used to describe mixture toxicity. The origin of these concepts lies in the work of Bliss (1939) and Hewlett and Plackett (1959), who thought in terms of mode of action and identified 4 types of possible combination mechanisms for the joint action of toxicants. Two of these 4 mechanisms, both assuming no interaction of the chemicals in the mixture, are simple similar action and independent action. These 2 concepts turned out to be easily described in mathematical terms (see Chapter 4), and therefore have found general acceptance in human and ecotoxicology. In both disciplines, instead of simple similar action, the term "concentration addition" (CA) has become generally accepted. This term therefore is used in this book. In ecotoxicology, the term "independent action" (IA) has widely been accepted, whereas in the field of human toxicology the term "response addition" (RA) seems preferred. To avoid inconsistency with the current literature, in this book we use both terms, assuming they have the same meaning.

This book focuses on environmental mixtures, but approaches described are also applicable to other mixtures, like those encountered during occupational exposure or in pharmacology. The book addresses 2 approaches to mixtures:

1) Predictive assessments that involve determining effects of chemicals in a mixture in relation to effects expected from the toxicity of the single chemicals. This generally concerns well-defined mixtures, often containing a few chemicals.
2) The analysis of complex mixtures with unknown composition and containing many different chemicals present in environmental samples, like effluents and waste materials. In this case, identifying which chemicals are responsible for the toxicity of the mixture is the main goal, but assessment may also focus on determining the best methods of risk reduction or quantifying potential effects associated with exposure to the mixture.

Each chapter of this book provides an essential overview of the state of the art in both human and ecotoxicological mixture risk assessment, focusing especially on the much excellent work that has been published in the intervening years between publication of this and the previous mixture volume. Each chapter, then, ends by identifying current possible crosslinks and recommendations for mutual developments that can improve the state of knowledge on mixture toxicity and ultimately lead to better and more integrated risk assessment. A glossary is added that provides definitions of common terms used throughout the book.

The in-depth way the book covers the state of the art for mixture toxicology and ecotoxicology principles means it serves well as a textbook on the subject. At the same time, the inclusion of the considerations on application of novel developments in these principles, and their integration across human and environmental mixture risk assessment, makes it an ideal tool for researchers, regulators, and other risk assessment practitioners as mixture considerations start to enter regulatory forums over the next years.

About the Editors

Cornelis A.M. (Kees) van Gestel studied environmental sciences at Wageningen University, the Netherlands. From 1981 to 1986, he was scientific advisor on the ecotoxicological risk assessment of pesticides and from 1986 to 1992 head of the Department of Soil Ecotoxicology at the National Institute of Public Health and the Environment (RIVM) in Bilthoven, the Netherlands. He obtained his PhD from Utrecht University in 1991. Since 1992, van Gestel has been associate professor of ecotoxicology at the Department of Animal Ecology of the Vrije Universiteit (VU) in Amsterdam. He is teaching undergraduate and postgraduate courses on various topics from basic biology to ecotoxicology and supervising undergraduate and PhD students working on various aspects of ecotoxicology. He is author or coauthor of more than 185 papers and book chapters, member of the editorial boards of *Ecotoxicology*, *Ecotoxicology and Environmental Safety*, and *Environmental Pollution*, and editor of *Environmental Toxicology and Chemistry* (2005–2010) and *Applied Soil Ecology* (since 2009).

Martijs J Jonker received his PhD in 2003 from Wageningen University, the Netherlands. His thesis, entitled "Joint toxic effects on *Caenorhabditis elegans* on the analysis and interpretation of mixture toxicity data," was to a large extent focused on the experimental design and statistical analysis of mixture toxicity studies. After his PhD, he did a 3-year fellowship at the Centre for Ecology and Hydrology at Monkswood, Cambridgeshire, UK. The aim of this project was to identify genes that are functionally linked to important parameters of the life histories of the invertebrate species *Lumbricus rubellus* and *Caenorhabditis elegans*. These genes would be candidate targets for the development of "biomarkers of effect." He is currently working at the Microarray Department and Integrative Bioinformatics Unit of the University of Amsterdam, the Netherlands, as bioinformatician/biostatistician.

Jan Kammenga studied biochemistry and toxicology at Wageningen University, the Netherlands. He holds a position as associate professor at the Laboratory of Nematology, Wageningen University, where he leads a research group on the genetics of stress biology in nematodes, in particular the well-studied biological model Caenorhabditis elegans. He has published more than 50 publications in peer-reviewed journals and books and has been coordinator of 3 different EU-funded projects relating to multiple stress biology. He served for 3 years on the Editorial Board of *Environmental Toxicology and Chemistry*. He teaches undergraduate and postgraduate courses on various topics from basic biology to ecotoxicology and supervises undergraduate and PhD students working on aspects of stress biology and genetics.

Ryszard Laskowski, completed his studies in biology in 1984 at the Jagiellonian University in Kraków, Poland. He is a professor at Jagiellonian University and Head of the Ecotoxicology and Stress Ecology Research Group at the Institute of Environmental Sciences. From 2002 to 2008 he was a deputy director of the Institute. He worked also at the Swedish University of Agricultural Sciences in Uppsala, the University of Reading, UK, and Oregon State University, USA. He is the coauthor of 5 books, including 3 major textbooks. He authored and coauthored more than 80 research, review, and popular articles. Ryszard Laskowski specializes in terrestrial ecotoxicology, population ecology, and evolutionary biology. He has lead a number of research projects on the effects of toxic chemicals on biodiversity of terrestrial invertebrates, microbial processes, biogeochemistry, and population dynamics of various species. He teaches, or has taught before, general ecology, ecotoxicology, soil ecology, terrestrial ecology, population ecology, tropical ecology, global ecological problems, and nature photography. In his private life, he is a badminton player, traveler, and nature photographer.

Claus Svendsen is a senior ecotoxicologist at the Centre for Ecology and Hydrology at Wallingford, UK. Svendsen studied chemistry and biology at Odense University (Denmark), gaining his BSc after thesis work at the University of Amsterdam in 1992 and his MSc after a year's thesis project developing and validating biomarkers at the Institute of Terrestrial Ecology, Monks Wood, UK, in 1995. He completed his PhD from the University of Reading, UK, in 2000 after investigating terrestrial biomarker systems and contaminated land assessment. Since 2000, he has worked on fundamental and applied environmental research at the Centre for Ecology and Hydrology's sites at Monks Wood and Wallingford, including the mechanistics and joint effects of contaminant mixtures and multiple stressors, identification and ecological risk assessment of contaminated land areas, and developing metabolic and microarray-based methods for assessing responses of soil invertebrate species for pollutant exposure. During this period, Svendsen was visiting researcher at National Research Canada's Biotechnology Research Centre, Montreal and Landcare Research, Christchurch, New Zealand, and was awarded the Society for Environmental Toxicology and Chemistry Europe's "Best First Paper on Environmental Research Award" in 1996." Svendsen is author or coauthor of more than 50 papers and book chapters. His current research includes comparative environmental genomics, bioavailability, nanoparticle ecotoxicity and environmental fate, and mixture toxicity is focusing especially on how effects at these mechanistic levels translate to population effects for terrestrial invertebrates and how these survive as populations in polluted habitats. He is also the coordinator of the EU FP7 project NanoFATE- Nanoparticle Fate Assessment and Toxicity in the Environment.

Workshop Participants[*]

Rolf Altenburger
Department of Bioanalytical
  Ecotoxicology
UFZ Helmholtz Centre for
  Environmental Research
Leipzig, Germany

Thomas Backhaus
Institute for Plant and Environmental
  Sciences
Göteborg, Sweden

Mieke Broerse
Institute of Ecological Science
Department of Animal Ecology
VU University
Amsterdam, The Netherlands

Christina E. Cowan
Environmental Science Department
Ivorydale Technical Center
The Procter and Gamble Company
Cincinnati, Ohio, United States

Jean Lou C. M. Dorne
Unit of Contaminants in the Food
  Chain
European Food Safety Authority
Parma, Italy

Almut Gerhardt
Schweizerisches Zentrum für
  angewandte Oekotoxikologie
Oekotoxzentrum EAWAG-EPFL
Duebendorf, Switzerland

John Groten
NV Organon
Toxicology & Drug Depostion
Oss, The Netherlands

Sami Haddad
University of Quebec in Montreal
  (UQAM)
Montréal, Québec, Canada

Fred Heimbach
RIFCON GmbH
Leichlingen, Germany

Geoff Hodges
Safety and Environmental Assurance
  Centre
Unilever Colworth
Sharnbrook, Bedford, United Kingdom

Tjalling Jager
Department of Theoretical Biology
VU University
Amsterdam, The Netherlands

Martijs Jonker
Microarray Department & Integrative
  Bioinformatics Unit
Faculty of Science
University of Amsterdam
Amsterdam, The Netherlands

Jan Kammenga
Laboratory of Nematology
Wageningen University
Wageningen, The Netherlands

Andreas Kortenkamp
The School of Pharmacy
University of London
London, United Kingdom

---

[*] Affiliations were current at the time of the workshop.

Paulina Kramarz
Department of Ecotoxicology
Institute of Environmental Sciences
Jagiellonian University
Krakow, Poland

Ryszard Laskowski
Department of Ecotoxicology
Institute of Environmental Sciences
Jagiellonian University
Krakow, Poland

Hans Løkke
Aarhus University
National Environmental Research
    Institute (NERI)
Department of Terrestrial Ecology
Silkeborg, Denmark

Susana Loureiro
Department of Biology
University of Aveiro
Aveiro, Portugal

Hana R. Pohl
Agency for Toxic Substances and
    Disease Registry (ATSDR)
Atlanta, Georgia, United States

Leo Posthuma
National Institute for Public Health and
    the Environment (RIVM)
Laboratory for Ecological Risk
    Assessment
Bilthoven, The Netherlands

Ad M. J. Ragas
Department of Environmental Science
Institute for Wetland and Water
    Research
Radboud University
Nijmegen, The Netherlands

Nissanka Rajapakse
Food Standards Agency (FSA)
Aviation House
London, United Kingdom

Martin Scholze
The School of Pharmacy
University of London
London, United Kingdom

David J. Spurgeon
Soil and Invertebrate Ecotoxicology
Centre for Ecology and Hydrology
Wallingford, Oxfordshire, United
    Kingdom

Claus Svendsen
Soil and Invertebrate Ecotoxicology
Centre for Ecology and Hydrology
Wallingford, Oxfordshire, United
    Kingdom

Linda K. Teuschler
US Environmental Protection Agency
Office of Research and Development
National Center for Environmental
    Assessment
Cincinnati, Ohio, United States

Cornelis A. M. van Gestel
Institute of Ecological Science
Department of Animal Ecology
VU University
Amsterdam, The Netherlands

Raymond S. H. Yang
Department of Environmental and
    Radiological Health Sciences,
    Physiology
Colorado State University
Fort Collins, Colorado, United States

WORKSHOP ASSISTANTS

Agnieszka Bednarska
Department of Ecotoxicology
Institute of Environmental Sciences
Jagiellonian University
Krakow, Poland

Renata Sliwinska
Department of Ecotoxicology
Institute of Environmental Sciences
Jagiellonian University
Krakow, Poland

Joanna Zietara
Department of Ecotoxicology
Institute of Environmental Sciences
Jagiellonian University
Krakow, Poland

Executive Summary

This book is the final outcome from discussions started at an international workshop on mixture toxicity, held 2–6 April 2006 in Krakow, Poland, the aim of which was to produce an updated review of the state of the art of mixture toxicity research and the potential for integrating its use in environmental and human health risk assessment. Since the previous key book on mixture toxicity (Yang 1994), there has been great progress in the development of concepts and models for mixture toxicity, both in human and environmental toxicology. However, due to the different protection goals of the 2 fields, developments have often progressed in parallel and with little integration. The workshop was, therefore, aimed mainly at developing mutual understanding, generating awareness across the fields, and investigating options for cross-fertilization and integration. In the time since the workshop, exchange of views and ideas has continued, resulting finally in this volume.

This book presents an overview of developments in both fields, comparing and contrasting their current state of the art to identify where one field can learn from the other. In terms of subject matter the book progresses from exposure through to risk assessment, at each step identifying the special complications that are typically raised by mixtures (compared to single chemicals). Five chapters are included, each addressing a specific step from exposure to risk assessment for mixtures:

1) exposure (how to quantify the amounts of chemicals that may enter the living organism);
2) kinetics, dynamics, and metabolism (how the chemicals enter and travel within the organism, how they are metabolized and reach the target site, and finally, the development of toxicity with time);
3) toxicity (its detrimental effects on the organism);
4) test design and complex mixture characterization (how chemicals interact, how to measure effects of mixtures, and how to identify responsible chemicals); and
5) risk assessment to man and the environment.

Each chapter provides an essential overview of the state of the art in both human and ecotoxicological mixture risk assessment, focusing especially on the excellent work that has been published in the intervening years between publication of this and the previous mixture volume. Each chapter then ends by identifying current possible crosslinks and recommendations for mutual developments that can improve the state of knowledge on mixture toxicity, and ultimately lead to better and more integrated risk assessment. The in-depth way the book covers the state of the art for mixture toxicology and ecotoxicology principles means it serves well as a textbook on the subject. At the same time, the inclusion of the considerations on application of novel developments in these principles, and their integration across human and environmental mixture risk assessment, makes it an ideal tool for researchers, regulators, and other risk assessment practitioners as mixture considerations start to enter regulatory forums over the next years.

EXPOSURE

Following the general principle of toxicology, dose determines the effect. As a consequence, any assessment of (eco)toxicological risk cannot do without a proper assessment and quantification of exposure. In this book, emphasis is on exposure to mixtures of chemicals, although principles will be the same as for assessing a single chemical exposure. Assessment of exposure may involve direct and indirect methods, the direct methods involving measurements of chemical concentrations at the point of contact or uptake, and the indirect methods using modeling and extrapolation methods to estimate exposure levels.

The first step in exposure assessment is the identification of the potential sources of emission. In the case of mixtures, emission may be due to a single source emitting a mixture of chemicals or a combination of sources producing a mix of chemicals. Composition of the mixture of chemicals emitted may vary in time. Several methods exist to estimate emission. Emission Scenario Documents (ESDs) and Pollution Release and Transfer Registers (PRTRs) may be helpful in identifying emission patterns and establishing emission factors for relevant chemicals. So far, these documents focus mainly on single chemicals rather than on mixtures. Nevertheless, they may be helpful also in assessing exposure to mixtures. There is, however, an urgent need to generate emission data that may help in identifying and quantifying mixture exposure. This need also begs for international collaboration and exchange of emission data and expanding the knowledge on emission scenarios for new and existing chemicals.

Exposure is determined by the fate of chemicals, sorption processes, degradation or transformation in the environment, and transportation from the site of emission to the places where organisms live. Different chemicals in a mixture may have different chemical properties, and as a consequence, their fate in the environment and final distribution over environmental compartments may differ. The composition of a mixture at the local scale may, therefore, be quite different from the one emitted. Chemical–chemical reactions may lead to the formation of new and potentially dangerous chemicals or mixtures. So far, little insight exists in these chemical–chemical interactions and their consequences for bioavailability, uptake, and toxicity. Multicompartment fate models, using information on physical–chemical properties of chemicals (such as partition coefficients air–water, octanol–water, and octanol–air) and characteristics of the receiving environmental compartments (such as pH and organic matter content), may be useful in assessing the fate of chemicals, single and in mixtures. Validation of such models, especially with respect to mixtures (composition and concentrations), is urgently needed.

Total concentrations of chemicals in the compartment of exposure are considered of limited relevance because, due to sorption and sequestration processes, often only a small fraction of the total amount is available for uptake by organisms. Properties of the chemical and of the environment determine sorption, chemical speciation, and bioavailability. Information on chemical concentrations therefore is much more useful when it goes along with data on the properties of the environment (e.g., pH, organic matter content). Bioavailability may also be species specific, making it more difficult to predict. In case of pesticides, formulation may have an impact on bioavailability of the active substances.

For any species, including humans, exposure is highly dependent on the life stage, with striking differences between, for example, adults and infants or juveniles, leading to large differences in individual exposure levels. Dietary intake and air are important routes of human exposure. Also human activity, for example, working or indoor and outdoor behavioral patterns, may lead to a significant exposure to (mixtures of) chemicals. There is a growing awareness that assessment of human exposure should take into account life stage, lifestyle, and activity patterns. Models available for this purpose need improvement to better account for these factors in predicting human exposures to mixtures.

Ecological exposure assessment often seems hampered by the fact that the routes of exposure are not well known and may be different for different species. The relative importance of a particular route of uptake may also depend on the exposure level. Behavioral aspects also are important in determining exposure. In addition, the fact that different life stages of organisms may live in different environmental compartments or have completely different physiologies and behaviors may affect exposure. So far, these aspects have received little attention.

Monitoring may be useful to assess exposure, although current monitoring activities, for both humans and ecosystems, mainly are focused on individual chemicals rather than mixtures. Monitoring data could be useful for identifying probable mixture scenarios, for example, by providing information on the combinations of chemicals most frequently encountered. The use of monitoring data might require refinement of existing monitoring programs, such as the Arctic Monitoring and Assessment Program (AMAP) and the European Monitoring and Evaluation Program (EMEP). As argued above, for a proper interpretation of monitoring data and assessment of available exposure concentrations, monitoring programs should also include characterization of the environments analyzed.

One of the most important knowledge gaps hampering a proper assessment of mixture effects is a lack of information about cumulative exposure scenarios, including simultaneous as well as sequential exposures and taking into account temporal and spatial aspects. There also is a need for integrated models that link exposure and toxicity, such as the biotic ligand model (BLM) for metals or the critical body residue (CBR) approach for organic chemicals. In addition to the modeling and monitoring programs mentioned above, application of whole mixture analysis methods, bioassay-directed fractionation (BDF) and toxicity identification evaluation (TIE) concepts may provide valuable information.

TOXICOKINETICS AND TOXICODYNAMICS

After uptake, a chemical may be absorbed, distributed, metabolized, and excreted (toxicokinetics (TK)), and once the biological target in the organism has been reached it may exert toxic effects (toxicodynamics (TD)). So, toxicokinetics can be considered what the body does with the chemicals, and toxicodynamics what the chemicals do to the body.

TK modeling within human toxicology is well established, and there are good data and models at the level of binary mixtures. The binary to multichemical extrapolation approach for modeling toxicokinetic interactions within mixtures has proven successful for chemicals interacting by affecting each other's metabolism, but remains to be challenged for interactions that are not of a metabolic nature. This approach is scientifically most appealing because it allows the use of the wealth of information that has already been generated on binary interactions. However, its major drawback is the large amount of experimental studies and resources needed to obtain missing information on multichemical mixtures. Chemical lumping is a potential way to reduce the information needed, and the use of in vitro studies has proven promising for obtaining mechanistic information and parameter values on binary interactions more efficiently.

Other emerging technologies may enhance our ability to determine mechanisms of binary interaction and hence facilitate the use of binary-interactions-based physiologically based toxicokinetic (PBTK) models of mixtures. These include the different in silico technologies that are being developed to predict ligand–enzyme interactions such as quantitative structure–activity relationship (QSAR) and 3D modeling of the different enzymes involved in the biotransformation of xenobiotics. Additionally, biochemical reaction networks (BRNs) are very promising tools that could help predict the rate of reactions and inhibition constants.

In ecotoxicology, there are only a few published examples for which a TK model has been used in the analysis of toxicity data for mixtures, often using a 1-compartment model without kinetic interactions. Few attempts have been made to develop more elaborate models, accounting for biotransformation and an additional state of damage (also without interaction). For single compounds, a broad range of models has been developed and successfully applied in ecotoxicology. In the near future, most of these TK models can probably be easily adapted to accommodate mixtures of compounds, although some further work is needed on interactions in uptake and excretion kinetics, metabolic transformation interactions, and binding interactions at the target site. For all these areas, considerable experience can be gained from the data generated in the mammalian toxicology area.

At present for many ecotoxicological applications, regarding the organism as a 1-compartment model is a good initial choice for whole body residues, and probably also for mixture TK. There are, of course, several situations where such simplicity will not apply and where more elaborate PBTK modeling may be required. These include larger organisms where internal redistribution is not fast enough to make a 1-compartment model reasonable, species with known tissues for handling and detoxification of chemicals where these are likely to interact, and also situations where the kinetics of receptor interactions or the accumulation of internal damage may need to be modeled explicitly as additional state variables. However, it is generally advisable to start with the simple 1-compartment model, and build more complex models when needed (in view of the limited data available). For single metals, there is growing evidence in aquatic ecotoxicology that advocates the concept of biodynamic modeling for linking internal metal concentrations to toxicity. This has been achieved by taking into account what proportion of the total metal is in free form and able to be biologically active. To get better estimates for the uptake and effects of organic chemicals, data based on membrane–water partitioning and chemical activity should be incorporated, and this would provide information to perform cross-species extrapolation. For mixtures of metals, approaches for addressing joint toxicology will have to account for speciation and competition between metals (in different forms) at the target organ level. The combination of multiple BLM approaches offers great possibilities for the interpretation of toxicity data for metal mixtures, but has not yet been applied to actual data. Nevertheless, this complexity may not be necessary for all mixture studies.

For wildlife ecotoxicology, PBTK modeling may be a viable choice, applying the knowledge available from human toxicology combined with data from the most relevant test species available. Examples of mixture TK models in this area are currently unavailable.

Toxicodynamic studies involving chemical mixtures are relatively scarce in the human and ecological arenas. Of the few available, a greater portion of such TD studies are relatively simple, without much mechanistic insight. Physiologically based toxicodynamic (PBTD) modeling of chemical mixtures holds great promise in various fields of human toxicology. For noncancer effects, the use of PBTD models has elucidated the fundamental mechanisms of toxicological interactions. Such mechanistic knowledge linked with Monte Carlo simulations has initially been employed in in silico toxicology to develop models that predict the toxicity of mixtures in time. A combination of PBTK and PBTD models for individual compounds into binary PBTK and PBTD models can be achieved by incorporating key mechanistic knowledge on metabolism inhibitions and interactions through shared enzyme pathways. Simulations of such models can then be compared to experimental data and allow conclusions to be reached about their pharmacokinetics and the likelihood of effects being dose additive.

In cancer research the resource-intensive chronic cancer bioassays originally needed to evaluate carcinogenic potentials of chemicals or chemical mixtures have led to the development of computer simulation of clonal growth of initiated liver cells in relation to carcinogenesis. This model describes the process of carcinogenesis based on the 2-stage model with 2 critical rate-limiting steps: 1) from normal to initiated cells and 2) from initiated cells to malignant states. Because this approach can incorporate relevant biological and kinetic information available, it usefully facilitates description of the carcinogenesis process with time-dependent values without the need for chronic exposures.

To stretch toward the goal of having the ability for predicting the toxicity of the infinite number of possible mixture exposure scenarios, it is clear that computer modeling must be employed in conjunction with experimental work. The latest development toward such abilities is the development of initial BRN modeling. While a seemingly insurmountable task, it provides a system for collating the ever-building TK and TD information in a way where the toxicology of chemical mixtures of increasing component numbers can be assessed and predicted.

Within the field of ecotoxicology, promising approaches for explaining and predicting effects as a function of exposure time, using biologically based models, such as the dynamic energy budget (DEB) model, show a lot of promise, but need developing and testing for mixtures. The problem with the application of any TD model in ecotoxicology to mixtures is that there is an extreme lack of data on toxic effects of mixtures measured as a function of time. Almost all studies focus on the effects of the mixture after a fixed exposure period only, which is of limited use for the application of dynamic approaches. For mortality, the individual tolerance concept (using a fixed critical body residue (CBR), or the more elaborate damage assessment model (DAM)) has been applied to mixtures at the 50% effect level, but more work needs to be done to validate its applicability, and to test it against the stochastic approach. For the stochastic approach, several mixture toxicity studies have been performed so far, and they look promising.

For sublethal responses, the level of resource allocation is essential. The DEB approach offers great promise as a TD model in ecotoxicology. It has been applied to the combination of a toxicant with another stressor (food limitation), but its application to mixtures of chemicals requires further work and a comparison to dedicated experimental data.

There are moves for the disciplines of human and ecological risk assessment to start closing the gap between considering effects at the individual and the population levels. Human risk assessment has undergone fundamental changes in recent years, in order to consider the population-level variation. This change has particularly involved PBTK modeling, which has gone on to population PBTK modeling using Bayesian statistics and Markov chain Monte Carlo simulations. Nevertheless, while including population-level variability, human risk assessment still aims to protect the individual. Within ecological risk assessment there are developments to include receptor characteristics, which when perfected would allow the latter to better employ the approaches of human risk assessment, but with the challenge of accounting for one extra level of biological organization (i.e., interspecies differences).

TOXICOLOGY OF MIXTURES

Very large numbers of substances are found simultaneously in ecosystems, food webs, and human tissues, however, all at quite low levels. This fact triggers the question of whether exposures to multiple chemicals are associated with risks to human health, wildlife, and ecosystems. Current assessment procedures focus mainly on dealing with environmental and human health risks on a chemical-by-chemical basis. It therefore is important to address the following questions: 1) Can the effects of mixtures be predicted from the toxicity of individual components? 2) Is there any risk to be expected from exposure to multiple chemicals at low dosages? 3) How likely is it that chemicals interact with each other, leading to effects that are larger than expected from the toxicities of the individual compounds (synergism), and which factors determine the potential for synergism?

A "whole mixture approach," investigating a complex mixture as if it were a single agent without assessing the individual effects of all the components, is useful for studying unresolved mixtures or specific combinations on a case-by-case basis. This approach, however, does not allow for an identification of the chemicals contributing to the overall mixture effect or how they work together in producing a joint effect. To understand how chemicals act together in producing combination effects, component-based analyses are required that aim at explaining the effect of a mixture in terms of the responses of its individual constituents. Thus, an attempt is made to anticipate joint effects quantitatively from knowledge of the effects of the chemicals that make up the mixture. The concepts of concentration addition (CA) and independent action (IA; also termed "response addition" (RA)) allow valid calculations of expected effects, when the toxicities of the individual mixture components are known.

An overview of the literature shows that in the majority of cases, CA did yield accurate predictions of combination effects, even with mixtures composed of chemicals that operate by diverse modes of action. The studies available were dealing with mixtures of chemicals having an unspecific mode of action (membrane disturbance or narcosis) or with pesticides, mycotoxins, or endocrine disruptors. In ecotoxicology, CA usually produced more conservative predictions than IA. There are indications that this is true also for mammalian toxicology, but more data are needed to come to more definitive conclusions. The validity of CA or IA was confirmed for individual-based endpoints like growth or reproduction, but also for effects at the cellular or subcellular level and for community-based endpoints.

Deviations from expected additive effects can be assessed in terms of synergism or antagonism (the mixture being less toxic than expected from the toxicity of the individual chemicals). In only a few cases, the mechanisms underlying such deviations are well understood. Interaction typically occurs when one substance induces toxifying (or detoxifying) steps effective for another mixture component, which in turn alters profoundly the efficacy of the second chemical. Interactions may also lead to changes in doses available at the target site or alterations in time to effects. In any case, it is obvious that disregard of mixture effects may lead to considerable underestimations of hazards from chemicals, and application of either CA or IA has to be seen as superior to ignoring combined effects.

There is good evidence that combinations of chemicals are able to cause significant mixture effects at doses or concentrations well below no-observed adverse effect levels (NOAELs), irrespective of perceived similarity or dissimilarity of the underlying modes of action. On the basis of the available experimental evidence as well as theoretical considerations, the possibility of combination effects therefore cannot easily be ruled out. On the other hand, this possibility cannot readily be confirmed either. Knowledge about relevant exposures, in terms of the nature of active chemicals, their number, potency, and levels, and simultaneous or sequential occurrence, is essential for a proper prediction of mixture effects. Demonstration of interactive effects of mixtures at low exposure levels requires proper experimental designs, with careful selection of exposure levels.

Human epidemiology strongly indicates that environmental pollutants may act together at existing exposure levels to produce health effects. However, more evidence is needed to substantiate the case. In ecotoxicology, the evidence is much stronger. In both fields, however, considerable advances are needed to quantify combination effects better.

The advances made with assessing the effects of multiple chemicals at low doses in laboratory experiments have yet to be fully realized in human epidemiology. At present, too much of human epidemiology is still focused on individual chemicals. Epidemiology needs to embrace the reality of mixture effects at low doses by developing better tools for the investigation of cumulative exposures. The application of biomarkers able to capture cumulative internal exposures is promising in this respect. Only an approach that fully integrates epidemiology with laboratory science can hope to achieve this task.

Toxicity testing of mixtures should move beyond the standard tests for deviations from the default models of CA and IA, toward a more mechanistic understanding of the process involved in mixture toxicity. These studies should focus not only on the processes and effects involved in concurrent exposure to multiple substances (i.e., cocktails), but also on those involved in sequential exposure to multiple substances.

TEST DESIGN

Mixture toxicity testing may have several aims, ranging from unraveling the mechanisms by which chemicals interact to the assessment of the risk of complex mixtures. Basically, 2 approaches can be identified: 1) a "component-based approach" that is based on predicting and assessing the toxicity of mixtures of a known chemical composition on the basis of the toxicity of the single compounds, and 2) a "whole mixture approach" in which the toxicity of (environmental samples containing) complex mixtures is assessed with a subsequent study in order to analyze which individual compounds drive the observed total toxicity of the sample. The first approach is also often used to unravel the mechanisms of mixture interactions. The experimental design highly depends on practical and technical considerations, including the biology of the test organism, the number of mixture components, and the aims of the study.

Fundamental for both the component-based and whole mixture approaches are 2 concepts of mixture toxicity: the concept of CA and the concept of IA or RA. CA assumes similar action of the chemicals in the mixture, while IA takes dissimilar action as the starting point. In practice, this means that CA is used as the reference when testing chemicals with the same or similar modes of action, while IA is the preferred reference in cases of chemicals with different modes of action.

The CA concept uses the toxic unit (TU) or the toxicity equivalence factor (TEF), defined as the concentration of a chemical divided by a measure of its toxicity (e.g., EC50) to scale toxicities of different chemicals in a mixture. As a consequence, the CA concept assumes that each chemical in the mixture contributes to toxicity, even at concentrations below its no-effect concentrations. The IA or RA concept, on the other hand, follows a statistical concept of independent random events; it sums the (probability of) effect caused by each chemical at its concentration in the mixture. In the case of IA, the only chemicals with concentrations above the no-effect concentration contribute to the toxicity of the mixture. The IA model requires an adequate model to describe the (full) dose–response curve, enabling a precise estimate of the effect expected at the concentration at which each individual chemical is present in the mixture. The concepts generally are used as the reference models when assessing mixture toxicity or investigating interactions of chemicals in a mixture. Interactions are defined as deviations from the expected additivity of effects based on using either CA or IA as a reference. Interactions may result in higher (synergism) or lower (antagonism) toxicity of the mixture than expected from the toxicities of the individual chemicals.

The component-based approach usually starts from existing knowledge on the toxicity of the chemicals in the mixture, either from the literature or from a range-finding test. Several test designs may be chosen to unravel the mechanisms of inter-action in the mixture or to determine the toxicity of the mixture, the CA and IA concepts serving as the reference. In addition to just testing for synergistic or antago-nistic deviations from the reference concepts, the focus may also be on detecting concentration-ratio- or concentration-level-dependent deviations. Experiments can be designed to determine the full concentration–response surface, often taking a full factorial design or a fixed-ratio design. When resources are limited or the question to be answered is more specific, the test design may be restricted to determining isobo-les. Isoboles are isoeffective lines through the mixture-response surface, defined by all combinations of the chemicals that provoke an identical mixture effect. Another alternative is fractionated factorial designs, such as Box-Behnken or central compos-ite designs, allowing for detection of interactions between chemicals and curvature of the response surface with a relatively low workload. Sometimes designs limited to "chemical A" in the presence of "chemical B" or point designs are also used, but they are not recommended because of the low resolving power and inability to detect curvilinear relationships for mixture effects. In all cases, it is desirable to combine tests on the mixtures with tests on the single chemicals in 1 experiment.

Several problems may hamper a proper interpretation of results, when effects of the mixtures are compared with effects of the individual chemicals. Measured con-centrations may be different from the initial (nominal) ones, and adsorption, chemi-cal-chemical interactions, and (bio)degradation may affect the bioavailability of the chemicals in the mixture. As a consequence, the exposure concentrations may be different from the starting point, and in fact, the experimental design has changed. This needs to be acknowledged while analyzing the data. It therefore is essential to investigate whether the concentration layout (experimental design) still supports the model parameters sufficiently. Hormesis, the finding of a stimulatory rather than an inhibitory response at low concentrations of a toxicant, raises all kinds of con-ceptual and technical issues in case of mixture toxicity and may lead to difficulties in estimating model parameters. Modeling mixture toxicity may also be hampered when responses to individual mixture components have different end levels at high concentrations, resulting in incomplete dose–response curves. Effect concentrations may be endpoint specific and dependent on exposure time. Also these aspects have consequences for the experimental design of a mixture toxicity study. In the latter case, test designs may benefit from a more detailed understanding of toxicokinetics and toxicodynamics.

The whole mixture approach generally consists of testing the complex mixture in bioassays (both in the laboratory and in situ), usually applying the same prin-ciples as used in the single chemical toxicity tests. By performing whole mixture tests on gradients of pollution or on concentrates or dilutions of (extracts of) the polluted sample, concentration–response relationships can be created. Bioassays and biosensors may be applied for that purpose. These tests will, however, not provide any information on the nature of the components in the mixture responsible for its toxicity. By using toxicity identification evaluation (TIE) approaches, including chemical fractionation of the sample, it is possible to get further insight into the groups or fractions of chemicals responsible for toxicity of the mixture. Also, comparison with similar mixtures may assist in determining toxicity of a complex mixture. Such a comparison may be based on the chemical characterization of the mixture in combination with multivariate statistical methods. Effect-directed analysis (EDA) and the 2-step prediction (TSP) model may be used to predict toxicity when full chemical characterization of the complex mixture is possible and toxicity data are available for all chemicals in the mixture. Such a prediction can, however, be reliable only when sufficient knowledge of the modes of action of the different chemicals in the complex mixture is available. In other cases, bioassays remain the only way of obtaining reliable estimates of the toxicity and potential risk of complex mixtures.

RISK ASSESSMENT

The risk assessment for mixtures shows much similarity with that for single substances, but there also are some important differences. In order to make accurate risk predictions, risk assessment should pay specific attention to all aspects of mixture exposures and effects. The establishment of a safe dose or concentration level for mixtures is useful only for common mixtures with more or less constant concentration ratios between the mixture components and for mixtures of which the effect is strongly associated with one of the components. For mixtures of unknown or unique composition, determination of a safe concentration level (or a dose–response relationship) is inefficient, because the effect data cannot be reused to assess the risks of other mixtures. One alternative is to test the toxicity of the mixture of concern in the laboratory or the field to determine the adverse effects and subsequently determine the acceptability of these effects. Another option is to analyze the mixture composition and apply an algorithm that relates the concentrations of the individual mixture components to a mixture risk or effect level, which can subsequently be evaluated in terms of acceptability.

There are many concepts in use for the assessment of risks or impacts of chemical mixtures, both for human and ecological risk assessment. Many of these concepts are identical or similar in both disciplines, for example, whole mixture tests, (partial) mixture characterization, mixture fractionation, and the concepts of CA and RA (or IA). The regulatory application and implementation of bioassays for uncharacterized whole mixtures is typical for the field of ecological risk assessment. The human field is leading in the development and application of process-based mixture models such as PBTK and BRN models and qualitative binary weight-of-evidence (BINWOE) methods. Mixture assessment methods from human and ecological problem definition contexts should be further compared, and the comparison results should be used to improve methods.

Most national laws on chemical pollution do account for mixture effects, but explicit regulatory guidelines to address mixtures are scarce. Only the United States has fairly detailed guidelines for assessing mixture risks for humans, for example, using the hazard index (HI), relative potency factors (RPFs), and toxicity equivalency factors (TEFs). Most of these regulations are applied for chemicals with similar modes of action, and make use of the concept of CA. Also, in ecological risk assessment, TEFs and RPFs are applied. Recently the concept of multisubstance probably affected fraction (msPAF) has been introduced as a method to estimate potential risk of chemical mixtures to ecosystems. This method may be part of the "TRIAD approach" of contaminated site assessment.

The multitude of different mixture assessment techniques is typical for the current state of the art in mixture assessment. There is a clear need for a comprehensive and solid conceptual framework to evaluate the risks of chemical mixtures. For that purpose, Chapter 5 outlines a system that can be considered a first step toward a conceptual framework for integrated assessment of mixture risks. The framework is proposed as a possible line of thinking, not as a final solution. Distinction is made between approaches for assessment of whole mixtures and component-based approaches. The most accurate assessment results are obtained by using toxicity data on the mixture of concern. If these are not available, alternatives can be used, such as the concept of sufficient similarity, (partial) characterization of mixtures, and component-based methods. Which method is most suitable depends on the situation at hand. A single mixture assessment method that always provides accurate risk estimates is not available. Tiering is proposed as an instrument to balance the accuracy of mixture assessments with the costs. When lower tiers do not provide sufficiently accurate answers for the problem at hand, the option exists to go to a higher tier, for example, by more detailed characterization of the mixture or application of more sophisticated mixture models. The general framework proposed for organizing problem definitions and associated mixture assessment approaches should be critically tested and improved.

In both human and ecological risk assessment, there is considerable scientific latitude to develop novel methods (e.g., those that exist in only one of the subdisciplines could be useful in the other one) and to refine approaches (e.g., by considering complex reaction networks and more specific attention for modes of action). The refinements are needed to improve the scientific evidence that is available for underpinning risk assessments. Several key issues in risk assessment of chemical mixtures were identified, that is, exposure assessment of mixtures (e.g., mixture fate and sequential exposure), the concept of sufficient similarity (requires clear criteria), mixture interactions, QSARs, uncertainty assessment, and the perception of mixture risks. Resolving these key issues will significantly improve risk assessment of chemical mixtures.

Tools should be developed to support the identification of mixture exposure situations that may cause unexpectedly high risks compared to the standard null models of concentration addition and response addition, for example, based on an analysis of food consumption and behavioral patterns, and the occurrence of common mixture combinations that cause synergistic effects. Criteria should be developed for the inclusion of interaction data in mixture assessments.

Finally, the review of risk assessment approaches for mixtures clearly showed the need for improved regulations. National authorities should develop legislation that enables the assessment and management of potential high-risk situations caused by sequential exposures to different chemicals and exposures through multiple pathways, with specific emphasis on a systems approach rather than on approaches focusing solely on chemicals, or on water or soil as compartments.

1 Exposure

David J. Spurgeon, Hana R. Pohl, Susana Loureiro,

Hans Løkke, and Cornelis A. M. van Gestel

CONTENTS

1.1   INTRODUCTION

In the environment, organisms including man are exposed to mixtures of chemicals rather than single compounds. Examples of mixtures are food and feedstuff, pesticide and medical products, dyes, cosmetics, and alloys. Many other commercial products, such as printing inks, contain a mixture of substances, possibly up to 60 individual chemicals in 1 formulation. Preparation of these chemicals may involve the use of several hundred other substances in upstream processes.

As a first step in the risk assessment of chemicals, it is essential to have an insight into the magnitude and duration of exposure. Following the toxicological principle that dose determines the effect, one may assume that no exposure means no risk. In the case of chemical mixtures, a proper assessment of exposure assists in adequately interpreting the interacting effects of chemicals. So, exposure assessment is an essential component of any risk assessment study of mixtures, since it can be used to reduce uncertainty and provide data.

The exposure of organisms includes man-made chemicals as well as natural compounds. Natural compounds are, for example, toxins in plants, ozone, or natural occurring metals. The total number of man-made chemicals is vast. To assess exposure, the ambient concentrations of chemicals resulting from man-made sources need to be known or estimated. Chemical Abstracts, covering more than 8000 journals since 1907, registers more than 20 million entries. This section focuses on man-made chemicals. In Europe, around 30,000 chemicals are commonly used and thereby emitted to the environment (EC 2001).

In human health risk assessment, "direct" and "indirect" methods of exposure assessment are distinguished. The direct method involves measurements of exposure at the point of contact or uptake, for instance, by monitoring chemical concentrations in humans or the environments they are exposed to (food, air, water). The indirect methods use modeling and extrapolation techniques to estimate exposure levels (Fryer et al. 2006). Also in environmental exposure assessment, these 2 ways to assess exposure may be applied.

Indirect exposure assessment, both human and environmental, starts with emission data and a prediction of the fate of chemicals in the environment and the resulting concentrations in different environmental compartments. Foster et al. (2005) outlined 5 steps in a strategy to conduct exposure assessment of complex mixtures, consisting of many different components, such as gasoline. These steps, as outlined below, are also relevant when assessing exposure to less complex mixtures.

1) Determination of mixture composition. Composition of the mixture may vary spatially and temporally. Measurements at the source (point of emission) may help in identifying (variations in) mixture composition.
2) Selection of component groups (optional). Within a mixture, different (groups of) components may be identified. These components may be grouped on the basis of properties that affect their fate in the environment.
3) Compilation of relevant property data for each group. This step consists of collecting properties relevant for predicting the fate of the different (groups of) components in the environment.
4) Assessment of the environmental fate of each group. Fate models may be used to predict environmental fate of mixture components on different spatial scales. Such models may yield a predicted distribution over air, water, soil, and sediment.
5) Assessment of environmental and human exposure. As a final step, concentrations can be calculated for each of the (groups of) mixture components in different exposure media (inhaled air, ingested water, food items) or environmental compartments (soil, sediment, air, and surface or groundwater). This information may not, however, represent the complete picture: often only part of the total concentration in an environmental compartment is biologically available for uptake by organisms. In addition, species habits and individual behavior may affect the nature of exposure. Finally, life-stage-specific aspects may be highly important in determining exposure to mixtures; this aspect is best studied for human exposure, but is also relevant to ecological assessment for some taxa.

For exposure assessment of ecosystems, direct exposure assessment involves taking field samples at the site and time of exposure and measuring chemical concentrations in these samples or in the organisms exposed at the site. Direct assessment of (potential) exposure also is possible by performing bioassays in which selected test organisms are exposed to the environmental sample, in the laboratory or in the field. The latter approach is discussed in more detail in Chapter 4.

In this chapter, the different steps in the assessment of mixture exposure are discussed. The chapter starts from emission scenarios and subsequently discusses transformation processes taking place in the environment and their effects on mixture composition. Next, bioavailability is discussed, and exposure scenarios for both humans and biota in the environment are described. These descriptions also consider methods to assess exposure to mixtures. Most data available on mixture exposure are restricted to North America and Europe, but we recognize that there are emerging problems in other regions of the world. We restrict our discussion to man-made chemicals and those natural chemicals subject to regulation (metals, polycyclic aromatic hydrocarbons (PAHs)), because these represent the most well-studied group and the current priorities for risk assessment.

1.2   EMISSION SCENARIOS

An emission is defined as the amount of chemical discharged or transferred per unit time, or it is the amount of chemical per unit volume of gas or liquid emitted. The emission can be characterized by the following attributes (OECD 2006):

1) Pollutant type
2) Release medium
3) Source type
4) Spatial scale
5) Temporal scale

Normally, an emission assessment deals with a single chemical or a group of chemicals, which have similar properties, such as PAHs, metals, ozone-depleting substances, or chlorinated biphenyls. In the environment, organisms and man are exposed to mixtures of chemicals with different properties rather than single chemicals or chemicals with similar properties. As an example, many commercial products (e.g., inks, oils, lubricants) contain a mixture of substances in a single formulation and so may be released simultaneously to environmental media, including land, surface water, groundwater, and indoor and outdoor air.

1.2.1   MAJOR EMISSION SOURCES

Emission sources are generally divided between point, diffuse, and mobile sources (OECD 2006). Point sources, such as industrial plants, power stations, waste incinerators, and sewage treatment plants, may play a major role as sources of mixtures of chemicals. Emissions from such sources are frequently of multiple chemicals; even in cases where the emission is dominated by a single chemical, overlap of plumes from other nearby point sources for different chemicals means that the surrounding areas are subject to combined exposure. Diffuse emissions from the application of pesticides and biocides and the domestic and widespread commercial use of chemicals can make a major contribution to the release of chemical mixtures into air, soils, and waters. In the case of pesticides, these biologically active compounds are applied as mixtures, or the application is repeated with other types of active ingredients within a short period, so that more than 1 chemical is present. As for local sources, even when there is not deliberate combined release, overlapping release and transport mechanisms in the atmospheric and the aquatic environments result in the widespread presence of chemical mixtures in different environmental compartments. Emissions from mobile sources, such as vehicles, may be regarded in effect as diffuse emission and in the same way can contribute to the widespread contamination of the environment with chemical mixtures. Thus, diffuse emission can comprise contributions from several emission sources and product emissions. In addition to exposure through environmental media, such as air, soil, and water, the indoor conditions of private households can be relevant for many airborne mixtures in relation to human health due to a large variety of products that are used indoors, where the ventilation can be limited. Also, food intake can be considered for potential relevant mixture exposures for humans and for species in the higher tier of ecological food webs.

For the terrestrial environment, waste sites may act as major emission sources of mixtures. In the United States, the Agency for Toxic Substances and Disease Registry (ATSDR) has performed a trend analysis to identify priority chemical mixtures associated with hazardous waste sites (De Rosa et al. 2001, 2004; Fay 2005). The information was extracted from the Hazardous Substance Release/Health Effects Database (HazDat) (ATSDR 1997). The HazDat contains data from hundreds of hazardous waste sites in the United States. A trend analysis was completed for frequently co-occurring chemicals in binary or ternary combinations found in air, water, and soil at or around hazardous waste sites (Fay and Mumtaz 1996; De Rosa et al. 2001, 2004). Table 1.1 gives an overview of frequently occurring substances at hazardous waste sites in the United States.

1.2.2 EMISSION ESTIMATION METHODS

In the work of OECD (2006) on assessment of emissions, a distinction is made between Emission Scenario Documents (ESDs) and Pollution Release and Transfer Registers (PRTRs). An ESD provides a description of activities related to emissions and methods to estimate these emissions. A PRTR is an environmental database of potentially harmful chemicals released to air, water, and soil (on-site releases) and transported to treatment and disposal sites (off-site transfers). PRTRs contain data on releases or transfers, by source, and are publicly available in many countries, including Australia, Canada, Japan, several European countries, and the United States. An OECD study identified the similarities and differences between the emission estimation methods used in ESDs and PRTRs, showing that PRTR mass balance and emission factor methods yielded more conservative estimates than the ESD fixation-based method (OECD 2006). The PRTR mass balance method was found to account for a more thorough analysis of parameters, such as substance sources and recycles, which could impact emissions. Both ESD and PRTR methods might be applied to complex chemical mixtures, although no studies are available at present.

The emission estimation methods of the PRTR approach are described by OECD (2002a, 2002b, 2002c) and include direct monitoring, mass balance, emission factor, and engineering calculations and judgment. These methods are all feasible for the estimation of mixture emissions. The mass balance approach is based on the principle of mass conservation. Emissions from a system can be estimated by knowing the amount of a substance going into the system and the amount that is created or removed (dissipated or released to other compartments, degraded, transformed, or bound):

$$\Sigma(\text{output}) = \Sigma(\text{input}) - \Sigma(\text{removed}) + \Sigma(\text{generated})$$

For mixtures of chemicals, this equation should be used to estimate the concentration of each component under steady-state conditions, or under dynamic conditions when data are available to describe temporal conditions. These calculations lead to the (constant or varying) composition of the mixture over time.

Table 1.1 Frequencies of single substances and combination of substances at hazardous waste sites in the United States

Rank	Percent of sites	Single substance	Percent of sites	Binary combination		Percent of sites	Ternary (tertiary) combinations		
Water									
1	42.4	TCE	23.5	TCE	Perc	11.6	TCE	1,1,1-TCA	Perc
2	38.4	Lead	18.9	Lead	Chromium	10.6	Lead	Benzene	Perc
3	27.3	Perc	17.9	1,1,1-TCA	TCE	10.6	Lead	Cadmium	Chromium
4	25.8	Benzene	17.3	TCE	Lead	9.8	TCE	1,1-DCA	1,1,1-TCA
5	25.8	Chromium	17.3	Lead	Cadmium	9.7	Lead	Arsenic	Cadmium
6	23.9	Arsenic	17.0	Benzene	TCE	9.7	TCE	Perc	TCE
7	20.8	1,1,1-TCA	16.3	Lead	Arsenic	9.6	Lead	Arsenic	Chromium
8	20.3	Toluene	14.5	TCE	Trans-1,2-DCE	9.4	Benzene	TCE	Toluene
9	19.8	Cadmium	13.6	TCE	Toluene	9.3	TCE	Perc	Trans-1,2-DCE
10	17.7	MeCl	13.5	Benzene	Lead	9.1	TCE	Lead	Chromium
Soil									
1	37.7	Lead	20.5	Lead	Chromium	12.0	Lead	Cadmium	Chromium
2	25.3	Chromium	17.8	Chromium	Arsenic	11.6	Lead	Arsenic	Chromium
3	23.0	Arsenic	17.6	Arsenic	Cadmium	10.9	Lead	Arsenic	Cadmium
4	19.7	Cadmium	13.3	Cadmium	Arsenic	8.4	Arsenic	Cadmium	Chromium
5	19.1	TCE	12.9	TCE	Chromium	8.1	Lead	Nickel	Chromium
6	16.0	Toluene	11.6	Arsenic	Cadmium	7.9	Lead	Chromium	Zinc
7	14.8	Perc	10.9	TCE	Perc	7.7	TCE	Copper	Chromium
8	13.6	PCBs	10.9	Lead	Zinc	7.6	Toluene	Lead	Copper
9	13.0	Xylenes	10.4	Ethylbenzene	Toluene	7.5	Ethylbenzene	Toluene	Chromium
10	12.8	Ethylbenzene	10.4	Nickel	Lead	7.5	Lead	Nickel	Cadmium

Air

1	6.0	Benzene	3.5	Benzene	Toluene	Benzene	2.2	TCE	Perc
2	4.7	Toluene	2.7	Benzene	TCE	Benzene	1.9	Ethylbenzene	Toluene
3	3.8	TCE	2.6	Benzene	Perc	Benzene	1.8	Toluene	Perc
4	3.4	Perc	2.6	TCE	Perc	TCE	1.8	TCE	Toluene
5	3.1	1,1,1-TCA	2.3	Toluene	Perc	TCE	1.8	Toluene	Perc
6	2.6	Lead	2.1	Ethylbenzene	Toluene	1,1,1-TCA	1.4	Toluene	Perc
7	2.5	Ethylbenzene	2.1	TCE	Toluene	1,1,1-TCA	1.4	TCE	Perc
8	2.4	MeCl	1.9	1,1,1-TCA	TCE	Benzene	1.3	1,1,1-TCA	Perc
9	2.4	Xylenes	1.9	Toluene	Xylenes	Benzene	1.3	Toluene	Xylenes
10	1.8	Chloroform	1.9	1,1,1-TCA	Perc	1,1,1-TCA	1.3	TCE	Toluene

Source: Adapted from De Rosa CT, El-Masri HE, Pohl H, Cibulas W, Mumtaz MM. 2004. J. Toxicol. Environ. Health 7:339–350.

Note: MeCl = methylene chloride, PCBs = polychlorinated biphenyls, Perc = perchloroethylene (tetrachloroethylene), 1,1,1-TCA = 1,1,1-trichloroethane, TCE = trichloroethylene, Trans-1,2-DCE = trans-1,2-dichloroethylene, 1,1-DCA = 1,1-dichloroethane.

An emission factor is defined as a constant that relates the intensity of an activity to an emission (OECD 2002a). Emission factors can be used to estimate releases from nearly any source that generates emissions with a strong proportional dependence on the extent. Emission factors are used for specific cases where no release information is available, or if the release is only given for 1 specific compartment. The complementary release estimates can be obtained using the OECD approach or from the European Technical Guidance Document (ECB 2003a). The release of a chemical to an environmental compartment a is calculated as

$$\text{Release}_a = F_a \times \text{Prodvol}$$

That is, the release to the compartment (e.g., freshwater or air) is equal to the product of the fraction of the produced volume released, F_a, for example, during production, and the produced volume of the chemical (Prodvol). F_a is the emission factor. Emission factors can be applied to essentially any pollution or source (OECD 2006). They may be derived by many different techniques, but often are developed by taking the average emission rate during a representative time interval and relating it to the extent of the activity in question (OECD 1999). Emission factors can be used to estimate emissions of mixtures when data on the components are available. If the ratio between concentrations of different chemicals in the emission is stable, periodic monitoring of certain pollutants can be used to represent other pollutants by applying average ratios (OECD 2006).

More complex calculations can be made based on mathematical relationships between variables within a system, the outcome being dependent on the quality of data and the validity of assumptions. No specific models are available for chemical mixtures; however, modeling tools might be useful to estimate the varying composition of mixtures in time and space.

In Europe, ESDs are commonly used to facilitate the risk assessment of single substances, and they often deal with groups of chemicals. The methods used in ESDs are designed to deal with broader emissions than those used in PRTRs by dealing with emissions from clusters of facilities. This provides information on data on a local or regional basis. Although the ESDs are directed toward single chemicals, they may provide data for estimation of the emission of mixtures. Recently such information on emissions of single chemicals has been collected by the European Chemicals Bureau (ECB 2003b). The data are currently available for the different industrial categories (ICs) and biocidal product types. This documentation has been developed by different competent authorities and by industry. In most cases the documents are based on in-depth studies of the environmental release of substances used in different industrial categories and of different biocidal product types. Some documents describe environmental releases from specific use categories under an industrial category. Data are not available for all industrial categories and biocidal product types; some documents are still under preparation. It is anticipated that this set of emission scenario documents will be expanded continuously in the future. For industrial chemicals 9 areas have been developed so far: chemical, leather processing, metal extraction, photographic, textile processing, rubber, and paint industries, as well as personal or domestic and public domains.

Data on the emission of (mixtures of) chemicals may also be obtained from the European Pollutant Emission Register (EPER), which is the first European-wide register of industrial emissions into air and water (http://eper.ec.europa.eu/ (last accessed November 2009)). EPER gives access to information on the annual emissions of approximately 9200 industrial facilities in the member states of the European Union as well as Norway mostly for the year 2001, and approximately 12,000 facilities for the year 2004. It has the option to group information by pollutant, activity (sector), air and water (direct or via a sewerage system), or country, and even gives access to data on individual facilities. Such information thus has value for developing realistic emission scenarios for diffuse release and also at the local scale.

Prediction of the composition of a chemical mixture that organisms are exposed to in a certain area requires considering all emission sources in that area or contributing to the input of chemicals in that area. The European Chemicals Bureau has described the relevant factors for estimating the release or emission of chemicals, including their intermediates and degradation products (ECB 2003a):

- the emission factor (release fraction) for processing of the intermediate,
- local production volume per time unit,
- the emission factor (release fraction) for production of the intermediate,
- the elimination in on-site treatment facilities, and
- the elimination in biological wastewater treatment facilities.

As an example, emission to surface water (in g/s) was estimated in a German scenario. The local concentration in rivers was calculated from that emission and the river flow (in m^3/s), taking into account adsorption processes in the surface water. This approach was based on a statistically evaluated database. Volumes of wastewater flows from the production or processing facility were not taken into account. Although for the time being the database is restricted to a set of 29 substances and to German conditions, it can be regarded as a realistic worst-case situation since it combines two 90th percentiles (discharge × river flow) (ECB 2003a).

1.2.3 PRIORITIZATION

As described, living organisms are constantly exposed to vast amounts of chemicals, that is, to 1 big chemical mixture. We lack the information on how to properly capture the entire exposure, on how to address the toxicity, and how to evaluate the associated risk. Therefore, the initial approach is to properly define mixtures on a smaller scale. These are the "mixtures of concern" and are usually associated with a specific exposure scenario and possible health implications. Exposures to chemical mixtures of concern can range from simple and well-defined mixtures to complex and poorly defined mixtures. For example, morphine in combination with other epidural anesthetics is used in hospital settings for pain relief. The mixture can be characterized as simple (<10 chemicals), and in some cases such mixtures can be well defined because it is easier to identify the chemicals involved and know their dose, toxicity, and potential interactions. In contrast, complex mixtures are composed of many (>10) chemicals. Their composition may be largely known or at least reproducible (e.g., fumes from a specific paint), or the mixture may not be fully characterized either qualitatively or quantitatively, and may even vary from one similar exposure scenario to another (e.g., diesel fuels or gasoline from different sources). The sheer number of interactions means that the complete set of individual and interactive effects can never be established. Such cases thus present a particular challenge, and Chapter 4 discusses tools available to assessing the toxicity of such complex mixtures.

In the prioritization of chemicals for setting scenarios that are the most realistic and probable for chemical mixtures, tools are needed to overcome the almost infinite number of chemicals or combinations of chemicals and of differing concentrations in emissions. The global production of chemicals has increased from 1 million tons in 1930 to 400 million tons today. There are about 100,000 different substances registered in the EU market, of which 10,000 are marketed in volumes of more than 10 tons (produced per manufacturer or imported per importer per annum), and a further 20,000 are marketed at 1 to 10 tons (EC 2001). In the selection of chemicals for mixture scenarios, the volume of marketing is an important parameter; however, other factors, such as emission pattern (spatial and temporal), degradation, and toxicity, are also very important.

1.2.4 VALIDATION STUDIES

In many countries, emissions are determined by direct monitoring. In these cases the measurements should be subject to quality assessment, and the sampling plan should be evaluated to estimate the uncertainty in all steps of the procedure. To assess emissions of mixtures of chemicals, concentrations of chemicals should be measured simultaneously, but the validation procedures would be the same as for single chemicals. When emissions are estimated from data on produced, processed, or used amounts of the individual chemicals in mixtures, the calculations should be validated by measurements in the field.

1.3 INTERACTIONS AFFECTING AVAILABILITY AND EXPOSURE TO CHEMICAL MIXTURES

1.3.1 CHARACTERISTICS OF THE MAJOR ENVIRONMENTAL COMPARTMENTS

Identification of the most probable mixture exposure scenarios is a first step to establishing the nature of multiple chemical exposures, but even when such information is available, the interaction of the chemical constituents of the mixture with the environment needs to be considered. The physical and chemical nature of different environmental compartments has a large influence on the magnitude, duration, and stability of exposure (Table 1.2). On the one hand, chemicals in air are highly mobile and can travel over long distances, but also can be subject to rapid dilution through mixing. On the other hand, soils and sediments are immobile, and so the chemical released into these media can remain patchily distributed and dilution through mixing (e.g., by bioturbation) occurs at a very slow rate. In soils, sediments, and waters, the chemical properties of the media (such as pH, percentage of total and dissolved organic matter, cation exchange capacity, and concentrations of some ionic species) can have a large

Table 1.2   Physical characteristics influencing the duration and magnitude of exposure for the major environmental compartments

Media or compartment	Characteristics of the exposure
Soil	Immobile; hard to dilute; exposure, especially to persistent compounds, can be temporally quite constant, but spatially patchy; soil properties (pH, organic matter content, cat ion exchange capacity) have large effects on exposure (bioavailability) and can show spatial and temporal variability
Sediment	Immobile; hard to dilute; exposure, especially to persistent compounds, can be temporally quite constant, but spatially patchy; sediment characteristics can affect exposure; anoxia common and can affect exposure
Water	Mobile; can be diluted; pollutants can disperse within water column over moderate distances, pulsed exposure common; chemical characteristics can affect exposure, although these are less variable than, for example, soil
Air	Highly mobile, so pollutants can travel long distances (transboundary issues arise); composition quite stable at given altitude, and so local effect on bioavailability not such an important issue; rapid dissolution through mixing can occur; pulsed exposure common
Food chain for higher organism	The physiology of the species involved (both predator and prey) strongly influences the nature of exposure; exact exposure of an individual depends on the home range and dietary composition of the species or individual involved

influence on chemical availability. In ecotoxicology, a great deal of research effort has been focused on understanding the relationships between chemicals and soil, sediment, and water properties. The models and methods developed from this work are likely to be broadly applicable to exposure assessments for mixtures, although work is needed to validate this assumption. The importance of food chain transfer as an exposure route is greatly influenced by the properties of the chemical involved (e.g., persistence) and the physiology of the species within the particular food web. For example, Hendriks et al. (2001) modeled bioaccumulation of organic chemicals and Hendriks and Heikens (2001) modeled metal accumulation and demonstrated the need to include both chemical properties (log K_{ow} in case of organic chemicals) and species characteristics (body size, metabolic rate, trophic position, route of exposure) in order to allow for an effective assessment of bioaccumulation.

1.3.2   Environmental Fate Affecting Mixture Composition

1.3.2.1   Single Compounds as Chemical Mixtures

Once released into the environment, selected single compounds can be subject to a range of transformations that may change the chemical identity of part or the entire amount of the original chemical released. Thus over time, even release of a single compound will result in the presence of a chemical mixture. The most obvious example of cases where a single chemical release results ultimately in the presence of a complex and changing mixture is in organic molecule (especially pesticide and biocide) degradation. There is an extensive literature that describes in detail many of the catabolic processes that can transform organic molecules in the environment, and it is not realistic to include anything other than a sample of such information.

When considering how such degradation affects exposure and potential effect, it is obviously tempting to assume that degradation of biologically active molecules such as pesticides will result in a lower toxicity of the metabolite-parent compound mixture than for the parent compound alone. While this is often the case, it certainly cannot be seen as a universal law. Thus, dichlorodiphenyl trichloroethane (DDT), which is already a highly persistent and bioaccumulative insecticide, is converted in organisms into the even more stable metabolites dichlorodiphenyl dichloroethylene (DDE) and dichlorodiphenyl dichloroethane (DDD). Thus, for this compound, the metabolite can represent an important component of the resulting mixture. The same principles apply for the pesticide aldrin, which is degraded in soil by microbial epoxidation to form the more persistent dieldrin (O'Halloran 2006). Such effects can be media specific depending on the prevalent processes. Thus, the organophosphate pesticide chlorpyrifos is degraded in air by reaction with $OH^\bullet$ radicals to form different products, including chlorpyrifos oxon, which has a higher toxicity than the parent compound. The oxon may subsequently be converted into other products by side chain oxidations. In water, soil, and sediment, degradation of chlorpyrifos by photolysis or hydrolysis results in the formation of 3,5,6-trichloro-pyridinol (TCP). The importance of the metabolite as a component of the mixture depends on the speed at which conversion reactions occur. In the case of water, soil, or sediment, hydrolysis of the oxon metabolite leading to the formation of TCP is expected to take place faster than that of the thiol (parent) molecule, meaning that the metabolite is only likely to be present at low concentrations, and so have only a limited contribution to overall toxicity compared to other components (Cahill et al. 2003).

As well as organic molecules, metals can also be the subject of chemical and biological transformations that can change chemical identity and result in altered toxicity. The most well-known and well-studied example of this is the conversion of mercury (Hg^0) into methylated and ionic species (such as Hg^{2+}, CH_3Hg^+, and $(CH_3)_2Hg$). Mercury is emitted to the environment from many different sources, usually as phenyl mercury, metallic mercury, and bivalent inorganic mercury. In anoxic sediments, the dominating form is sulfide (HgS), which is almost insoluble in water. Different bacteria (especially methane bacteria) as well as several species of fungi are capable of forming methyl mercury (CH_3Hg^+) and dimethyl mercury ($(CH_3)_2Hg$), which are soluble in water (Regnell 1994; King et al. 2002; Bisinoti and Jardim 2003).

These examples show that also in the case of single chemicals, such as pesticides and metals, exposure assessment should not only focus on the parent chemical but also include the metabolites and transformation products produced either in the environment or upon biotransformation in the organism.

1.3.2.2 Chemical Fate Effects on Mixture Composition

Once a mixture of chemicals is introduced into the environment, processes may take place that affect the composition of the mixture in space and time (see, e.g., Foster et al. 2005; Haws et al. 2006). When (complex) mixtures are released, differing fate properties of the individual chemicals can lead to a change in the exposure over time, and exposure in different environmental compartments and through different routes. An example is the distribution of gasoline over different environmental compartments:

depending on the compartment of first introduction and the properties of the different (groups of) components in the gasoline mixture, their distribution over the environment may differ (Foster et al. 2005). Properties of single chemicals determining their distribution and fate in the environment include physical–chemical properties, such as molar mass, boiling point, density, vapor pressure, water solubility, Henry's law constant, partition coefficients like the octanol–water partition coefficient (K_{ow}) and the octanol–air partition coefficient (K_{oa}), and degradation half-lives in different environmental compartments (air, water, soil, and sediment). In addition, bioaccumulation factors may be needed to determine partitioning in biological compartments (Foster et al. 2005).

The residence time of a compound in a mixture is not only determined by its fugacity (Mackay et al. 1992a), but also by its susceptibility for degradation. Persistency (usually indicated by its half-life) not only depends on the properties of the compound, but also on those of the environmental compartment (Haws et al. 2006). Since persistency of most organic compounds in soil, water, and sediment mainly relies on biological activity, it is important to focus on processes affecting biodegradation. It is assumed that biodegradation only takes place in the bulk aqueous phase of soil (or sediment). Processes affecting the sorption of chemicals (see above) determine the availability of chemicals in solution, and therefore play a key role in biodegradation. In addition, biological factors, such as microbial abundance and activity, and affinity for the contaminant, determine biodegradation rate. Such interactions can have both negative and positive effects on the degradation rates for particular compounds.

Two types of interaction with the potential to negatively affect degradation rates of chemicals in mixtures compared to single chemicals are competition and toxicity. When substrates are present, this can lead to the inhibition of biodegradation simply because more substrate molecules are competing for the same (and limited) number of active enzyme sites. This inhibition can be competitive in case of homologous chemicals, but also noncompetitive when 2 chemicals independently bind to the same enzyme, leading to a reduction of its overall utilization rate. It is also possible that the second chemical (inhibitor) binds to the enzyme complex but not to the free enzyme (Haws et al. 2006).[1] Toxicity to the microbial community can change biodegradation potential in cases where the species responsible for the catabolism of one compound are sensitive to the presence of another that is present in a mixture. Such interactions can potentially affect the ratio at which metabolites are formed or change the nature of degradation pathways (Haws et al. 2006).

As well as inhibition, some interactions in mixtures have the potential to stimulate degradation processes for a particular chemical. One example of an increased biodegradation rate in a mixture is as a result of increased biomass growth. This can occur, for example, when an easily metabolized substrate is present in conjunction with a more recalcitrant chemical. The rapid breakdown of the easily metabolized chemical supports the expansion of the degradative community, and this larger community is then better able to metabolize the less readily catabolized compound. This

---

[1] Similar principles of competitive, noncompetitive, and uncompetitive inhibition of metabolism will also be discussed in Chapter 2 in relation to toxicokinetics.

was the case with a mixture of naphthalene, phenanthrene, and pyrene. Compared to single compound studies, the mineralization rate of the easily degraded naphthalene was decreased compared to rates found for the single chemical alone, while that of the more persistent phenanthrene and pyrene was increased (Guha et al. 1999). The presence of solubilizing agents, such as biosurfactants, is a further means by which the degradation of chemicals present in combination with such chemicals may be increased. This effect has been shown for a number of PAHs (Guha et al. 1998). These biosurfactants, which enhance the bioavailability of the PAH, may, however, be toxic to the microorganisms responsible for PAH biodegradation, leading to a reduced degradation rate. This has been shown for phenanthrene (Shin et al. 2005) and demonstrates the potential complexities of the interactions that may occur.

1.3.3 AVAILABILITY

1.3.3.1 Availability and Bioavailability

Bioavailability is an important issue, not only regarding the transfer of contaminants within food chains, but also for a robust effect assessment of chemical mixtures in the environment. Pathways and mechanisms on how chemicals interact with each other within a mixture, how they enter the organism, and how they accumulate under defined circumstances need to be unraveled (CSTEE 2000). It is generally accepted that the total concentration of chemicals and chemical mixtures in all environmental compartments is not enough to predict biological or ecosystem effects. Bioavailability can be defined as the fraction of a chemical compound in a specific environmental compartment that, within a given time span, is or can be made available for uptake by organisms or is made available at the site of physiological activity (cf. Peijnenburg and Jager 2003). Within the same scope other words are often used in human health issues, such as availability and bioaccessibility; the latter one usually is defined as the fraction of a chemical that is capable of being used by a living organism. These 3 concepts can be considered synonyms, and from now on in this section the word "bioavailability" is used.

For mixtures, like for single chemicals, bioavailability is a key factor governing the magnitude and duration of exposure. Processes affecting the sorption of chemicals in, for example, soils and waters are vital in determining the availability of chemicals that enter the solution phase. Physical and chemical characteristics of the sorbent (particle size, organic matter content, pH) and sorbates (K_{ow}, ionization) are the major determinants of the bioavailability of single chemicals. Additionally, in cases where mixtures of chemicals are found, the presence of other chemicals may affect bioavailability (Haws et al. 2006). Competitive sorption especially may occur when chemicals have an affinity for similar sorption sites. This has been reported for phenanthrene and pyrene, with phenanthrene desorption increasing in the presence of pyrene (White and Pignatello 1999). Also, in case of metal mixtures, competitive sorption may occur: nickel sorption to soils decreased when other cations (including H^+) were present and was highest at neutral pH (Staunton 2004). And desorption of cadmium was increased by the presence of zinc, but zinc desorption was not increased by the presence of cadmium (Van Gestel and Hensbergen 1997).

In principle, sorption of chemicals in a mixture can be predicted from isotherms for sorption of the single chemicals, but such predictions may become less accurate for mixtures with significant interactions of the chemicals (Haws et al. 2006). Another aspect hampering a proper prediction of the sorption of single chemicals, as well as of mixtures, is the fact that sorption may change with time. Due to the process of aging, sorption of organic chemicals increases with time and is stronger than expected on the basis of physical–chemical properties (such as K_{ow}) or laboratory sorption experiments (Alexander 1995; Hatzinger and Alexander 1995; Kelsey and Alexander 1997). This also is the case for metals (Smit and Van Gestel 1998; Lock and Janssen 2003). Also, in case of changing redox conditions, sorption, of especially metals may no longer be directly predictable, due to the formation of sulfides that have a very low solubility (Lee and Lee 2005).

Although the effects of chemical mixtures in the environment on humans are thought to have important health implications, they are not widely studied or especially well understood. Their potential risks, therefore, have to be considered as a crucial environmental health problem requiring clear and rigorous future investigation. Examples pertaining to human exposure to chemical mixtures can be associated with emissions from point sources and diffuse and mobile sources, although because they often represent the most severe absolute exposure concentrations, the majority of published work is focused on local scale studies associated with industrial and mining releases. One such example is the study of Pereira et al. (2004) in an abandoned mine in the southeast of Portugal (S. Domingos mine) on integrated human and environmental risk assessment. Hair samples from the scalp were analyzed for metal presence (As, Cd, Cr, Cu, Mn, and Zn) in the human population living nearby the mine. High concentrations of Cd, Cu, and As were recorded in individuals living near the mine compared to individuals that live apart. The concentrations reported in the hair of this group of people were above the reference values. It was also concluded in this study that metal concentrations were related to concentrations in soil, probably related to the consumption of milk and cheese from cattle from the region. In a further local scale study around an industrial facility, Cui et al. (2005) showed that Cd intake through vegetables was higher in samples from one village closer to a factory (500 m) than in samples from another village located at 1500 m. Cd concentration in urine and serum from residents was, however, much lower in the village closer to the factory. The authors suggested that the possible reason for this was the higher intake of Fe, Ca, and Pb from vegetables of the far-away village, suggesting also that high intake of these ions could lead to a decrease in Cd body burdens. This therefore indicates clearly that, like for environmental species, it is not the absolute concentration of a chemical that is present in the exposure medium, but rather the concentration that is actually available for uptake that is the primary determinant of exposure.

Oomen et al. (2003) defined 4 steps in the oral bioavailability of chemicals present in contaminated soils to man: soil ingestion, mobilization from soil during digestion (i.e., bioaccessibility), absorption from the intestinal lumen, and first-pass effect. An in vitro model of the human digestive system was used to study the uptake of chemicals from ingested soil. When an artificial soil, spiked with a mixture of polychlorinated biphenyls (PCBs) and lindane, was introduced into the system, approximately 35% of PCBs and 57% of the lindane appeared to be accessible. Bioaccessibility was explained from the distribution of these chemicals over the soil solid phase, bile salt micelles, and proteins (Oomen et al. 2000). It turned out that the first step in the uptake process, the mobilization of the chemicals from soil, was most important in determining the uptake flux, and therefore the bioavailability of these contaminants in ingested soil particles (Oomen et al. 2001). Other studies demonstrated that from an artificial soil dosed at 530 mg Pb/kg, only 23% of the Pb was bioaccessible, and that only a part of the bioaccessible soil-borne lead was actually taken up (Oomen et al. 2003). These studies demonstrate that the composition of a mixture of chemicals originally present in ingested soil particles may change in the course of the uptake process, affecting exposure.

This example suggests that organisms may also affect the sorption equilibrium of chemicals in soil or sediment. Another example of that was found in the uptake of cadmium in *Hyallela azteca*. In water-only exposure, cadmium uptake and toxicity in this organism were not affected by phenanthrene. When exposed in sediment, however, cadmium uptake was increased when the animals were simultaneously exposed to phenanthrene. This most likely is a result of associated feeding, causing alterations in ingestion or digestive processes, leading to an increased cadmium uptake. So, from a toxicological point of view, the synergism seen cannot be explained from an interaction in the animals but rather is exposure related (Gust and Fleeger 2005).

1.3.3.2 Influence of Medium Physical–Chemical Properties on Chemical Availability

In case of metal mixtures in soils and sediments, bioavailability is controlled by the strength of binding of metal ions to the soil or sediment particles. The partitioning of metals to the soil or sediment reduces the availability for mobilization and uptake by plants, animals, and microbes. The binding strength depends highly on properties that control the partitioning process, such as pH, dissolved organic matter (DOM), and other organic ligands, calcium, organic matter content, inorganic ligands, and the solid-phase metal oxide (Allen 2002).

The partitioning of metal ions can be modified in the digestive track of soil-dwelling organisms or in the rhizosphere. Depending on the organism, bioavailability can be related to several key factors. In plants and microorganisms, it has been correlated to the activity of the free metal ion and its presence and diffusion in the soil pore water. Bioavailability of metals in invertebrates can be influenced by the organic matter content that is digested in their guts, and might be physiologically moderated by pH and competing cations (e.g., Ca^{2+}).

Soil or water pH is one of the most important parameters when dealing with metal bioavailability. Also for ionizable organic chemicals, like chlorophenols, pH may affect bioavailability. Changes of conditions in environmental compartments may depend on their surrounding environments and climatic changes, and pH is one of those properties that can show larger changes.

In aquatic systems, in addition to the complexation of metal ions by natural organic matter, metal bioavailability, bioaccumulation, and toxicity are highly affected by water hardness and alkalinity (Banks et al. 2003). This is also applicable to metal mixtures where complexation of metals can occur even at higher rates than when single chemical compounds are present.

Bioavailability of organic chemicals is strongly dependent on aqueous solubility. The equilibrium partitioning theory has been applied to sediment toxicity studies, and it was concluded that uptake from sediment as well as from (pore) water is possible at the same time; however, the exposure route in equilibrium is not necessarily important. For substances with log $K_{ow} < 5$, the equilibrium partitioning theory is considered acceptable to assess the risk. For substances with log $K_{ow} > 5$, a safety factor of 10 is applied, in order to include the additional uptake by sediment ingestion (Loonen et al. 1997).

Bioavailability can vary with the contact time of chemicals with soil or sediment particle constituents. Chemicals in newly deposited sediments may become more bioavailable than older buried materials. In soils the same can happen. In a study where a 70-year-old and a freshly copper-contaminated soil were compared, in the soil contaminated 70 years ago no copper toxicity to *Folsomia fimetaria* was observed for concentrations of copper as high as 2911 mg/kg. The newly spiked soil, however, caused a 10% decrease in reproduction at 337 mg Cu/kg (Scott Fordsmand et al. 2000). In the work of Smit and Van Gestel (1998), aging was also shown to be important to understanding the toxicity of Zn to *F. candida*. Within time soil pH can change, which also induces a change of Zn toxicity by altering Zn sorption to soil particles (e.g., an increase of soil pH leads to an increased Zn sorption). For chemical mixtures the same may happen due to the effect of soil characteristics on chemical adsorption to soil particles and also to chemical interactions through aging. For organic chemicals, it was suggested that molecules slowly become sequestered within the soil matrix and therefore become less available to organisms. Besides abiotic parameters, biotic interactions also may play a role. Bacteria, fungi, and soil invertebrates may alter the behavior of chemicals in soils, altering also their persistence and sorption to soil particles. These facts are crucial when evaluating soil toxicity and risk assessment for decision makers and cleanup purposes (Alexander 1995).

When exposed to mixtures, chemicals in the exposure medium may affect each other's uptake by humans in a manner that is analogous to some of the bioavailability effects outlined here for environmental species. This was, for instance, shown for the neurotoxicity of EPN (O-ethyl-O-4-nitrophenyl phenylphosphonothionate), which was enhanced by aliphatic hydrocarbons due in part to increased dermal absorption (Abou-Donia et al. 1985). It was also shown that dietary zinc inhibits some aspects of lead toxicity, which could in part be explained by decreasing dietary lead absorption (Cerklewski and Forbes 1976). Other examples of interactions of chemicals at the uptake phase in humans, which may in part be related to bioavailability interactions, are summarized in Table 1.3.

1.3.3.3 Metal Speciation Determines Bioavailability

To understand chemical exposure and bioavailability the biotic ligand model (BLM) was developed for single (cationic) metal species, assuming that the amount of a

Table 1.3   Examples of interactions of chemicals at the uptake and absorption level in humans

Antimicrobial agent	Other drug	Mechanism
Decreased absorption		
Lincomycin	Kaolin-pectin	Irreversible adsorption to kaolin-pectin
Lincomycin	Cyclamate	Possibly complexation
Ritampicin	PAS "granulates"	Adsorption to bentonite granules
Tetracycline	Sodium bicarbonate, cimetidine, bi- and trivalent metal cat ions (including anatides), atropine	Alkaline pH inhibits dissolution, probably reduced dissolution, chelation, slowed gut peristalsis
Pivampicillin PAS	Diphenhydramine	Slowed gut peristalsis
Neomycin	Digoxin	Induced malabsorption
Neomycin	Warfarin	Induced malabsorption
Neomycin	Penicillin V	Induced malabsorption
Increased absorption		
Tetracycline	Metoclopramine	Increased gut peristalsis
Pivampicillin	Metoclopramine	Increased gut peristalsis

Source: Based on Calabrese, EJ. 1991. Multiple Chemical interactions, Part 4: Drugs; Part 5: The drug–pollutant interface. Chelsea (MI): Lewis Publishers, p 389–578.

metal binding to a sensitive biological ligand determines its toxicity. This fraction binding to the biological membrane is considered the bioavailable fraction. In addition to metals, other cationic elements may bind to the same target sites on the biological membrane. On the other hand, concentration and activity of free metal ions is determined by the presence of organic and inorganic ligands. The BLM incorporates the competition and affinity to the site of toxic action on the organism of the free metal ion, other naturally occurring cations, and also the possible complexation by abiotic ligands (e.g., dissolved organic matter, chloride, carbonates, sulfide). On this basis, the model includes ion uptake pathways that can quantify chemical characteristics like metal affinity and capacity in vivo. In general, the greater the affinity to the binding sites, the higher the toxicity of a particular metal. The BLM was also applied to mixtures to evaluate how it would respond, using the classic toxic unit's concept of additivity. Ion competition has been included from the beginning in the BLM because cat ions like Ca^{2+} and H^+ are known to decrease metal accumulation at the ligand (Playle 2004).

As an example, in the study of Sanchez-Dardon et al. (1999) exposure of the rainbow trout (*Oncorhynchus mykiss*) to Zn decreased Hg and Cd toxicity, probably by competition at the entry sites to gills' cells, through competition at intracellular binding sites or through induced synthesis of metallothionein. Another example is with *Daphnia magna* exposed to Zn that increased their tolerance toward Cd, probably through a competition at the ligand site in the gut or through an induction of metallothionein synthesis (Barata et al. 2002). These studies are 2 examples of possible competition between ions to bind in the uptake and intracellular sites in organisms. See also Chapter 2 for a discussion of BLM-type interactions in relation to toxicokinetics and toxicodynamics.

1.3.3.4 Species Specificity

Bioavailability is also considered species and organ or tissue specific because what is available to one species might not be the same for the other, and the same counts for different organs or tissues inside a living organism. Barahona et al. (2005) found that physiological differences among oat, wheat, and sunflower roots, such as root waxes, might be responsible for differences in permeability for nonpolar compounds and therefore for differences in sensitivity of the root elongation process.

Another example for species-specific differences in bioavailability of chemical mixtures is the study of Loureiro et al. (2005), where avoidance behavior of earthworms and isopods was studied in 2 soils from the vicinities of an abandoned gold mine. Isopods (*Porcellionides pruinosis*) turned out to show avoidance behavior, whereas earthworms (*Eisenia andrei*) did not, suggesting isopods to be more sensitive than earthworms for these soils that contained a mixture of metals. It remains unclear which factors explain this difference, but routes of exposure and bioavailability difference may have played a role.

Since bioavailability is an integral factor in the estimation of the internal dose (or dose at target tissue) of the chemical, it is important in human studies to consider the environmental and physiological characteristics of uptake sites and their interactions since these are clearly important for defining the extent of exposure. When exposure is through food, the gastrointestinal tract and its physiology has an important effect on the amount of different chemicals that are taken up. It has been suggested, for example, that gut uptake of 2,3,7,8-tetrachlorodibenzodioxin (TCDD) and related compounds is variable, incomplete, and congener and vehicle specific, and that more lipid-soluble congeners, such as 2,3,7,8-tetrachlorodibenzofuran, are almost completely absorbed, while the extremely insoluble octachlorodibenzodioxin is less well absorbed, depending on the dosing regimen. The fact that high doses may be absorbed at a lower rate, whereas low repetitive doses may be absorbed at a greater rate, thus has the potential to alter internal exposure from the assumptions based on external exposure concentrations. To date, the only study of TCDD bioavailability in humans was reported by Poiger and Schlatter (1986) and was based on a single male in which the gastrointestinal absorption was >87% when TCDD was administered in corn oil. Laboratory data suggest that there are no major interspecies differences in the gastrointestinal absorption of dioxins and dibenzofurans. However, absorption of TCDD is dependent on conditions and characteristics of the soil medium. In animals, absorption of TCDD from different soils ranged from 0.5% (Umbreit et al. 1986a, 1986b) to 50% (Lucier et al. 1986). Absorption from a diet was 50 to 60% in rats (Fries and Marrow 1992). Therefore, exposure with food rather than with oil as a vehicle relates more closely to exposure from soil. Bioavailability has to be considered when calculating the hypothetical ingestion dose.

1.3.3.5 Formulating Agents

Another important issue on the bioavailability of chemical mixtures is the use of formulating agents for several purposes (e.g., agriculture, veterinary, human health). Some studies have been performed to compare the toxicity of pure compounds and formulations. Azadirachtin, a plant-derived extract, is often used as a repellent, anti-feedant, and molt regulator for several insect species and can be applied as several formulations, like Azantin-EC, Bioneem, or Neemix. When exposed to the pure compound and 2 formulations, the cladoceran *Daphnia pulex* showed clear differences in sensitivity. LC50 values for the formulations Bioneem and Neemix were 0.07 and 0.03 µg/L, respectively, while a 50- to 100-fold higher value was found for the toxicity of the pure compound (0.382 µg/L). The higher toxicity of both formulations suggests that azadirachtin is not the only active compound in the formulations, or that the inert ingredients significantly enhanced the toxicity of azadirachtin. However, it remains unclear whether this is due to a higher bioavailability in the formulation, although it is suggested that the formulation may also affect fate (sorption, biodegradation) of azadirachtin (Goktepe and Plhak 2002).

Garcia-Ortega et al. (2006) reported conflicting results for the fate and effect of pure (Pestanal) propetamphos and its formulation Ectomort Centenary (8% active ingredient) in sediments. Sorption of propetamphos to sediments was stronger in the formulation. Biodegradation rate decreased with increasing sorption but was not affected by the formulation. Toxicity to sediment microbial communities was, however, significantly greater for the commercial formulation than for the pure compound. This cannot be explained from a higher availability of propetamphos in the formulation. Garcia-Ortega et al. (2006) therefore assume that ingredients of the formulation enhance toxicity of the active ingredient.

These studies do not confirm the effect of formulations on the bioavailability of compounds, but they do emphasize the need to test formulations (next to or instead of active ingredients) when performing a risk assessment of commercial pesticides (Garcia-Ortega et al. 2006).

1.3.3.6 Analytical–Chemical Procedures

Several methodologies have been developed for bioavailability assessment using chemical-analytical procedures. To determine the bioavailable fraction of metals or organic compounds in soils and sediments, several methods can be applied, such as extraction techniques. Usually desorption procedures are applied, using an extracting agent to desorb chemicals from the soil or sediment solid phase. An aqueous extraction might be considered the simplest alternative, although with some limitations. Deionized water has a very low ionic strength compared to natural (pore) water. That is why a weak salt extraction, for example, using 0.01 M $CaCl_2$ (Houba et al. 1996), is sometimes preferred, because it better simulates the soil solution. Such water and neutral salt extractions can be used to determine partition coefficients (K_d).

Specific extractants and sequential fractionation are also widely used procedures to estimate metal or nutrient availability to plants (see, e.g., Houba et al. 1996). Several extractants can be considered, like diethylene triamine penta acetate (DTPA), ethylene diamine tetra acetic acid (EDTA), acetic acid, HNO_3, HCl, or other mineral acids (Allen 2002). Each reagent might be considered specific for extracting a certain fraction of metals. As an example, EDTA and DTPA are often used as extractants of exchangeable and organically bound trace metals and also to dissolve metal precipitates. These procedures can also be considered as chelating extractions, showing a correlation between water and total digestion extractions. Another technique also used for the identification of the solid phase associated with metals is an x-ray absorption fine structure methodology (EXAFS) (Manceau et al. 2003).

Recently, attempts have been made to develop biomimetic methods, simulating plant uptake of metals. An example of such a method is DGT (diffusive gradients in thin films), developed by Zhang et al. (2001), for measuring metal availability to plants. In this case, metal accumulation in a chelex layer is measured. By taking into account thickness of the diffusive layer covering the chelex layer and contact time with the soil sample, it is possible to estimate the available metal concentration in the soil solution. The DGT method may also be used to estimate metal speciation in surface water (Zhang 2004).

Similar biomimetic methods have been developed for assessing the (bio)availability of organic chemicals in water, sediments, and soil (Mayer et al. 2003; Ter Laak et al. 2006). The main advantage of these nondepletion techniques, such as the solid-phase microextraction methods (SPMEs), is that they may be used without disturbing the chemical distribution in surface water or soils or sediments.

1.3.4 CHEMICAL–CHEMICAL INTERACTIONS IN MIXTURES

Another form of interaction that may change the nature of exposure for chemicals in mixtures is through direct chemical-chemical interaction. One example is the formation of nitrosamines (which are carcinogenic) from noncarcinogenic nitrates and amines in the stomach (Klaassen 1996). The formation of N-nitrosoatrazine from atrazine and nitrite has been demonstrated in vitro in human gastric juice (pH 1.5 to 2.0) during 1.5 to 12 hours of incubation at 37 °C (Cova et al. 1996). The percent formation peaked at 3 hours, and gradually declined thereafter, due to degradation of N-nitrosoatrazine to atrazine. Peak formation of N-nitrosoatrazine was 2% from 0.05 mM atrazine and 0.5 mM nitrite, 23% from 0.05 mM atrazine and 3 mM nitrite, and 53% from 1 mM atrazine and 3 mM nitrite. The formation of N-nitrosoatrazine from atrazine and nitrite also has been demonstrated in vivo, the amount formed being dependent on the ratio of atrazine and nitrite concentrations and on pH (Krull et al. 1980).

1.4   ENVIRONMENTAL FATE MODELING

Environmental fate models make use of chemical properties to describe transfer, partitioning, and degradation (Mackay et al. 1992a; Cahill et al. 2003). For organic chemicals, quantitative structure-property relationships (QSPRs) may be used to predict partitioning from physical–chemical properties, such as K_{ow} and K_{oa}. Such properties may also allow for a prediction of the transfer of chemicals between compartments. Recently, some successful attempts have also been made to predict persistency of chemicals (Raymond et al. 2001), although this mainly concerns degradation rate under standardized conditions (Posthumus et al. 2005). But in most cases, specific knowledge on degradation pathways is required to predict the formation of metabolites. OECD (2004) presented an overview of multimedia fate models to predict overall environmental persistence (P_{ov}) and the potential for long-range transport (LRTP) of organic chemicals. P_{ov} and LRTP derive from both chemical properties and environmental conditions. Multimedia fate models can be used during exposure assessment to identify spatial extent of exposure, environmental partitioning (media of concern), and residence time of chemicals in a certain environmental compartment. Four levels of model complexity were identified, ranging from closed systems at equilibrium to dynamic open systems. OECD (2004) identifies generic multimedia models, region-specific multimedia fate models, and multizone multimedia models (Figure 1.1). These models may be helpful in describing distribution on a global scale (Toose et al. 2004), on a regional scale (Mackay et al. 1992a), or on a smaller scale, such as in surface water (e.g., the Exposure Analysis Modeling System (EXAMS); Schramm 1990).

The input required by multimedia fate models includes properties of the chemicals (such as distribution over compartments air, water, and soil or sediment), properties of the environment or landscape receiving the contaminants, and emission patterns and mode of entry of chemicals into the environment (OECD 2004) (Figure 1.1). Fenner et al. (2005) compared the outcome of 9 multimedia fate models by applying them to a set of 3175 hypothetical chemicals covering a range of 25 half-life combinations (in water, air, or soil or sediment) and 127 combinations of partition coefficients (air-water (K_{aw}), K_{ow}, and K_{oa}). Results show great similarities between the model outputs for P_{ov} predictions, but less for LRTP. P_{ov} and, to a lesser extent,

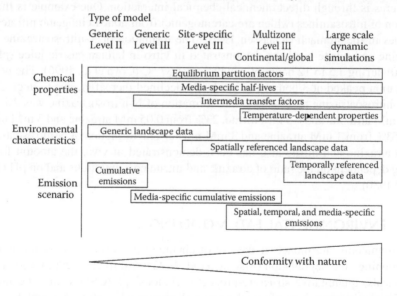

Figure 1.1 The continuum of multimedia fate models available for estimating overall persistence (P_{ov}) and potential for long-range transport of chemicals (Redrawn from OECD [2004]).

LRTP depend mainly on chemical properties. Models show significant differences in certain regions of the chemical space, with model uncertainty being higher than parameter uncertainty in case of LRTP predictions, and vice versa for P_{ov}. On the basis of this analysis, Fenner et al. (2005) conclude that it always is best to not rely on just 1 model but apply more than 1 model. Model selection is not an arbitrary task but requires careful consideration of the question and context of the assessment. Fenner et al. (2005) therefore provide guidance for selecting the most suitable model for a certain task.

So far, the main focus of multimedia fate models has been on single chemicals, but extensions may become available to include fate of transformation products. This may open the way to making the models applicable to mixtures (OECD 2004). Initially such development may simply be made through the serial analysis of the fate of individual chemicals, and from this a derivation of probable concentrations of each, assuming no interaction. Such analysis is, for example, feasible for many of the most widely used "down the drain" and is at present being extended to other product types, such as personal care chemicals and human pharmaceuticals. Such combined analysis would in fact represent a considerable step forward in addressing the nature of likely mixture exposures; however, if the interactions with the environment and between chemicals as outlined above are to be considered, then this would require a considerable effort to understand and include the major processes involved within existing models.

1.5 EXPOSURE SCENARIOS AND MONITORING

An important issue in pollution management is the identification of exposure scenarios that pose relatively high risks to humans or the environment (Thomsen et al. 2006). The regulation of these high-risk scenarios should be given priority in order to realize an effective reduction of the overall risks. An exposure scenario can be broadly defined as a set of parameters that together determine the risk level. This may include parameters such as the emission load, the emission compartment, fate parameters (persistence, biodegradability, vapor pressure), and toxic properties. The exposure scenario is thus the set of parameter values that needs to be known as a condition to estimate the risk. Thomsen et al. (2006) have described a new methodology for the identification of exposure scenarios that cause high risks for the environment or human health. They define a scenario type as the set of criteria that is considered important after analysis of a problem tree. A criterion has to be described by data, and thus empirical knowledge, before it can be taken into account as a condition for scenario selection. A data type that describes a specific criterion is denoted a descriptor. A specific scenario is defined as a combination of realistic descriptor values. In this procedure, descriptors for emission are consulted (descriptors for different product groups), as this can be seen as first screening for potential mixtures of relevance for further studies.

According to this new system, a harmful mixture consists of substances that together reach the target on 1 side and also have a toxic action that has potential to yield harmful effects. Realistic mixtures can thus be defined based on 1 of 2 principal general approaches:

1) toxicity-driven approach watching for common harmful toxicity of mixtures and thus based on the mode of action, where the combination of substances is identified that yields a high combined toxicity as a result of either
   additive or synergistic mixture toxicity mechanisms; and
2) exposure-driven approach watching for simultaneous exposure, where substances that can cause simultaneously exposure are investigated for combined toxicity.

It is not possible to claim one of these approaches as best because they are complementary to each other and a combination of them is ideal if possible. The method
takes concentration addition (CA) as the starting point for mixture toxicity, so that
the same risk descriptors can indicate toxic actions that add to each other between
different substances (Thomsen et al. 2006).

1.5.1 HUMAN EXPOSURE

The mere presence of any single chemical or chemical mixture in the environment
does not indicate that a health threat exists. An important step of mixture risk assessment is the evaluation of completed exposure pathways. Completed exposure pathways link together the source of contamination, environmental medium, point of
exposure, route of exposure, and a receptor population. It means that without the
potential for chemicals actually entering (or contacting) the human body, no threat
is present.

Personal behavior and practices also add a further level of complexity to the
estimation of individual total exposure for humans. Many publications pointed out
harmful effects of high-dose alcohol consumption, smoking, drugs, or the use of
mercurials in religious practices, just to name a few. All these chemicals contribute to overall exposures and may affect the toxicity of other chemicals entering the
human body (Calabrese 1991). Such personal exposure patterns can also be overlaid
on top of more regionally based environmental exposure resulting from diffuse environmental pollution. In this section, first, different pathways of human exposure are
considered as well as life-stage-related exposures. Second, the use of monitoring
data to assess mixture exposure is discussed.

1.5.1.1 Environmental Exposures Excluding Food

Environmental exposures are present through the human lifetime. However, they may
vary considerably over time at the same location, for example, because of the local
or global changes in emission and environmental pollution levels. Environmental
exposures of humans consist of exposures outdoors and indoors as well as at work-
places; these environments may significantly differ. The exposure media include air,
water, and soil and dust. Historically, research on human exposures to chemicals
and associated health effects has been conducted mostly on single chemicals. In
addition, several studies have dealt with complex mixtures, such as diesel fuel and
gasoline, by-products from coal combustion, and tobacco smoke. A common problem of complex mixtures is that the composition may vary from one exposure to
another and, as a result, the associated toxicity may vary. For a better understanding

of joint toxic action of chemicals and their effects on human health, it is important to identify combinations of chemicals that represent the most frequently occurring simple mixtures.

Some studies have attempted to address this need. For example, further analyses of the ATSDR data on mixtures at and around hazardous waste sites (see Section 1.2) considered completed exposure pathways (Table 1.4) (De Rosa et al. 2004; Fay 2005). The results show that the number of sites with mixtures in completed exposure pathways is lower than the number of sites for which only the frequency of co-occurrence was analyzed. Data from 1706 hazardous waste sites indicated that completed exposure pathways exist at 743 (44%) of the sites (Fay 2005). Of these, 588 had 2 or more chemicals in the completed exposure pathway. That means that exposure to mixtures occurred at 79% of the sites with exposure. As indicated in Table 1.4, mixtures of inorganic chemicals were found predominantly in soil, mixtures of organic chemicals were detected in air, and combinations of both in water.

A study of the US Geological Survey identified mixtures of chemicals in groundwater used for drinking water in the United States (Squillace et al. 2002). Samples were analyzed from 1255 domestic drinking water wells and 242 public supply wells between 1992 and 1999. Water in 11.6% samples did not meet current drinking water standards or human health criteria established by the US Environmental Protection Agency (USEPA). Volatile organic compounds (VOCs) were detected in 44% of the samples, pesticides in 38%, and nitrate in 28%. Many mixtures (i.e., possible combinations of chemicals) were found in the samples; however, only 402 mixtures were detected at least 15 times (>1% detection frequency). From all the samples, 47% contained at least 2 analyzed compounds and 33% contained at least 3 compounds. A list of the top 25 most frequently detected mixtures is shown in Table 1.5. Since the study was done on drinking water, human exposure can be assumed. In addition to these compounds in groundwater, drinking water may also contain a complex mixture of products resulting from disinfection, including trihalomethanes, haloacetic acids, haloacetonitriles, and bromate (Teuschler et al. 2000).

The USEPA's Total Exposure Assessment Methodology (TEAM) studies found levels of about a dozen common organic pollutants to be 2 to 5 times higher inside homes than outside, regardless of whether the homes were located in rural or highly industrial areas (USEPA 2006b). Evidence for the contribution of indoor air pollution to human exposure to mixtures of chemicals has also been obtained for other areas in the world, for instance, through the project Towards Healthy Air in Dwellings in Europe (THADE), which was sponsored by the European Union (Franchi et al. 2006). While the consumers are using products containing organic and inorganic chemicals indoors, not only can they expose themselves to high chemical concentrations, but increased concentrations can also persist in the air long after the activity is completed. Among the chemicals often found inside are carbon monoxide, nitrogen dioxide, formaldehyde, methylene chloride, and tetrachloroethylene. Carbon monoxide is generated as a product of incomplete combustion from sources, which include home furnaces and fireplaces. Similarly, nitrogen dioxide may be found in houses with poorly vented fireplaces and furnaces. Formaldehyde is found in many products used around the house, such as antiseptics, medicines, cosmetics, dishwashing liquids, fabric softeners, shoe-care agents, carpet cleaners, glues, adhesives, lacquers,

Table 1.4 Chemical mixtures in completed exposure pathways at and around hazardous waste sites in the United States

Rank	No. sites	Binary combinations		Rank	No. sites	Binary combinations	
Water							
1	120	TCE	Perc	10	45	Chloroform	TCE
2	64	1,1,1-TCA	TCE	12	42	TCE	1,2,-DCA
3	58	1,1,-DCE	TCE	13	40	1,1,1-TCA	1,2,-DCA
4	55	Benzene	TCE	13	40	1,1,1-TCA	1,2,-DCE
5	54	TCE	Lead	13	40	TCE	Trans-1,2-DCE
6	51	1,1,1-TCA	Perc	16	39	Lead	Cadmium
7	49	1,1-DCA	TCE	16	39	Perc	Lead
8	47	1,1,-DCE	Perc	18	38	Vinyl Chloride	TCE
9	46	TCE	Toluene	19	37	1,1-DCA	Perc
10	13	Lead	Arsenic	19	37	MeCl	TCE
Soil							
1	60	Lead	Arsenic	10	34	Lead	Nickel
2	56	Lead	Chromium	12	33	Copper	Zinc
3	52	Lead	Cadmium	13	32	Arsenic	Zinc
4	47	Arsenic	Chromium	13	32	PCBs	Lead
5	46	Arsenic	Cadmium	15	30	Cadmium	Copper
6	44	Lead	Zinc	16	29	Nickel	Chromium
7	39	Cadmium	Chromium	17	28	Antimony	Arsenic
8	38	Cadmium	Zinc	17	28	Arsenic	Copper
9	36	Lead	Copper	17	28	Chromium	Copper
10	34	Chromium	Zinc	27	28	Lead	Antimony
Air							
1	18	Benzene	Toluene	11	10	1,1,1-TCA	Toluene
2	16	Benzene	TCE	11	10	1,1,1-TCA	TCE
3	15	Benzene	Perc	13	9	Benzene	Xylenes
4	15	TCE	Perc	13	9	Ethylbenzene	Perc
5	14	Benzene	Ethylbenzene	15	8	1,1,1-TCA	Perc
6	13	Ethylbenzene	Toluene	15	8	Benzene	Chlorobenzene
7	12	Benzene	1,1,1-TCA	15	8	Benzene	MeCl
7	12	TCE	Toluene	15	8	Ethylbenzene	Chlorobenzene
9	11	Toluene	Perc	15	8	TCE	Ethylbenzene
10	11	Toluene	Xylenes	20	7	1,1,1-TCA	Ethylbenzene

Source: Adapted from De Rosa CT, El-Masri HE, Rohl, H, Cibulas W, Mumtaz, MM. 2004. J. Toxicol. Environ. Health 7:339–350.

Note: Binary combinations at the 1188 sites surveyed. MeCl = methylene chloride, PCBs = polychlorinated biphenyls, Perc = perchloroethylene (tetrachloroethylene), 1,1,1-TCA = 1,1,1-trichloroethane, TCE = trichloroethylene, Trans-1,2-DCE = trans-1,2-dichloroethylene, 1,1-DCA = 1,1-dichloroethane, 1,1-DCE = 1,1-dichloroethene.

Table 1.5 Top 25 most frequently detected mixtures in groundwater used for drinking water in the United States

Rank	Compounds in mixture				No. samples (out of 1497) with mixture
1	Atrazine	Deethylatrazine			284
2	Deethylatrazine	Nitrate			214
3	Atrazine	Nitrate			198
4	Atrazine	Deethylatrazine	Nitrate		179
5	Atrazine	Simazine			138
6	Deethylatrazine	Simazine			127
7	Atrazine	Deethylatrazine	Simazine		120
8	Nitrate	Simazine			111
9	Atrazine	Metolachlor			103
10	Deethylatrazine	Metolachlor			99
11	Deethylatrazine	Trichloromethane			97
12	Atrazine	Prometon			96
13	Atrazine	Deethylatrazine	Metolachlor		95
14	Atrazine	Nitrate	Simazine		92
15	Deethylatrazine	Nitrate	Simazine		92
16	Deethylatrazine	Prometon			90
17	Atrazine	Deethylatrazine	Prometon		87
18	Nitrate	Trichloromethane			86
19	Tetrachloroethene	Trichloromethane			86
20	Atrazine	Deethylatrazine	Nitrate	Simazine	86
21	Atrazine	Trichloromethane			78
22	Metolachlor	Nitrate			76
23	Nitrate	Prometon			73
24	Deethylatrazine	Metolachlor	Nitrate		71
25	Atrazine	Metolachlor	Nitrate		70

Source: Adapted from Squillace PJ, Scott JC, Moran MJ Nolan T, Koplin DW. 2002. Environ. Sci. Technol 36:1923–1930.

plastics, and some types of wood products. Methylene chloride is widely used as an industrial solvent and as a paint stripper and can also be found in certain aerosol and pesticide products, some spray paints, and automotive cleaners. Tetrachloroethylene may be found in the home environment as a result of dry cleaning of textiles. Another important group of indoor contaminants consists of pesticides. For example, chlorpyrifos, an organophosphorus insecticide, is the most widely used insecticide for indoor and outdoor residential applications in the United States (ATSDR 1997). Based on the monitoring of different outdoor and indoor media, a study indicated that indoor dust and air were the primary exposure media for the residents (Whyatt et al. 2002). Similar results were obtained in a large group of pregnant women; the chlorpyrifos body burden that did not exhibit any seasonal variations was thought to come primarily from indoor exposures (Berkowitz et al. 2003). Metals can be also found in the households; for example, lead is generated from deteriorating lead paint and is a major concern in regards to children exposed to lead-contaminated house dust (ATSDR 1999). Further, radon can be found in many homes all over the United States (ATSDR 1998b).

1.5.1.2 Food

Another major exposure route for humans is via contaminated food. For example, North America's Great Lakes, which are the largest body of freshwater in the world, are polluted with about 362 contaminants that were found in quantifiable amounts in the water, sediment, and biota (IJC 1983; USEPA 1994). The critical pollutants were identified as PCBs, DDT, dieldrin, toxaphene, mirex, methyl mercury, benzo(a) pyrene, hexachlorobenzene, polychlorinated dibenzodioxins (PCDDs), polychlorinated dibenzofurans (PCDFs), and alkylated lead. Some of these pollutants biomagnify in the aquatic food chain and can be detected in increased levels in cooked Great Lakes fish. Consequently, the blood serum levels of these chemicals are significantly increased in consumers of contaminated Great Lakes sport fish compared to people who do not eat such fish (Humphrey 1983; Fiore et al. 1989; Sonzogni et al. 1991).

Another example is human exposure to mixtures of PCDDs, PCDFs, and PCBs. The primary route of exposure for the general population is the food supply (ATSDR 1998a). When the USEPA and US Department of Agriculture (USDA) completed the first statistically designed survey of the occurrence and concentrations of PCDDs and Fs in beef fat (Ferrario et al. 1996; Winters et al. 1996), pork fat (Lorber et al. 1997), poultry fat (Ferrario et al. 1997) and US milk supply (Lorber et al. 1998), the total TEQ values[1] were the highest for pork and the lowest for chicken and milk. This exposure results in a background body burden of about 5 ng TEQ/kg body weight for a mixture of dioxin-like PCDDS, PCDFs, and PCBs (USEPA 2000a). Analysis of fish oil dietary supplements showed exceedance of the WHO-TEQ limit of 2 ng/kg for dioxins in approximately 30% of the samples. When combined with whole diet intake, exposure to dioxin-like compounds was estimated at 1.8 to 8.9 pg WHO-TEQ per kilogram body weight per day for adults and 1.4 to 14 pg WHO-TEQ per kilogram body weight per day for children (Fernandes et al. 2006).

1.5.1.3 Human Exposure in Different Life Stages

Most environmental exposures are similar for a given population at a given location and time period; however, human exposures to chemicals also have some distinctive characteristics related to the life stages (Figure 1.2).

---

[1] Toxicity of dioxin-like chemicals is expressed in toxicity equivalents (TEQs). TEQ is defined as the product of the concentration, C_i, of an individual "dioxin-like compound" in a complex environmental mixture and the corresponding TCDD (2,3,7,8-tetrachlorodibenzo-p-dioxin) toxicity equivalency factor (TEF$_i$) for that compound. TEFs are based on congener-specific data. The TEF scheme compares the relative toxicity of individual dioxin-like compounds to that of TCDD, which had been traditionally assigned a toxicity of one (ATSDR 1998a). In 1998, the World Health Organization (WHO) released an updated system where a TEF of one was assigned not only to TCDD but also to 1,2,3,7,8-pentaCDD (Van den Berg et al. 1998).

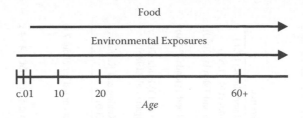

Figure 1.2 Exposures in life stages. c = conception.

1.5.1.3.1 Fetuses

In utero exposures represent the first contact of the developing organism with environmental pollutants. Less than half of all human conceptions result in the birth of a completely normal, healthy infant. For example, approximately 60% of spontaneous abortions are thought to be related to genetic, infectious, hormonal, and immunological factors (Bulletti et al. 1996). However, the role of the environment in the etiology of spontaneous abortion remains poorly understood. The placenta plays a key role in influencing fetal exposure by helping to regulate blood flow, by offering a transport barrier, and by metabolizing chemicals (Shiverick et al. 2003). However, the placenta never acts as a complete barrier; virtually any substance present in the maternal plasma is transported to some extent by the placenta. There is a distinction between direct and indirect developmental toxicants. Direct chemicals such as thalidomide and retinoids induce developmental toxicity without maternal toxicity. Indirect chemicals such as ethanol and cocaine mostly affect the fetus at levels also toxic to the mother. Epidemiological studies indicated that environmental exposures to low concentrations of some chemicals may cause subtle neurodevelopmental changes (Jacobson and Jacobson 1996) or may disrupt the endocrine system (Kavlock et al. 1996).

1.5.1.3.2 Infants

Breast-feeding is recognized as providing the developing infant the benefits of balanced nutrition and passive immunization against infections; however, exposure of infants to anthropogenic chemicals via breast milk is of concern. Many chemicals persist in the environment for a long time; they bioaccumulate in the organism and biomagnify through the food chain. Among the environmental pollutants found in breast milk are PCDDs, PCDFs, PCBs, metals, and pesticides. Considering the relatively short period of breast-feeding and relatively high daily intake, the exposure may be substantial. For example, Schecter and Gasiewicz (1987a, 1987b) estimated the daily intake of PCDDs and PCDFs in TEQs by nursing infants (10 kg) in the United States to be 83 pg TEQs/kg body weight/day. Later, Schecter et al. (1994) estimated lower intakes of 35 to 53 pg TEQs/kg body weight/day for infants (7.3 kg) (see exposure changes under Section 1.5.1.2). In contrast, intake for infants who were fed soy formula was substantially lower, ranging from 0.07 to 0.16 pg TEQ/kg body weight/day. In comparison, Schecter et al. (1994) estimated the daily intake of PCDDs and PCDFs for adults (70 kg) as 0.26 to 2.75 pg TEQs/kg body weight/day. Koppe (1995) reported that daily dietary intake of PCDDs during lactation represents only 14% of the daily secretion of PCDDs in breast milk; the rest (about

Table 1.6 Levels of various chemicals in human breast milk samples from general populations

Chemical	Range of mean or median concentrations (ng/g lipid)	Newborn intake via breast milk[a] (µg/kg/day)	Region	Reference
PCDDs and PCDFs	0.013–0.028[b]	0.00009–0.00057[c]	United States, Canada, Germany, New Zealand, Japan, Russia	Pohl and Hibbs (1996)
Mercury (total)	130–793[c]	0.922–5.625	Japan, Germany, Sweden	Abadin et al. (1997)
Hexachlorobenzene	5–63	0.035–0.447	New Zealand, Brazil, Arkansas, Australia, Canada, Mexico, Quebec Caucasians	Pohl and Tylenda (2000)
Hexachlorobenzene	100 to >1000	0.709–7.094	France, Spain, Quebec Inuits, Slovak Republic, Czech Republic	Pohl and Tylenda (2000)
p,p'-DDE	300 to >3000	2.128–21.281	New Zealand, Brazil, France, Australia, Quebec Caucasians and Inuits, Arkansas, Canada, Slovak Republic, Czech Republic, Germany, North Carolina	Pohl and Tylenda (2000), Rogan et al. (1986)
PCBs	167–1770	1.185–12.556	Japan, Quebec Caucasians and Inuits, New York, Michigan, the Netherlands, Poland, Finland, Croatia, North Carolina	DeKoning and Karmaus (2000)

Source: Adapted from Pohl HR, McClure P, De Rosa CT. 2004. Environ Toxicol Pharmacol 18: 259–266.

[a] Converted from 0.6 to 3.6 µg Hg/dL, using a conversion factor of 45.4 g lipid/10 dL milk (DeKoning and Karmaus 2000). Organic forms accounted for about 7 to 50% of total mercury (Abadin et al. 1997).

[b] Measured in 2,3,7,8-TCDD toxicity equivalents (TEQs).

[c] Calculated, based on assumptions of 3.2 kg body weight, 45.4 g fat/L milk, and 0.5 L milk/day (DeKoning and Karmaus 2000), as follows: 5 ng/g fat × 45.4 g fat/L × 0.5 L/day × 1/3.2 kg × 1 µg/1000 ng = 0.035 µg/kg/day.

86%) is derived from PCDDs stored in adipose tissue. Levels of selected chemicals found in breast milk around the world are presented in Table 1.6.

1.5.1.3.3   Children

There are many differences between children and adults. The first obvious difference is in size; children consume more food and water per kilogram of body weight, they have higher inhalation rates, and they have larger surface-area-to-volume ratios than adults. For example, Schecter and Li (1997) conducted a congener-specific analysis of PCDDs, PCDFs, and dioxin-like PCBs in US fast foods. They reported TEQ values from 0.03 to 0.28 pg/g wet weight for McDonald's big mac, 0.03 to 0.29 pg/g for Pizza Hut's personal pan supreme pizza with all toppings, 0.01 to 0.49 pg/g for Kentucky Fried Chicken's 3-piece original recipe meal, and 0.3 to 0.31 pg/g for Häagen-Dazs' chocolate–chocolate chip ice cream. Daily TEQ consumption per kilogram body weight assuming a 65 kg adult, from 1 serving of each of the fast foods tested, ranged between 0.046 and 1.556 pg/kg. The daily intake from 1 serving of each of the fast foods tested, assuming a 20 kg child (6-year-old), ranged from 0.15 to 5.05 pg TEQs/kg. A child on average consumes 3 times more TEQs on a per kilogram body weight basis than adults eating any one of the fast foods tested.

Children may also be more sensitive to harmful environmental chemicals because of differences in absorption, excretion, and metabolism (see also Chapter 2). Immaturity of some systems (e.g., immune, nervous systems) also contributes to children's vulnerability to chemicals.

Another difference between children and adults is in their behavior. Children spend more time outside and play in the dirt. Associated with this activity is soil ingestion, where hand-to-mouth ingestion has been recognized as a major exposure route (Clark et al. 1996; Hemond and Solo-Gabriele 2004). The soil ingestion value for children is based on a number of studies estimating the average soil ingestion in populations of normal children (Binder et al. 1986; Clausing et al. 1987). One of the reports suggested that an average child ingests only about 25 to 40 mg of soil daily (Gough 1991). However, about 1 to 2% of children are geophagic ("pica children") and ingest from 5 to 10 g of soil daily (USEPA 1989a). Known child-specific exposure factors have been reviewed by the USEPA (2002a).

1.5.1.3.4   Adults

Occupational exposures play an important role during adulthood. Often the exposures last for many years and exposure levels are much higher than those of the general population. Some of the common complex mixtures regulated in the workplace include coal tar pitch volatiles, mineral oil mist, petroleum distillates, and Stoddard solvent[1] (Hearl 2005). There are numerous possible combinations of chemicals found at the workplace; that is, most exposures are to chemical mixtures.

---

[1] Stoddard solvent is a colorless, flammable liquid that smells and tastes like kerosene. It will turn into a vapor at temperatures of 150 to 200 °C. Stoddard solvent is a petroleum mixture that is also known as dry cleaning safety solvent, petroleum solvent, and varnoline; its registered trade names are Texsolve S® and Varsol 1®. It is a chemical mixture that is similar to white spirits. Stoddard solvent is used as paint thinner; in some types of photocopier toners, printing inks, and adhesives; as a dry cleaning solvent; and as a general cleaner and degreaser.

Additive and synergistic effects of chemicals with common targets of toxicity are of great concern in occupational settings. For example, coexposure to acetone, sec-butyl acetate, and methyl ethyl ketone causes increased skin irritation; coexposure to heptane, methyl chloroform, and perchloroethylene increases central nervous system effects (Hearl 2005). In addition, personal behavior can affect the outcome of occupational exposures. For example, smoking increases the development of cancer in occupational exposures to asbestos (Selikoff et al. 1980) and to radon (Lundin et al. 1969; Archer 1985); alcohol increases liver effects of occupational exposure to hepatotoxicants such as carbon tetrachloride (Manno et al. 1996). Inadequate dietary protein enhances the toxicity of pesticides such as heptachlor (Boyd 1969; Shakman 1974). Household members, including children, can be exposed to workplace chemicals by coming into contact with contaminated work cloths (e.g., off-gassing of tetrachloroethylene).

1.5.1.3.5 Seniors
Elderly people are obviously not the only age category exposed to pharmaceuticals, but they are by far the most exposed. It is beyond the scope of this chapter to describe all the exposure scenarios and possible interactions. Interested readers are encouraged to consult other detailed literature on the topic (Calabrese 1991).

1.5.1.4 Modeling and Measuring Human Exposure
Several exposure models are available for human health risk assessment, some of which are summarized in a review by Fryer et al. (2006). They categorize exposure models according to the route of exposure:

   1) environmental, distinguishing environmental concentration models (see Section 1.4) and human intake models;
   2) dietary;
   3) consumer product;
   4) occupational;
   5) aggregate, including multiple exposure pathways; and
   6) cumulative, including multiple chemical exposure.

Fryer et al. (2006) conclude that the use of human exposure models still is fragmentary, with different organizations using different models for very similar exposure assessment situations. The main problem in the use of models is the lack of input data and the lack of validation of both the input data and the output of the model. They therefore recommend the development of an overall framework for human exposure (and risk) assessment.

One problem encountered when assessing exposure of human populations to contaminated land is spatial heterogeneity of pollution. To overcome this problem, Gay and Korre (2006) propose the combinations of spatial statistical methods for mapping soil concentrations, and probabilistic human health risk assessment methods. They applied geostatistical methods to map As concentrations in soil. Subsequently, an age-stratified human population was mapped across the contaminated area, and the intake of As by individuals was calculated using a modified version of the Contaminated Land Exposure Assessment (CLEA) model. This approach allowed a determination of sites with clearly elevated human exposure, and may also be used to determine exposure to mixtures of chemicals.

Weis et al. (2005) reported the results of the ad hoc Committee on Environmental Exposure Technology Development. They identified a toolbox of methods for measuring external (environmental) and internal (biological) exposure and assessing human behaviors that influence the likelihood of exposure. The toolbox of environmental exposure methods includes environmental sensors, such as in vitro sensors, like personal dosimeters, to detect and quantify priority environmental exposure, and Geographical Information System (GIS) technology to map and link environmental and personal exposure. The latter technique may also be used to identify populations at risk. The internal exposure methods comprise biological sensors, toxicogenomic measurements, and body burden analyses. The latter also include determination of biomarkers of exposure, such as DNA adducts. All these methods may be used to determine exposure to single chemicals and mixtures.

1.5.1.5 Human Biobanks and Human Volunteer Monitoring of Exposure

Monitoring body burdens of chemicals in human populations may contribute to a better understanding of what chemicals and at what concentrations they get into the body. For chemicals with known toxicity levels, it may be possible to learn the prevalence of people with concentrations exceeding those toxicity levels. However, it should be noted that most biomonitoring studies in the general population are designed as survey studies. Many of the studies do not take into account exposure history. Therefore, the results often represent a snapshot at 1 point in time.

In the United States, the Centers for Disease Control and Prevention (CDC) released the third National Report on Human Exposure to Environmental Chemicals (CDC 2005). The report presents blood and urine levels for 148 environmental chemicals (or their metabolites) found in the general civilian US population over the 2-year period 2001–2002. The study is part of the National Health and Nutrition Examination Survey (NHANES), designed to provide an insight into the health and nutritional status of the US population. Future follow-ups are planned to encompass 2-year periods with information on trends of exposure in population groups defined by age, sex, and race. The latest report monitored 32 more chemicals than the second report, which encompassed the 2-year period from 1999 to 2000 (CDC 2003). The major chemical groups monitored included metals, phytoestrogens, PAHs, PCDDs, PCDFs, coplanar and mono-ortho-substituted biphenyls, non-dioxin-like PCBs, carbamate pesticides, organochlorine pesticides, pyrethroid pesticides (added in the third report), phthalates, organophosphate insecticides, and herbicides. Cotinine, a major metabolite of nicotine, was measured as an indicator of smoking. Urine creatinine was analyzed as a continuous variable for chemicals measured in urine to adjust for urinary dilution. Sample sizes varied under each of the categories from several hundreds to several thousands.

The large-scale monitoring studies enable researchers to track, over established time periods, trends in levels of exposure in human populations and to assess the effectiveness of public health efforts to reduce exposure to specific harmful chemicals. Aylward and Hays (2002) summarized recent trends in dioxins intake in the United States and in Western Europe. The intake estimates show clear decreases of dioxin exposures in these countries. For example, USEPA's 2000 estimate of 0.6 pg TEQ/kg/day is 66% lower than the 1994 estimate of 1.7 pg/kg/day for PCDDs/Fs. In the United Kingdom, the intake levels for PCDDs/Fs were 4.6, 1.6, and 0.9 pg TEQ/kg/day in 1982, 1992, and 1997, respectively. Similarly, PCDDs/Fs intakes were estimated as 4.2, 1.8, and 0.5 pg TEQ/kg/day in 1978, 1984, and 1994, respectively, in the Netherlands. Similar trends were reported for PCB intakes (Aylward et al. 2002). These decreases in intake are reflected in decreases in human body burdens. A large number of studies in the general population in the United States, Canada, Germany, and France during 1972–1999 show a trend of substantial (almost 10-fold) decreases in human TCDD-only body burden over that time period (Aylward and Hays 2002). Considering the long half-life of TCDD, a 1-compartment pharmacokinetic model estimated that the decrease in intake must have been more than 95%. A recent retrospective time-trend study that analyzed levels of major halogenated aromatic hydrocarbons in human serum concluded that PCB and polybrominated biphenyl (PBB) levels are also decreasing since their phaseout in the 1970s (Sjodin et al. 2004). In contrast, concentrations of polybrominated diphenyl ethers (PBDEs) have been increasing in recent years in the United States, because of their use as flame retardants. Such a very substantial increase has not been observed in Europe, where PBDE levels in human serum are about 10 times lower than in the United States (Thomas et al. 2006).

Large-scale monitoring studies are also under way in Europe (Pohl et al. 2005). In 2003, the EU commission launched an initiative called SCALE, which stands for Science, Children, Raising Awareness, Legal Instruments, and Evaluation. The objectives are to reduce disease burden caused by environmental factors, to identify and prevent new environmental threats, and to strengthen the EU policy-making capacity. The EU commission asked for the cooperation of all stakeholders in identifying and addressing the most relevant children's environmental health issues. Biomonitoring was specifically addressed in a document that called for unified testing approaches all over Europe so that the study results are comparable (Pohl et al. 2005). From 2003 to 2005, the Robert Koch Institute monitored 18,000 children aged from <1 to 18 years at 150 different places in Germany over a 3-year period. Detailed interviews on pregnant women, parents, and children—together with medical examinations—focused on health risks in modern life, relevant environmental pollution, psychological health, and motoric development in childhood. Similarly, the Erasmus Medical Centre in Rotterdam monitors about 10,000 subjects over a 20-year period: from early fetal life until young adulthood. The project started in 2002. Physical examinations, questionnaires, interviews, and ultrasound examinations are performed; biological samples are collected as well. The focus is on pediatric growth, neurobehavioral development, pediatric diseases, and preventive health care for mother and child.

1.5.2 EXPOSURE IN ECOSYSTEMS

Estimating exposure through measurement of chemical concentrations in tissues or body fluids (e.g., blood, urine) of environmental species (or a suitable surrogate) is a long-standing concept going back many years. In recent times, interest in the "biomonitoring" approach has grown to the point where the subject was the focus of a *Nature* editorial that looked explicitly at application in developed and developing regions (Whitfield 2001). Some current monitoring programs have been measuring residue levels of multiple chemicals in biological samples over extended periods, while others are targeted studies addressing particular sites or issues. The principal driver behind each of the studies is not usually to quantify the exact nature of multiple chemicals exposure, but instead to assess spatial and temporal changes in exposure level for particular contaminant groups of concern (e.g., metals, PCBs, organochlorine insecticides). Despite this focus, many of these monitoring schemes can provide valuable information on the nature of complex exposure scenarios, and a few even explicitly report on the potential mixture effects that may result from a multiple chemical exposure.

The nature of the different environmental compartments means that the species selected for exposure monitoring may vary between different environmental types (e.g., air, soils, sediment, freshwaters, and marine waters). Similarly, the characteristics of the chemicals being investigated (often persistent pollutants) can also influence species choice. For some environmental compartments, such as soils and sediments, it is possible to select species for monitoring that are exposed through different routes, such as via pore water or by ingestion of contaminated particulate matter. Because of the potential for pulsed exposure in mobile media such as air and running waters, sampling for exposure biomonitoring in these ecosystems may need to be repeated to avoid missing potentially significant exposures to mobile or ephemeral pollutants. In soil and sediment, which act as contaminant sinks, such temporal repetition may be less of an issue, although this depends on the persistence of the compounds being assessed. Further in these environments, spatial heterogeneity can be more of an issue. Approaches for different environmental compartments and details of the finding and outcomes of particular schemes associated with each are outlined in detail below.

1.5.2.1 Air

Exposure through air is particularly important for gases, volatile compounds and chemicals associated with the surfaces of small airborne particulates. Since plants are in intimate contact with the air through leaf cuticles and stomatal pores, this group has been most widely used for airborne contaminant monitoring. The first widespread application of plants for biomonitoring was to support policy implementation on issues relating to acidification and eutrophication (Cape et al. 1990; Bobbink et al. 1992). Building on this work on sulfur and nutrients, metals have also been the subject of numerous studies measuring both spatial and temporal aspects of accumulation in plant leaves. Burton et al. (1986) reported that many surveys involving metal analysis of lichen thalli reflected the spatial variability of metal deposition. Bargagli et al. (2002) compared metal concentrations in 2 common biomonitor species *Hypnum cupressiforme* (a moss) and *Parmelia caperata* (lichen) around an intensive mining area. While both moss and lichen were able to indicate the nature of the complex atmospheric emission occurring in the region, each accumulated different concentrations of different metals. This emphasizes the difficulty in reading across exposure scenarios between species.

This problem is, however, not unique to plants, but is in fact applicable to many taxa (Hopkin et al. 1993; Morgan and Morgan 1993; Newton et al. 1993). One area where the analysis of mosses and lichens has made a significant contribution to understanding of present and past pollution trends has been through the analysis of samples stored in herbaria. These samples provide useful information on temporal trends in deposition that can be used in the development of dynamic models to describe pollutant loading and fate in terrestrial environments (Hassanin et al. 2005). This issue may be particularly relevant in describing past exposure that may lead to current environmental and human health concerns (e.g., past carcinogen exposure).

An obvious issue regarding the use of plants for airborne pollutant biomonitoring is that they are also potentially exposed through the soil. This confounding effect is one reason why many monitoring studies of wet and dry deposition have focused on mosses and lichens, which lack roots and so rely on atmospheric deposition to surface cuticles to obtain adequate nutrients. For PAHs, and probably also for other lipophilic organic compounds, uptake from soil may also be limited by the strong adsorption to soil and lipid materials associated with root cells (Watts et al. 2006). For such compounds, therefore, air accounts for most of the burden on or in plant leaves and other aboveground tissues. This makes mosses and lichens excellent potential monitors for organic pollutants, with the physiology of the leaf, such as the presence and nature of leaf hairs and the form of any extracuticular wax present, being an important influence on the rate of uptake (Bakker et al. 1999; Jouraeva et al. 2002).

1.5.2.2 Water
The potential for sustained and pulsed exposure in riparian, lake, estuarine, and coastal systems has established biological monitoring as a potentially useful tool for characterizing the chemical status of these habitats. For monitoring contaminants in the water column, filter feeders and species with an extensive gill system are frequently favored. This is because the extensive contact between biological membranes and the water column in these species ensures that there is significant exposure. An example of the kind of coordinated approach that can be used for monitoring waterborne pollutants using a filter feeder species is the mussel watch program run by the National Center for Coastal Ocean Science in the United States. This scheme measures concentrations of over 100 contaminants in the tissues of marine mussels from almost 300 US sites covering the Atlantic, Pacific, and Gulf coasts, and also the American shores of 3 of the Great Lakes (Michigan, Huron, and Erie). Chemical groups measured include 18 elements (17 metals plus Si), over 50 PAHs, 31 PCBs, 31 organochlorinated compounds, organometals (such as butyl-tin compounds), PCDDs, and PCDFs. Major conclusions from the survey to date include evidence of a widespread, but declining exposure to multiple organochlorines; a widespread exposure to multiple PCBs that is reducing over time (but not as quickly as for the organochlorines), a widespread exposure to PAHs that has remained largely unchanged, except where there has been a specific accidental exposure (Page et al. 2005), a decreasing exposure to organo-tin compounds, and a very prominent exposure to metals that has reduced for some (e.g., lead) but remained unchanged for others (e.g., copper and zinc) (National Oceanic and Atmospheric Administration 2002).

The large-scale nature of the US mussel watch program has, of course, led to the development of many smaller-scale schemes in other countries and regions that have mirrored the use of mussel species as subjects for biological monitoring (Sole et al. 2000; Kim et al. 2002; Rainbow et al. 2004; Mendoza et al. 2006). Other schemes also used other bivalve species, such as oyster and scallop, as alternatives to mussels (Daskalakis 1996; Silva et al. 2003; Norum et al. 2005). These biomonitoring studies range from small-scale surveys of country or regional coastlines conducted for particular contaminant groups (e.g., metals, PAHs, PCBs) to major regional surveys that rival the US program (e.g., Tanabe 2000). Being also subject to the vagaries of the research funding system, it is often difficult to obtain sufficient funds to repeat surveys, meaning that in some cases, the temporal element that is so important in the mussel watch scheme is lost. The large-scale surveys set up for coastal waters in Asia are, however, beginning to generate data suitable for the derivation of temporal trends (Sudaryanto et al. 2002; Monirith et al. 2003).

A second major group of invertebrate species recommended for marine biomonitoring is crustaceans. The barnacle *Balanus improvisus* has been used as a biomonitor of the metals Cu, Zn, Cd, Fe, Pb, Mn, and Ni in the Gulf of Gdansk (Baumard et al. 1998). Insects have also been used in freshwater (Fialkowski et al. 2003). As well as invertebrates, fish have also been recommended as potential monitors of exposure to multiple contaminants. Despite particular issues with the use of fish for monitoring organic pollutants (see below), biomonitoring studies have been able to separate individuals from sampling regions (coastal sea vs. oceanic) with different levels of prevalent pollution (Stefanelli et al. 2004). Since fish are both abundant and important components of aquatic ecosystems (at intermediate to high levels within aquatic food webs), and also because fish are an important food source, monitoring of contaminant levels in fish can provide an important link between environmental exposure to multiple compounds and human exposure to the same compounds through diet (Meili et al. 2003).

While biomonitoring studies undoubtedly provide useful information of mixture exposure assessment, the presence or, more importantly, absence of a compound in a particular species does not provide the full picture. Physiology may have a major influence on the tendency of an organism to accumulate certain chemicals. For example, fish have a potential to metabolize organic molecules that is higher than in many of the invertebrate taxa that are the most common subject of invertebrate monitoring. This means that more recalcitrant persistent organic pollutants, such as PCBs and organochlorine insecticides, are accumulated at low or very low concentrations in fish, and easily biodegradable compounds, such as PAHs and chlorinated phenols, do not tend to accumulate (van der Oost et al. 2003). A further issue in water is the pulsed nature of potential exposure, particularly in rivers. In cases where conditions, such as surface flow, cause pollutant runoff (e.g., pesticides from an arable field), this exposure can cause detrimental effects to receiving stream ecosystems. Later chemical biomonitoring studies may, however, fail to identify this exposure if the compound is both degraded from stream sediment and rapidly metabolized by the chosen monitoring species. This problem can be seen as part of a wider issue with the use of biomonitoring to detect complex exposure to easily metabolized compounds, which is described later.

1.5.2.3 Sediment

As sediments act as pollutant sinks in aquatic systems, they can be important sources of exposure, and so of the entry of chemicals into aquatic food chains. Sediments are the ultimate residence location for many pollutants released to water. The widespread presence of complex mixtures of contaminants in sediment is thus likely to occur in any location where multiple localized and diffuse contaminant sources contribute to the overall chemical load within natural waters. The role of sediment in the receipt and resupply of the chemical to the water phase means that there is interest in monitoring sediment chemical pollutant load over both different spatial and temporal scales. Because the process of sediment deposition and chemical adsorption on the one hand and solubilization and resuspension on the other link the pollutant loads of the sediment and water column, many of the species that can be used to sample the environment for waterborne pollutants (e.g., filter feeders such as mussels) can also describe the pollutant load present in sediments (Baumard et al. 1998).

An alternative to the use of filter feeders as proxies for sediment exposure is to utilize fully sediment-dwelling species. Two groups have been suggested. The first are the annelids. In freshwaters, the most commonly studied is the sludge worm, *Tubifex tubifex* (Egeler et al. 1999), while in marine sediments 2 polychaetes, *Nereis diversicolor* and *Arenicola marina*, have been used as biomonitors of single and multiple chemical exposures with varying degrees of success (Kaag et al. 1998; Poirier et al. 2006). The second sediment-dwelling group used for biomonitoring is insects, and in particular chironomids (Bervoets et al. 2004; Martín-Díaz et al. 2005a). Overall, the use of sediment-dwelling species is less well established than for bivalves such as freshwater and marine mussel species and, as a result, this later group remains the preferred taxa for multiple chemical exposure assessment in aquatic systems.

1.5.2.4 Soil

As soil is an important sink for pollutants in terrestrial ecosystems, there is interest in assessing the exposure of humans and ecological species through measurement of concentrations in organisms exposed through this medium. Such measurements can reflect the direct application of chemicals to soils as pesticides or solid waste or by aerial deposition of contaminants to the soil surface (Van Brummelen et al. 1996b; Filzek et al. 2004). Unlike air and surface waters, soil is immobile and hard to dilute, so there is a potential for temporally more stable long-term exposure than in mobile media, although it is also more likely that exposure is more heterogeneous, even at relatively small spatial scales. Since true soil organisms, such as earthworms, nematodes, and springtails, have low mobility, this means that tissue residue concentrations in these species can provide a more reliable estimate of the local exposure than more active surface-dwelling species, such as beetles, centipedes, spiders, and small mammals, for which tissue residue concentrations provide a picture of the average of contaminant concentrations over the full home range. Thus, depending on the question being asked by selecting different soil species for monitoring, it is possible for direct analysis to both include and partially ignore local scale spatial heterogeneity. For example, by using earthworms as the subject species, Marinussen and Van der Zee (1996) were able to model the spatial patterns of exposure in heterogeneously contaminated soils at an industrial facility. Such an analysis would, however, have been even more complex for very mobile species or animals with a complex life history for a single compound, let alone if multiple chemicals and their combined effects were to be considered.

Despite issues relating to chemical heterogeneity and the subsequent choice of monitoring species, the success of the mussel watch program for marine waters has prompted the initiation of schemes in soils that mirrored the marine approach. As in marine work, it is important to consider the physiology of the species when designing and interpreting the outputs of any monitoring scheme. For example, in laboratory experiments, slugs have been shown to be sensitive to metal pollution and were thus suggested to be useful for biological assessment of soil exposure (Marigomez et al. 1998). Work to assess the potential of slugs as biomonitors was, therefore, conducted at a copper-contaminated site, concluding that exposure and effect biomarkers recorded in sentinel slugs could be sensitive, quick, and cheap indices of metal pollution in soils (Marigomez et al. 1998). Similarly using woodlice, Hopkin et al. (1986) were able to conduct reliable exposure maps for the presence of multiple metals from diffuse and localized industrial sources over a medium-sized town (Reading) of 250,000 people in the United Kingdom. Jones and Hopkin (1991) also looked at the biomonitoring potential of woodlice and mollusks, concluding that both groups had potential and that each sampled exposure in a similar manner. This was indicated by the high correlation between tissue concentrations for each metal found between the different measured species. The studies outlined above have demonstrated the feasibility of using soil species to provide a picture to environmental exposure. A review of the published literature, however, shows that while this line of research was certainly the fashion 10 years ago, it is losing favor, with few recent papers focusing on this kind of spatially based tissue concentration assessments.

1.5.2.5 Monitoring of Food Chain Transfer

The potential for some chemicals to transfer through food chains has resulted in the development of a set of long-term, large-scale monitoring studies that measure exposure of top predators (predatory birds and mammals) through quantification of chemical concentrations in tissues (e.g., liver) and eggs. Monitoring of top predators arose in response, first, to recognition of the effects of organochlorine insecticides on bird populations (Newton and Wyllie 1992; Walker et al. 2001) and, second, to the realization that these pollutants have the potential to circulate across regions and ultimately around the globe (Wania and Mackay 1996; Gouin et al. 2004). The focus on top predators, which can range across large territories, means that the residue levels measured represent a cumulative exposure across the landscape, rather than the specific regions in which they were sampled. For some bird species in particular, exposure can also occur in different regions as a result of migration, and this can result in different patterns of tissue concentration depending on specific regional chemical usage patterns (Minh et al. 2002).

Although there are numerous fairly small-scale academic studies of the concentration of persistent pollutants in avian or mammalian predators (Kenntner et al. 2003a, 2003b; Berger et al. 2004; Jaspers et al. 2005; Hela et al. 2006), the 2 most important long-term large-scale schemes for quantifying wildlife exposure through the food chain are the UK Predatory Bird Monitoring Scheme (UK-PBMS) and the Arctic Monitoring Program of the USEPA (US-AMP). To briefly summarize both schemes, the UK-PBMS has run since the mid-1960s and measured chemical concentrations in the tissues of birds collected as dead carcasses by volunteers and non-hatching (sterile) eggs collected under license from nest sites. Like many monitoring schemes, the focus has been on quantifying exposure, including spatial (Alcock et al. 2002; Broughton et al. 2003) and temporal (Newton et al. 1991, 1993) trends. Current analytes measured include organochlorines and their metabolites, PCBs, second-generation rodenticides, mercury, and in a more limited set of samples, poly-brominated compounds and PAHs.

Analyses in the UK-PBMS concentrate on the use of single methods to measuring concentrations of particular subgroups of chemicals rather than on the application of diverse methods to detect the full range of residues that may be present. Reporting also focuses on trends for single chemicals in isolation. The exception to this is for PCBs, where the scheme expressly considered the combined chemical dose that is present in the tissues of the birds using the TEQ method of Ahlborg et al. (1994) and Van den Berg et al. (1998). Sum totals of concentrations may also have been reported for other chemical subgroups, such as the organochlorines and PAHs. While relatively coarse, these summed values provide a relatively simple means of defining the changes in concentrations of the measured entirety of these contaminant groups in tissue.

The US-AMP is probably the largest-scale pollutants monitoring initiative currently undertaken. The scheme focuses on measuring chemical concentrations in soils, water, and biota at higher latitudes. The focus on polar latitudes can be linked to concerns regarding the potential for movement of persistent organic pollutants (POPs) over time to higher regions of the globe (Wania and Mackay 1996). This process occurs because of the fact that warmer temperature at lower latitudes favors volatilization of POPs, while the colder temperature at higher latitudes favors deposition. Such cycles of volatilization at lower latitudes and deposition at higher latitudes (often termed global distillation) may, therefore, lead to the accumulation of higher concentrations and an increasing contribution from the more volatile compounds in polar regions.

Biota samples measured in US-AMP include arctic plants such as mosses, large terrestrial herbivores such as caribou, waterfowl, and top predators including birds of prey and mustelids (mink and marten). Aquatic invertebrates and some fish species are also measured. Like UK-PBMS, the focus in US-AMP is on the spatial and temporal trends for individual contaminant groups, rather than holistic assessments of multiple chemical exposures. Overall, schemes such as the US-AMP and UK-PBMS and similar small-scale national programs (Sørensen et al. 2004) can provide an excellent summary of mixture exposure scenarios for top predators exposed through the food chain. The databases generated by these programs could become essential data resources for further data mining for historic and current mixture exposure assessment for wildlife and possibly even humans.

While monitoring of chemical residue levels can provide a useful snapshot of the range and extent of current exposure to some compound groups, the approach does have some limitations. Birds and mammals, like fish, have a high metabolic potential for some organic contaminants. This means that some chemicals that are subject to rapid degradation and metabolism are not easy to reliably detect. This favors measurement of more recalcitrant contaminants, such as PCBs and organochlorinated insecticides, and also some metals such as mercury and cadmium, for which modeling (Romijn et al. 1993a, 1993b, Spurgeon and Hopkin 1996) and measurements (Hunter et al. 1987, 1989; Read and Martin 1993; Kooistra et al. 2005) have indicated the potential for concentrations to reach potentially harmful levels in tissues (Nicholson et al. 1983). For organic compounds that rapidly decay or for metals that are subject to strong homeostatic regulation, analyses for top predators may provide only a limited summary of current exposure, even though such compounds may make a substantive contribution to toxic effects.

1.5.2.6 Multimedia Exposure Scenarios

The exposure scenarios detailed above focus on separate assessment using biomonitoring organisms applicable to the particular environment under consideration. This separation of exposure by environmental compartment does not reflect the true nature of exposure for many species. For example, higher plants are likely to be exposed to airborne contaminants through the leaf surface and through the soil solution; sediment-dwelling filter feeders can be exposed through the interstitial water and also through the main water column; and top predators can be exposed through air, water, and their food supply. Elucidating the principal exposure routes for different species is an extremely active area of research in both aquatic and terrestrial environments (Vink et al. 1995; Irving et al. 2003; Jager et al. 2003), and there are obvious implications for the assessment of exposure in different exposure routes, and as a result, careful consideration needs to be given to the assessment. Such assessments are likely to be specific for each taxon and need to consider the major life history and behavioral characteristics of the organism.

Even more complex than the situations of multiple exposure routes are situations where species move between different environments during different stages of the life cycles. At their simplest, such scenarios can simply be due to feeding in different places at varying times of the year. More complex scenarios occur in species that spend separate parts of their life cycle in different environments. A nice example is given by Linkov et al. (2002). In their model, PCB exposure and bioaccumulation in winter flounder is described, taking into account spatial and temporal exposure characteristics. Other examples include many species of insect (e.g., Ephemeroptera, Tricoptera, Chironomidae) that spend their larval stages fairly sedentary in water and their adult stages as mobile wide-ranging species in terrestrial ecosystems. For such species, it is difficult to establish the true complex nature of exposure, and so measurement may be the only appropriate method to evaluate the exact scenario. A promising attempt to model spatially explicit ecological exposure of terrestrial organisms, taking into account spatial, temporal, and multiple stressor interactions, and addressing landscape heterogeneity, has been described by Hope (2001, 2005). Hope (2005) identified physical (loss of habitat) and biological (lack of adequate food) stressors, in addition to chemical stressors. Other examples of studies that took into account spatial exposure patterns by using geostatistical or GIS-based methods include Clifford et al. (1995) for dieldrin and Kooistra et al. (2005) for Cd accumulation in terrestrial food chains.

1.5.2.7   Critique on Biomonitoring Studies for Complex Exposure Assessment

Biological monitoring is one of the best ways to provide a picture of current exposure to environmental mixtures. However, monitoring programs have to be carefully designed and results reviewed with caution. Even the largest and most comprehensive studies are limited in their scope to assess the true nature of complex environmental exposure simply by the fact that it is not feasible to measure the full range of potential pollutants. In most cases schemes are designed to meet particular policy objectives (i.e., characterized predator exposure to POPs) or for particular site-specific scenarios (i.e., metal levels in mosses around a metal smelter). Even when chemicals can conceivably be measured, the potential for metabolism can present a significant problem. This relates not only to compounds that are rapidly metabolized, and so difficult to detect even after a significant exposure event, but also to interindividual variability, due, for example, to enzyme polymorphism, that may introduce variability into the system that may mask time-dependent or spatial exposure trends.

As outlined in the discussion for particular sections above, an important fact to consider when reviewing biological monitoring data is the behavioral characteristics and lifestyles of the subject species. In the simplest case, monitoring of sessile species may be useful to provide a local scale view of exposure, while measurement of highly mobile species may provide a region scale exposure summary. Even when species have broadly similar lifestyles (e.g., birds of prey), behavioral and prey differences mean that biomonitoring can show different results for different species. While direct measurement in monitoring studies can account for these differences, this presents a particular problem for reading across exposure scenarios for species with different habitats, food sources, and behavior. Only in cases where the variables and differences that have the greatest influence on the nature of exposure are known is it possible to make interspecies predictions regarding a particular exposure scenario.

1.5.2.8   Effect-Directed Assessment

A final means of assessing combined exposure is through the direct application of biological testing for effects-based assessment of complexly polluted media (e.g., effluents, soils, sediments). For the use of bioassays for direct assessment of complex mixtures, the reader is referred to Chapter 4.

Matching the development of bioassays for complex exposure assessment has been the development of the use of biomarkers. This area has been the subject of a number of detailed reviews, some wholeheartedly recommending the approach and others being more critical (Decaprio 1997; Kammenga et al. 2000; Gagne and Blaise 2004; Forbes et al. 2006). Whatever the pros and cons of biomarkers, the use of effects-based analysis clearly has an appeal in assessing exposure to mixtures and its consequences. Such approaches have already been used in human exposure monitoring. For example, metabolite monitoring is regularly used for occupational human exposure in regulatory settings, and it is easy to envisage such approaches being used for exposure assessment in environmental species.

A final form of direct effect assessment that can be used for exposure monitoring is through monitoring the community composition for microorganisms, meso- and macrofauna, and plants. This approach is based on the fact that the exposure to specific pollution mixtures may be expected to result in the elimination of sensitive taxa and species. Physiological and community diversity-based profiling methods for microorganisms are becoming increasingly routine, and for microinvertebrates, the development of community-based monitoring systems, such as the River Invertebrate Prediction and Classification System (RIVPACS) scheme in the United Kingdom (Hawkins et al. 2000; Wright et al. 2000) and analogous schemes in other countries, is becoming increasingly widely used for assessing the ecological condition in regulatory regimes such as the EU Water Framework Directive. In these cases, the challenge lies practically in providing the required level of taxonomic resolution in "difficult taxa" and interpretationally in making causal links between the observed community change and the complex nature of the combined exposure to multiple stressors.

1.6   SUMMARY AND CONCLUSIONS

Estimation of mixture exposure requires an assessment of all steps from emission of chemicals, fate in the environment, bioavailability in different environmental compartments, and interactions at the uptake level. In addition, behavioral aspects and life-stage-specific exposures have to be considered.

Emission estimation methods are available for several sources and chemicals (or groups of chemicals), but focus is generally on single chemicals rather than on mixtures. Multimedia fate models may be used to predict or estimate the fate and distribution of chemicals in the environment. Such models are available for different scales, but the precision of the prediction usually increases with increasing level of details required. Physical–chemical properties of the chemicals and characteristics of the environment determine the composition of a mixture ending up at a certain site or in a certain place. Exposure is determined by factors affecting bioavailability of the chemical, such as binding strength to soil or sediment particles. Chemical–chemical interactions may also affect availability. In addition, individual behavior of organisms, including man, may influence exposure. Methods to determine exposure mainly include residue analysis, either in exposure media like air, water, soil, or food, or in tissues of organisms being exposed. Also, these monitoring methods usually are focusing on single chemicals rather than on mixtures. These long-term and large-scale data sets can provide essential information that can be used to validate the outcome of emission scenario and environmental fate models.

1.7   RECOMMENDATIONS

Although several emission registrations exist, these mainly focus on single (groups of) chemicals rather than on assessing mixture exposure. In addition, emission estimations or registrations seem mainly to take place on an ad hoc basis, with little international coordination.

1) Generate emission data that may help better estimate mixture exposure.
   a) Collection of data on existing chemicals should be evaluated to deter-
       mine if they are fit for the purpose of mixture scenario prediction.
   b) International collaboration for exchanging emission data.
   c) There is a special need to consider emission of new and emerging
       chemicals, for example, nanoparticles.

One of the reasons why current emission estimation methods do not focus on
mixtures might be that there are so many possible combinations of chemicals. With
some guidance on the combinations of chemicals most relevant or most likely to
cause problems, it would become easier to focus.

2) Prioritize approaches to emission estimation to focus on most common and
   most relevant mixture emission scenarios.
   a) Further development of desk-based methods for identifying most
       probable mixture scenarios and widespread release of modeling
       outcomes.
   b) Refine large-scale monitoring programs, for example, global programs,
       such as the Arctic Monitoring and Assessment Program (AMAP) and
       the European Monitoring and Evaluation Program (EMEP) to focus
       on inclusion of the most relevant old and new chemicals alone and in
       mixtures.

Although several recent studies demonstrate awareness of the fact that chemical
fate processes may have a large influence on the composition of mixtures, models
usually only focus on single chemicals. Nevertheless, such models may be very useful to predict composition of mixtures, but validation is needed.

3) Studies on fate in the environment should include aspects that cause a
   change in mixture composition from emission until exposure.
   a) Models that describe distribution of mixtures in the environment need
       to be validated.

Several factors and processes may lead to interactions between chemicals in the
environment. Such interactions not only determine fate and transport in the environment, but may also play a role in determining uptake. More insight into such interactions is highly needed in order to enable a more accurate exposure assessment of
mixtures.

4) Research on potential interactions between individual chemicals that might
   affect the exposure, availability, or toxicity of the mixture and inclusion of
   outcomes into developing multimedia fate models for mixtures.

Monitoring programs mainly focus on measuring (single) chemicals in the environment or in organisms as an indication of exposure. In many cases no additional
information is provided, hampering a proper interpretation of such data.

5) Monitoring programs measuring total concentrations should include measures of parameters (environmental characteristics, for example, clay, organic carbon content, pH, dissolved organic carbon (DOC)) and physical chemical properties of compounds that help to evaluate bioavailability.

Several models have been developed to link bioavailability or uptake of chemicals in organisms to their speciation in the environment. Again, focus generally is on single chemicals, while extension for use on mixtures is desired.

6) Research is needed to assess if it is possible to extend integrated models that link exposure-toxicity (like the BLM) for use with mixtures.

Like many data, emission and exposure data are presented as constant values, often a mean with standard deviation. In environmental risk assessment, however, awareness is growing that a stochastic or probabilistic approach is more suitable to obtain insight in the possible risk of chemicals. This also requires expressing exposure data as statistical, probabilistic distributions. Also in this case, the focus should be extended to mixtures.

7) Improve methods for identification of the probabilistic distribution of short- and long-term exposure of possible chemical mixtures for ecosystems and humans.

An adequate assessment of exposure to mixtures may require development of improved tools for measurement or detection of chemicals, but also for assessing temporal aspects of exposure.

8) Generate data that may help better estimate mixture exposure.
   a) Analytical methods should be available for new high-production-volume chemicals coming onto the market (e.g., as for pesticides).
   b) Differentiate between simultaneous and subsequent exposure
      i) Long-term trends at the same locations
      ii) Spatial sampling at the same time
   c) Better understanding of routes of exposure, potential entry and exposure of humans to new chemicals, and their contribution to mixtures.

Assessing human exposure is quite complex, because exposure is dependent on life stage and may be influenced by behavioral patterns. Prediction of exposure may be improved by accounting for these aspects.

9) Improve prediction of exposure for different life stages of humans for chemical mixtures accounting for behavior patterns.

ACKNOWLEDGMENTS

Thanks are due to Fred Heimbach for his valuable contribution to the discussions at the International SETAC/NoMiracle Workshop on Mixture Toxicity in Krakow that led to this chapter. Ad Ragas is acknowledged for his critical and valuable review of this chapter.

2 Toxicokinetics and Toxicodynamics

*Claus Svendsen, Tjalling Jager, Sami Haddad,
Raymond S. H. Yang, Jean Lou C. M. Dorne,
Mieke Broerse, and Paulina Kramarz*

CONTENTS

2.1   INTRODUCTION

The previous chapter dealt with how chemicals in a mixture may interact and affect
each other's availability for uptake by organisms. This chapter covers the approaches
dealing with 1) how such bioavailable fractions are absorbed, distributed, metabolized, and excreted in the organism (toxicokinetics), and 2) how the mixture toxicity
is caused once the biological target in the organism has been reached (toxicodynamics). Using plain English, toxicokinetics (TK) can be considered what the body does
to the chemicals, and toxicodynamics (TD) can thus be considered what the chemicals do to the body. It is often difficult to distinguish where toxicokinetics ends and
toxicodynamics starts.

This chapter is subdivided into 4 main parts. The first part introduces the fundamental principles of toxicokinetics and toxicodynamics, and their importance for
risk assessment, with particular emphasis on mixtures. The second part describes
the state of the art in toxicokinetics, and the third part in toxicodynamics. Both
later sections focus on how each aspect is dealt with in human toxicology and ecotoxicology. The differences and limitations in each field are addressed. Finally, a
general discussion, conclusions, and recommendations for future research explore
the application potentials of these approaches, with particular attention to cross-fertilization between human toxicology and ecotoxicology, and where they may hold
some promises.

2.2   FUNDAMENTAL PRINCIPLES

2.2.1   ABSORPTION, DISTRIBUTION, AND EFFECTS OF
CHEMICAL MIXTURES IN ORGANISMS

The effect cascade for a chemical exposure involves uptake, internal distribution,
and the effects at the site of action. Here we introduce the fundamental principles
of the effect cascade, while more in-depth discussions are covered in later sections.

Chemicals typically enter organisms through ingestion, inhalation, or permeation through the skin, and depending on their physicochemical properties, their uptake is assumed to be passive or active (mediated) by transporters. For neutral organic compounds, the assumption is that uptake is governed by passive uptake, while charged organic compounds and metals most often involve active uptake and carriers.

Once inside the organism, organic chemicals and metals are dealt with differently. Organic chemicals generally distribute based on their chemical properties (e.g., molecular size, lipophilicity, stereochemistry) and are eliminated through metabolism (phases I and II) or excretion (for example, renal excretion in mammals) of either the parent compound or the metabolites. Metals, on the other hand, can be split into essential and nonessential elements. Biochemical mechanisms and pathways have evolved to regulate essential metals with physiological functions. However, nonessential metals due to physicochemical similarities also use some of these pathways and thus affect the homeostasis of essential metals.

It is generally the free unbound form of organic chemicals and metals within the organism that may interact with biological systems such as metabolic pathways. These interactions can potentially cause adverse effects once certain critical threshold concentrations are exceeded. The mechanisms by which the chemicals may cause toxic effects are diverse. At given concentrations, the reaction rates of enzymatic reactions may be affected, and such effects can arise via several routes, which in principle can be related to simple Michaelis-Menten kinetics. Some chemicals interact with the organism via very specific interactions, for example, organophosphorous pesticides (OPs), which inhibit the enzyme acetylcholinesterase (AchE). Other chemicals are—often after biotransformation—very reactive and can bind to macromolecules such as lipids, proteins, or DNA and cause cytotoxicity or mutations (e.g., Hermens et al. 1985c). Because even a single organic chemical or metal may interact with several biological targets, the complexity of possible interactions at the level of disposition (toxicokinetics) or the expression of toxicity (toxicodynamics) following exposure to a mixture can be great.

2.2.2 Toxicokinetics and Toxicodynamics—Opening the Black Box

Historically, most mixture toxicity studies have treated the organism as a "black box." Test species are simply exposed to the mixture, and the toxic effects are observed for a single endpoint at the end of the exposure period. Subsequently, effects are related to the external concentration and compared to the expectations from the reference models of concentration addition (CA) (or similar action (SA)) or independent action (IA). Interactions between the mixture components are then inferred from this comparison. However, if the results do agree with CA, can we be confident that no interactions are occurring? It is possible that the effects of individual chemicals or interactions may cancel each other out, particularly for mixtures with a large number of compounds. And even if we observe a certain deviation from CA (say, "antagonism" or a "ratio-dependent interaction"), what is the underlying cause? Very simple kinetic or dynamic processes may already result in very complex dynamic behavior. For example, a mixture of 2 compounds with different toxicokinetics already results in a time-varying mixture within the body that may show specific patterns in

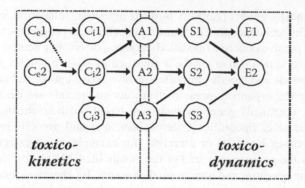

Figure 2.1 Hypothetical scheme for toxicokinetics and toxicodynamics for a mixture of 2 components. Ce = external concentration of chemical 1-2, Ci = internal concentration of chemical 1-3, where chemical 3 is a metabolite of chemical 2, A =target site, S = subsystem, E = final toxic effect. See text for further explanation.

the dose–response surface (even without any actual interaction between the mixture components). Furthermore, metabolic transformations may turn even a single compound into a complex time-varying mixture inside the body. In such cases, it makes little sense to relate the observed toxic effects to the external concentration in the medium; what affects the organism is the concentrations at the target site, organ, or tissue in time (see, e.g., McCarty et al. 1992).

To illustrate the processes involved in going from external exposure to effects, one has to borrow a concept from pharmacology (Ashford 1981). The effect of toxicants on a particular endpoint (E) can be considered using 3 main concepts: the stimulus (here, the toxicant C), the site of action (A), and the affected subsystems (S) (see Figure 2.1).

To fully understand the toxicology of a chemical or chemical mixture, we must consider both toxicokinetics and toxicodynamics. However, Figure 2.1 clearly illustrates that TK and TD modeling can already be quite complex for a single metabolized chemical (C2), and that mixtures provide an even greater challenge. Chemicals may interfere with each other's uptake (C1 on C2) or transformation (C1 on C2 to C3), and they may affect more than 1 site of action (C2 on A1 and A2). They may affect each other's potency (C1 on C2 affecting A2), and a site of action may affect more than 1 subsystem (A3 on S2 and S3). More than 1 subsystem may be involved in the observed effect (S1, S2, and S3 on E2). With the possibilities of interactions occurring at any step of this general scheme, a detailed TK/TD model easily becomes too complex for most practical applications. At the same time, many of the physiological and biochemical pathways responsible for accumulation or degradation need to be explained, even for well-studied organisms. This raises the question of justifying the effort of TK and TD modeling. The answer is that while mixtures may exhibit characteristics that are never fully recognized or understood, a sound understanding of the general physiological and biochemical pathways is of scientific interest and eventually also improves risk assessment of mixtures and potentially makes them more cost-effective.

2.2.3 APPLICATIONS OF TK AND TD MODELING IN RISK ASSESSMENT

Improving risk assessment can possibly be attained from TK and TD modeling, where it is physically impossible or not practically or ethically feasible to make direct measurements. Biologically based modeling may allow for informed extrapolation (e.g., between species). The most obvious example related to infeasible direct testing is the ethical limitations of human toxicological studies. Most mixture toxicity considerations for humans can only be undertaken by extrapolation from data obtained in experiments on other species. Especially, (suspected) carcinogenic chemicals cannot be administered to humans, but toxicological endpoints and models can be based on data from surrogate species (e.g., rats) or gained from subjects that are exposed to these chemicals on a daily basis from their own consent (e.g., smoking) or after an accident (e.g., radioactivity). In wildlife toxicology, similar issues apply, as it is impossible or impractical to test protected organisms. Furthermore, TK/TD insights provide a foundation for intraspecies extrapolation to cover certain sensitive subpopulations (e.g., children) or physiological states (e.g., organisms undertaking long-distance migration or hibernation).

Additionally, some risk assessment applications require complex exposure scenarios, such as pulse exposure (e.g., pesticide application, or use of a consumer product), time-varying mixture composition (e.g., differential degradation of a petroleum mixture), and sequential exposure to chemicals (e.g., chemicals present in different food items). Clearly, we cannot perform in vivo tests for all these scenarios and all organisms, and thus have to resort to modeling. Similarly, the number of chemicals in the environment is already very large, and the number of possible mixtures approaches infinity, and the use of the best available knowledge to construct informed TK/TD models provides a good starting point.

In order to perform sound risk assessments, an understanding of the mechanisms leading to mixture toxicity is required. While external concentrations of mixture components may often be all the data available, these are not always sufficient to understand or predict toxic effects, so either the measurement of the internal concentrations or the use of the best possible data set to estimate them would already improve predictions. Another aspect to understand is how mixture effects change in time and why results can differ between endpoints in the same test. For example, Van Gestel and Hensbergen (1997) observed mainly "concentration additive" effects of Cd and Zn on reproduction of the springtail *Folsomia candida*, but mainly "antagonism" on growth. Furthermore, the mixture effect became more antagonistic with exposure time. To understand these observations, we need to qualitatively and quantitatively investigate the TK and TD aspects.

Importantly, major differences exist between the state of knowledge and approaches in human toxicology and ecotoxicology, and these arise from the diverging aims of the 2 disciplines. Readers should refer to Chapter 5 for an extensive section on differences between human and ecological risk assessment. On the one hand, human risk assessment aims to protect the well-being of individuals, whereas ecological risk assessment aims to protect ecosystems (individual effects vs. ecosystem effects). The 2 fields therefore operate at different levels of biological organization (individual organisms vs. populations and ecosystems). Hence, their starting points are different. Human toxicology focuses on receptor characteristics to incorporate mechanistic information (e.g., metabolism, mechanism of toxicity) in 1 specific species (humans), while ecotoxicology mainly uses substance characteristics to deal with a high diversity of receptors within the range of species in which substances can have different modes of action (Dorne et al. 2006, 2007a). For humans, the interest is usually in very specific endpoints, such as enzyme induction or inhibition, tumor induction and growth, liver necrosis or any target organ toxicity, including reproductive toxicity, and changes on the birth weight of the offspring. In ecotoxicology, the endpoints of interest are most often integrated at higher levels of biological organization, such as total reproductive output, mortality, and population growth. Larger and rare or endangered wildlife constitute a particular case under which the protection goals and methodologies often need to be closer to those used for humans than those of general ecotoxicology. Finally, the exposure pathways generally considered for humans, mostly from the oral, dermal, and inhalation routes, are often much better defined than those having to be considered in ecotoxicity experiments where the organisms live in the exposure media (water, soil, or sediment) (Dorne et al. 2006, 2007a).

2.3 TOXICOKINETICS

The ultimate goal of TK modeling is to predict the time course of toxicant concentrations within organisms (internal dose), preferably at the target sites, starting from the concentrations outside the organism (external dose). This includes considerations on absorption, distribution, metabolism, and excretion, which are often referred to as ADME (see Section 2.3.1). In principle, these goals are the same for all studied species, whether humans, other vertebrates, invertebrates, plants, or microbes. The main difference lies in the available knowledge for each group of species; for vertebrates, and especially mammals, information is available, or can be obtained, and detailed TK modeling can be performed at the tissue level (see Section 2.3.2.2). For invertebrates, the metabolic and physiological information at the species level is often not available, and it is often difficult or even impossible to measure concentrations at the tissue level. Hence, invertebrate TK modeling may be restricted to simple relationships between external and total internal concentrations (see Section 2.3.2.1). Where steady state is reached, one may be able to use bioconcentration or bioaccumulation factors (BCF/BAF). While these might not technically qualify as TK models due to the lack of time course information, these are at times the best that can be achieved (e.g., species where experimentation is impossible and only randomly available field observations exist).

2.3.1 ABSORPTION, DISTRIBUTION, METABOLISM, AND EXCRETION (ADME)

2.3.1.1 Absorption

For neutral organic chemicals, the main assumption has been that the process of passive diffusion was the mechanism governing their uptake, and that their accumulation was based essentially on their octanol-water partitioning coefficient (K_{ow}). The theory is that because such chemicals have lipophilic and hydrophobic properties, they can diffuse across the lipid bilayer of biological membranes, being driven from the water phase of the environment to the lipid phase of the organism. Metals and polar organic compounds, on the other hand, generally occur as free and charged entities, which cannot cross biological membranes and require mediated transport, such as ion channels and specific carrier proteins or enzymes. When considering uptake of chemicals in mixtures, organic compounds have been assumed to have less potential for interacting directly with each other's uptake rates than metals, as metals may compete for the routes of mediated uptake. While present data show little reason to doubt that passive uptake applies to neutral organic substances, the last 10 years of research have characterized transporter proteins that are heavily involved in the transport of many organic chemicals (Dorne et al. 2003). In a recent publication primarily addressing drug uptake and bioavailability, Dobson and Kell (2008) emphasized that carrier-mediated cellular uptake of drugs could be more widespread than is assumed at present. Their principal argument is that the types of biophysical forces that determine the interaction of drugs with lipids are no different from those involved in their interactions with proteins. Dobson and Kell (2008) indicated that diffusion alone is too simplistic to explain the drug uptake into human cells, and they offered an impressive list of implications of a more prominent role for carrier-mediated drug uptake. Initial findings with environmental pollutants appeared to support the claim of Dobson and Kell (2008). Lohitnavy et al. (2008) provided evidence of the involvement of multidrug resistance-associated protein 2 (Mrp 2) in the pharmacokinetics of PCB126 and demonstrated an overall improvement of PBPK model simulations against experimental data from a number of laboratories with the incorporation of Mrp 2 binding in the liver. Thus, in considering the toxicokinetics of environmental pollutants, singly or as mixtures, we must bear in mind the importance of transporters. Additionally, even without direct interaction, exposure to a mixture of 2 chemicals at constant external concentrations, but with different uptake kinetics, results in a time-dependent internal mixture composition, as shown in Figure 2.2. Good information on uptake kinetics is therefore critical for understanding what effects may occur and, in cases where it is not possible to follow such effects in time, for estimating when steady state is reached and observations of such long-term effects may best be made.

2.3.1.2 Distribution

While in the systemic circulation within an organism, chemicals may end up in metabolically inactive parts of the body. Liphophilic chemicals may simply end up "stored" in fatty tissues, and remain as such, unless the fatty tissues are mobilized (e.g., during periods of starvation, or long-distance migration in birds). In contrast, metals may be actively excreted or stored in inert forms due to compartmentalization within the body and the presence of active regulation and detoxification mechanisms. Some organisms can store an excess of metals in inert forms within specialized organelles where they are not harmful (e.g., granules in isopods). The "biologically active" proportion, not the total internal concentration of metals, leads to toxic effects. Recent work by Rainbow (2002) and Luoma and Rainbow (2005) addresses how differences in physiology help explain how metal accumulation comes to vary

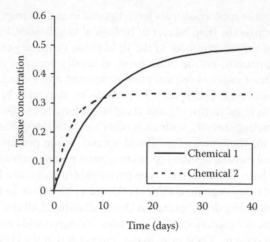

Figure 2.2 Development of body concentrations for 2 chemicals having different toxicokinetics, leading to an internal mixture composition that varies with time and differs from the external mixture.

so much between species and advocates the concept of biodynamic modeling to help understand how to link internal metal concentrations to toxicity. For the fraction of any chemical that is not yet stored in an inactive form or tissue, and therefore circulates within the body, the rate of overall accumulation in specific tissues can be influenced by elimination through processes such as biotransformation (e.g., metabolism in liver or sequestration) or excretion (e.g., renal excretion in urine). Here the situation for organic chemicals and metals has many of the same complications when considering mixtures, as 1 mixture component may affect the biochemical reaction rates for another, be it enzymatic transformation for organic chemicals or the binding to proteins for metals. However, biotransformation to metabolites in the case of organic chemicals does add to the mixture complexity over that of metals, as such metabolites may have a different toxicological profile than their parent chemical or could simply affect the conversion of other organic chemicals in the mixture.

2.3.1.3 Metabolism and Excretion

Over the years, research efforts in biomedical sciences from academia, industry, and government institutions have underpinned a wealth of detailed knowledge regarding metabolism and physiology in humans and vertebrates, and many TK models have been developed. A critical aspect for the development of such models is the identification of the specific enzymes involved in the metabolism of a particular compound. For vertebrates these enzymes are well characterized, at least in terms of structure, if not also in terms of function, but such detailed knowledge is not available for invertebrates. In humans, metabolic routes can be split into phase I, phase II, and renal excretion. However, the relatively recent characterization of transporters such as P-glycoprotein has introduced them in the system as phase 0 or phase III because they can transport the parent compound or the metabolite. Major metabolic routes include phase I enzymes responsible for initial oxidation, reduction, and hydrolysis (e.g., CYP1A2, CYP2A6, CYP2E1, CYP3A4, CYP2C9, CYP2C19, CYP2D6, alcohol dehydrogenases, and esterases) and phase II enzymes responsible for conjugation of phase I metabolite with endogenous substance usually containing an ionized group to aid water solubility and excretion (e.g., glucuronidation, glycine conjugation, sulfate conjugation, methyltransferases, N-acetyl transferases, and glutathione-S-transferases). Interindividual differences in these metabolic routes are large when considering subgroups of the population, such as extensive and poor metabolizers (due to polymorphic CYP2C9, CYP2C19, CYP2D6, and others), interage (e.g., neonates, children, and elderly), and interethnic differences. The integration of such differences is becoming of increasing importance for the validation and reliability of TK studies and models used in human risk assessments (Dorne et al. 2005; Dorne 2007).

When dealing with chemical mixtures, such assessment of metabolic variability is also critical, but is often difficult since the molecular interactions between mixture components need to be assessed at the TK and TD level to determine whether these interactions may potentiate toxicity. Recently, data for the CY2D6, CYP2C9, and CYP2C19 polymorphic enzymes were analyzed from the literature to quantify the magnitude of interaction between probe substrates of each isoform and inhibitors (competitive and noncompetitive) and inducers. Overall, inhibition or induction would increase or decrease exposure to chemicals in extensive metabolizers (EMs) (and poor metabolizers (PMs) for induction). EMs would be susceptible to toxicity if the compound was activated to a toxic species, and PMs would be at risk if the parent compound was the toxicant. This database was built using in vivo data quantifying human variability in TK interactions using therapeutic doses of pharmaceuticals (Dorne et al. 2007a, 2007b; Dorne and Papadopoulos 2008). A number of pesticides and environmental contaminants are known to either inhibit or induce CYP isoforms in animals and humans, and more research is needed to characterize their potential in vivo effects at the current level of exposure using recombinant technology and toxicokinetic assays (Hodgson and Rose 2005).

2.3.2 Data-Based and Physiologically Based Toxicokinetic Models

Fundamentally, 2 levels of toxicokinetic modeling approaches are commonly used for single compounds within the human and ecotoxicology arena: 1) data-based toxicokinetic (DBTK) modeling and 2) physiologically based toxicokinetic (PBTK) modeling. This naming should not give the impression that PBTK modeling is not data based. In fact, PBTK modeling is much more rigorous scientifically than DBTK, also known as "classical TK." The 2 types of modeling differ essentially in the fact that the mathematical equations in DBTK models, also commonly referred to as classical TK models, are purely descriptive of the experimental kinetic data (e.g., blood, tissue, or whole body concentration vs. time profiles), whereas in PBTK modeling the equations describe the mechanistic processes of ADME. It is interesting to point out that values for parameters in DBTK modeling are derived from model fitting to experimental data (e.g., overall uptake and clearance rates). On the other hand, the parameters used in PBTK refer to organism-specific or compound-specific properties that are available or can be estimated independently from the toxicokinetic data. Both types of modeling approaches, however, use the principle of mass balance differential equations to describe the time course of concentrations in the compartments of the model. It is worth mentioning up front that this distinction becomes difficult for some of the models that are discussed in the following text (e.g., a model based on empirically derived K_{ow} values could be argued to be mechanistic). Moreover, mixed models exist, and some of the relationships derived from such models are based on statistical regression of empirical data (i.e., no mechanistics) and others on mechanistic modeling.

2.3.2.1 Data-Based Toxicokinetic Models

The simplest DBTK model is the well-known 1-compartment model (see Figure 2.3). In this approach, the entire organism is treated as a single well-mixed compartment. Assuming first-order kinetics, the time course of the internal concentrations can be described by 2 parameters (e.g., an uptake and an elimination rate constant). Despite its simplicity, the 1-compartment model is generally very successful in describing uptake and elimination kinetics in a wide range of organisms. Provided the internal redistribution of the chemical is fast, relative to the uptake and elimination kinetics, this simple approach may suffice for many applications. The parameters are generally obtained by fitting the model to body residue data, but for organic chemicals, these parameters generally show predictable relationships with how hydrophobic a chemical is (i.e., octanol–water partitioning coefficients [K_{ow}]) (for overview see Barber 2003). Similarly, the liposome–water distribution ratio at pH 7 (D_{lipw} [pH 7]) has been proposed as a predictor for ionizable chemicals (Escher and Sigg, 2004). Mayer and Holmstrup (2008) recently found that springtail lethality correlates poorly with the octanol–water partition coefficients of the PAHs used in the exposures, whereas parameters closely related to the PAH's chemical activity provided much better correlations with the observed effects.

When measured internal concentrations are lacking, but a time series of effect data is available, then a 1-compartment model may still be useful to describe and analyze the toxicokinetic behavior of a substance. The effects pattern in time provides information on the toxicokinetics as well as toxicodynamics. Jager and Kooijman (2009) demonstrate that the relevant TK parameter can generally be derived from survival data with reasonable accuracy, and discuss how these scaled toxicokinetics relate to the whole body kinetics.

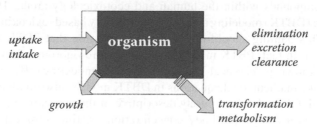

Figure 2.3 An example of a data-based toxicokinetic model, in this case a 1-compartment model. The concentration of chemical in an organism may increase by intake and uptake but decrease due to growth dilution, transformation or metabolism and elimination, excretion or clearance.

The 1-compartment model is easily extended to deal with body growth (e.g., Kooijman and Bedaux 1996), metabolism (by adding a first-order rate constant to the elimination rate, e.g., Mackay et al. 1992b), or more compartments. In some cases, additional descriptive compartments are added to accommodate observed patterns in the elimination of toxicants (e.g., Janssen et al. 1991; Steen Redeker and Blust 2004), often denoted the "central" and "peripheral" or "fast" and "slow" compartment. In other cases, compartments are added to provide physiological detail. Examples include the addition of a gut compartment to deal with uptake from food (Gobas et al. 1993; Jager et al. 2003; Steen Redeker et al. 2004), to describe storage of metals to a tightly bound fraction (Vijver et al. 2005), or to split up the organism in different tissues. In TK models for higher plants, a distinction is sometimes made between roots, stem, fruit, and leaves (Trapp and Matthies 1995; Trapp and McFarlane 1995). Some TK model approaches include more complex kinetics by including specific interactions with receptor proteins (Legierse et al. 1999; Jager and Kooijman 2005) or introducing damage to cell components as an additional state variable (Lee et al. 2002a). Even though most of these approaches are quite descriptive, we can still sometimes apply mechanistic knowledge or allometric scaling to go from one species to another, or from a small individual to a large one (see Sijm and Van der Linde 1995; Kooijman 2000).

The approach of DBTK modeling in human toxicology is well explained in book chapters (Renwick and Hayes 2001) and books dedicated to that subject (Rowland and Tozer 1995; Gabrielsson and Weiner 2000; Tozer and Rowland 2006). For more detailed information on TK models in ecotoxicology, the reader is referred to reviews (Mackay and Fraser 2000; Barber 2003), books (Nagel and Loskill 1991; Van Leeuwen and Vermeire 2007), and Section 2.3.4.

2.3.2.2 Physiologically Based Toxicokinetic Models

Apart from the descriptive DBTK models for vertebrates, PBTK approaches have been developed with numerous human and rodent PBTK models and, to a limited extent, within the ecotoxicology arena with examples in fish (Nichols et al. 1990; Law et al. 1991) and marine mammals (Hickie et al. 1999). There is also an increasing interest in using PBTK modeling for extrapolation to wildlife species (National Health and Environmental Effect Research 2005), and for this purpose, candidate species have been identified and include the mallard duck, Japanese quail, pheasant, and chicken. There is also growing research interest to develop models for other mammalian wildlife species and amphibians to address specific research and risk assessment needs.

One of the reasons for this growing interest in PBTK modeling is the fact that the equations are descriptions of the different mechanisms involved in ADME and incorporate relevant understanding and information on physiology, anatomy of the organism, and the physicochemical properties of the studied chemical. The biochemical interactions between the organism and the chemical are mathematically described with as much realism as possible or necessary for the purpose of concern. The following paragraphs describe how a PBTK model is developed and of what it consists. PBTK model development can be divided in the following steps: model representation, model parameterization, and model simulation, validation, and refinement. These steps have been well described in the exhaustive review by Krishnan et al. (2001).

2.3.2.2.1   PBTK Model Representation

In PBTK modeling, the whole organism is represented as different compartments of tissues or organs that are connected by blood circulation (Figure 2.4). The chemical typically enters the organism by absorption through the gut wall, inhalation in lungs, or permeation through skin. Once absorbed, the chemical reaches the tissue compartments by blood circulation, where it can accumulate as a result of its rate of entry by arterial blood and the rate of exit by venous blood (Figure 2.4). Chemicals must cross biological membranes from plasma to the organ cells. This diffusion rate dictates whether distribution in compartments can be described by perfusion limitation or diffusion limitation. For example, the PBTK model for pyrene in the rats (Haddad et al. 1998) had diffusion-limited distribution described

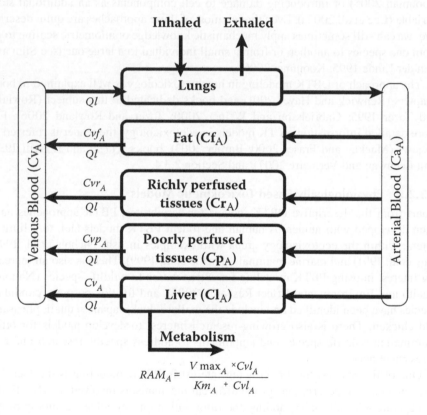

Figure 2.4   A typical conceptual representation of a PBTK model for a volatile organic chemical A. Each box represents a tissue compartment and arrows depict arterial and venous blood circulation. RAM refers to the rate of the amount metabolized. V_{max} and K_m refer to the maximal rate of metabolism and the Michaelis-Menten affinity constant, respectively. C is concentration in blood (V_A), fat (F_A), richly perfused tissues (RA), poorly perfused tissues (P_A), liver (L_A), and arterial blood (a_A). Ql is the blood flow.

for the muscles and the fat compartments, whereas other tissues were described by perfusion limitation. A plausible explanation for this difference is the porosity of the capillary wall, which is very different with other tissues. The rate of accumulation in tissue can also be influenced by some form of elimination, such as biotransformation (e.g., metabolism in liver) or excretion (e.g., renal excretion in urine or fecal depuration).

2.3.2.2.2 PBTK Model Parameterization

The parameters needed for PBTK models can be separated into 3 categories (Table 2.1): physiological, physicochemical, and biochemical. Whereas physiological parameters are specific to the organism of interest, the physicochemical and biochemical parameters are specific to the chemical and species.

The sources for the values of these parameters may be diverse. Physiological parameters are usually well known for humans and other mammals that have been extensively used in toxicology and pharmacology laboratories, such as rodents, dogs, and some primates, and can be found in the literature (Brown et al. 1997). Information is available on how physiological parameters change with respect to age (e.g., Price et al. 2003ab; Ginsberg et al. 2004; Bjorkman 2005; Nong et al. 2006) and during special physiological conditions such as pregnancy (Clewell et al. 2003; Gentry et al. 2003). Physiological parameter values are also available for some ecotoxicological relevant species and have been used to develop PBTK models in species such as trout (Law et al. 1991; Nichols et al. 2004a, 2004b), starry flounder (Namdari 1998), salmon (Brocklebank et al. 1997), channel catfish (Albers and Dixon 2002), and beluga (Hickie et al. 1999). When the values are not available, allometric scaling techniques can be used for their estimation (Young et al. 2001); otherwise, measurements or estimates (see below) must be done in the species of interest before PBTK modeling can be used.

In practice, because the physicochemical and biochemical parameters are chemical as well as species specific, they are seldom available (unless determined in previous studies), and therefore must be estimated. These estimations can be made by in vivo, in vitro, or even in silico methods. Reviews on estimation of these parameters are available (Krishnan et al. 2001; Theil et al. 2003; Nong et al. 2006). Briefly,

Table 2.1 The parameters most frequently used in human and mammalian PBTK models

Physiological	Physicochemical	Biochemical
Cardiac output	Partition coefficients	Rate constants (V_{max}, K_m) for
Body weight	Tissue-blood	Metabolism
Tissue volumes	Blood-air	Urinary excretion
Tissue blood flow	Blood-water (fish)	Biliary excretion
Glomerular filtration rate	Skin-air	Macromolecular binding constants
Tissue blood content	Skin-water	B_{max}
Tissue lipid and water content	Permeability coefficients	K_d
Skin surface area		

partition coefficients can be estimated using in vitro approaches such as ultracen-
trifugation or equilibrium dialysis, or in silico approaches that are mechanistically
based (Haddad et al. 2000a; Poulin et al. 2001; Poulin and Theil 2002; Rodgers et
al. 2005; Rodgers and Rowland 2006) or that use QSAR approaches (Fouchecourt
et al. 2001; Beliveau et al. 2003, 2005; Beliveau and Krishnan 2005). The mac-
romolecular binding parameters, which can greatly influence the distribution of
a chemical in blood and tissues, can usually be determined by the same in vitro
techniques. Many in vitro models are used to determine the values of biochemical
parameters; for example, metabolic rate constants (i.e., intrinsic clearance, V_{max},
K_m) can be determined using isolated hepatocytes, microsomal fractions, cytosolic
fractions, or postmitochondrial fractions preparation. Many of these approaches
for determining values for hepatic clearance parameters in vitro have been successfully extrapolated to in vivo (Mahmood 2002; Ito and Houston 2004, 2005;
Riley et al. 2005; Skaggs et al. 2006). There are also in silico methods that are currently being developed and which are very promising for predicting ligand-protein
interactions and rates of enzymatic metabolism. These in silico methods can be
based on the known chemical structures (QSAR) and their affinity to the proteins
or enzyme, and more recently with the more powerful computational technologies,
methods are being developed based on enzyme primary structure and 3D compu-
tations (De Groot and Ekins 2002; Ekins 2003; Balakin et al. 2004; De Groot et
al. 2004; De Graaf et al. 2005; Ekins et al. 2006). The use of such techniques and
knowledge, of course, presuppose that it is known which enzymes are involved in
the metabolism.

2.3.2.2.3   PBTK Model Simulation, Validation, and Refinement

Another attractive feature of PBTK modeling is that it can be used as a hypothesis
testing tool for mechanistic research. Because these models are constituted of mechanistic representations of the different mechanisms involved in the toxicokinetics of
chemicals, it is possible to make a hypothesis of a particular mechanism and test the
model by comparing model simulations to experimental data. When there is a dis-
agreement, the model can be further modified to test other proposed toxicokinetic
mechanisms until the simulations agree with the experimental data. Such mechanisms can then be confirmed through in vitro or in vivo tests. Such PBTK modeling
renders it very useful in orienting toxicological mechanistic research.

2.3.3   APPLICATION OF MIXTURE TK MODELING IN HUMAN AND MAMMALIAN TOXICOLOGY

There have been considerable efforts in modeling the toxicokinetics of mixture constituents in mammalian toxicology. This has mainly been done using PBTK modeling approaches. The reason is that, because of their mechanistic basis, PBTK models
have the advantage of facilitating different types of extrapolations: from one route
of exposure to another, from one species to another, from high-dose exposure to low
dose, from one exposure scenario to another, and as described below, from binary
mixtures to multichemical mixtures.

2.3.3.1　PBTK Modeling of Chemical Mixtures

A key feature of PBTK modeling is that it can be used to integrate information on toxicological interactions. Ideally, to develop PBTK models for interacting chemicals, one would like to use single chemical PBTK models that have been validated for the chemicals present in the mixture of interest. The interactions-based PBTK model can then be developed by linking the single chemical models by known or hypothesized mathematical descriptions of their mechanism of toxicokinetic interactions. For example, chemical A may inhibit the hepatic metabolism of chemical B, and vice versa, by competing for the active enzyme (competitive inhibition). The rates of metabolism (RAMs) of chemicals A and B are then adjusted by accounting for the concentration of the other chemical in the liver compartment. For chemical A this would be done multiplying the Michaelis-Menten constant $K_{m,A}$ by the term $(1 + C_{vl,B}/K_{i,AB})$, where $K_{i,AB}$ is the inhibition constant for chemical B on the metabolism of chemical A. There are numerous examples of interactions-based PBTK models for binary mixtures (see Krishnan et al. 1994; Yang and Andersen 2005), and in all cases these mathematical descriptions of toxicokinetic interactions can be expressed as a modification of a parameter of the ADME processes of one chemical as a function of the presence of another chemical. Although numerous nonbiochemical interactions have been observed (e.g., Krishnan and Brodeur 1991), the toxicokinetic interactions so far described by PBTK modeling have all dealt with modifications of biochemical parameters. The reason for this is simply that they are, by far, the most frequently observed toxicokinetic interactions. Among all the known toxicokinetic interactions, metabolic inhibition is the most frequently observed mechanism and can be placed in 2 categories: irreversible and reversible inhibition. Among other types of TK interactions that are observed at the biochemical level are enzyme induction (i.e., increase in enzyme levels) and competitive binding to plasma or tissue proteins (e.g., chemical displacement on plasma albumin or on uterine oestrogen receptor).

2.3.3.1.1　Reversible Metabolic Inhibition

Among all the interactions-based PBTK models published to date, reversible metabolic inhibition is by far the most frequently encountered type of interaction. There are 3 types of reversible enzyme inhibition: competitive, noncompetitive, and uncompetitive (Table 2.2), and examples of all are listed in Table 2.3. A large number of examples of such metabolic inhibition in humans and laboratory animals are available for specific CYP enzymes and therapeutic drugs (Dorne et al. 2007b).

Competitive inhibition occurs when chemicals compete for the space at another enzyme's active site and this results in decreased apparent affinity (i.e., increased K_m), and therefore reduces the rate of metabolism at lower substrate concentrations. Whether the competitive inhibitor is another substrate or strictly an inhibitor, the mathematical description is the same (Table 2.2). The inhibition constant (K_i) of a competing substrate should have the same value as its K_m for the enzyme involved (Segel 1975) since both parameters reflect the chemical's affinity to the enzyme site of action.

Noncompetitive inhibition occurs when a chemical binds to the enzyme (whether free or complexed with a substrate) at a binding site that is away from the catalytic

Table 2.2   Description of the effect of reversible metabolic inhibition on the parameters V_{max}, K_m, and intrinsic clearance

	Competitive	Noncompetitive	Uncompetitive
Equilibrium equation[a]	$E + S \leftrightharpoons ES \rightarrow E + P$ + I $\Updownarrow$ EI	$E + S \leftrightharpoons ES \rightarrow E + P$ + I $\Updownarrow$ EI+S	$E + S \leftrightharpoons ES \rightarrow E + P$ + I $\Updownarrow$ EIS
		$\leftrightharpoons$ EIS	
Apparent K_m	αK_m	K_m	K_m/α
Apparent V_{max}	V_{max}	V_{max}/α	V_{max}/α

Note: $\alpha = 1 + \dfrac{[I]}{K_i}$, E = enzyme, S = substrate, P = product, I = inhibitor, K_i = the constant describing competitive inhibition of metabolism, K_m = Michaelis constant for metabolic elimination, V_{max} = maximum metabolic rate.

[a]   Segel IH. 1975. Enzyme kinetics: behavior and analysis of rapid equilibrium and steady-state enzyme systems. Toronto (CA): John Wiley & Sons.

active site. This binding changes the conformation of the enzyme and results in a decreased catalytic activity (i.e., decreased V_{max}).

The less frequently encountered uncompetitive inhibition occurs when a chemical binds to the enzyme-substrate complex. The catalytic function is affected without interfering with substrate binding. The inhibitor causes a structural distortion of the active site and inactivates it (Voet and Voet 2004). This has the effect of reducing the available enzyme for the reaction, and hence reduces V_{max}, and also drives the reaction (E + S $\leftrightharpoons$ ES) to the right, hence decreasing K_m.

It is noteworthy that these types of reversible metabolic inhibitions can also be applied to active transport processes. In fact, Sugita et al. (1982) have described noncompetitive inhibition as the mechanism of binary interaction between sodium tolbutamide and 3 other drugs (sulfaphenozol, sodium sulfadimethoxazole, and sodium sulfadimethoxine) (Table 2.3).

Metabolites may also interact with the biotransformation of their parent compound. This phenomenon is called "product inhibition" and can use a competitive, noncompetitive, or uncompetitive mechanism. For example, *n*-hexane is metabolized to methyl *n*-butyl ketone (MnBK). MnBK is then further biotransformed to 2,5-hexanedione, the neurotoxic metabolite, by ω-1 oxidation, and to pentanoic acid by α-oxidation and decarboxylation. To simulate the kinetics of 2,5-hexanedione, Andersen and Clewell (1983) had to take into account multiple interactions occurring between n-hexane and metabolites (Figure 2.5), all of which were competitive inhibition.

2.3.3.1.2   Irreversible Metabolic Inhibition

When an inhibitor binds irreversibly to the enzyme at the active site, this usually inactivates the enzyme's catalytic activity. This type of inhibition decreases the concentration of the functional enzyme and therefore results in a decrease of V_{max} (i.e., an irreversible loss of enzyme activity). Briefly, the modeling of irreversible

Table 2.3 Interactions-based PBTK models developed for reversible inhibition in binary mixtures

Substance 1	Substance 2	Type	Reference
Warfarin	Bromosulfophthalein	C	Luecke and Wosilait (1979)
Sodium tolbutamide	Sulfaphenozol	N	Sugita et al. (1982)
	Sodium sulfadimethoxazole	N	
	Sodium Sulfadimethoxine	N	
Dibromomethane	Isofluorane	C	Clewell and Andersen (1985)
Trichloroethylene	1,1-Dichloroethylene	C	Andersen et al. (1987)
Phenolsulfonphthalein	Probenecid	C	Russel et al. (1987)
	Saliuric acid	C (2 sites)	
Saliuric acid	Phenolsulfonphthalein	N	Russel et al. (1987)
Iodopracet	Probenecid	C	Russel et al. (1989)
Benzene	Toluene	N	Purcell et al. (1990)
Toluene	m-Xylene	C	Tardif et al. (1993)
1,3-Butadiene	Styrene	C	Filser et al. (1993)
1,3-Butadiene	Styrene	C	Bond et al. (1994)
1,3-Butadiene	Benzene	C	Bond et al. (1994)
1,1-Dichloroethylene	Vinyl chloride	C	Barton et al. (1995)
Toluene	Xylene	C	Tardif et al. (1995)
Toluene	Dichloromethane	C	Krishnan and Pelekis (1995)
Trichloroethylene	1,1-Dichloroethylene	C	El-Masri et al. (1996a,b)
1,3-Butadiene	Styrene	C	Leavens and Bond (1996)
Toluene	Dichloromethane	C	Pelekis and Krishnan (1997)
Ranitidine	Cimetidine	C	Boom et al. (1998)
	Probenecid	C	
Toluene	n-Hexane	N	Yu et al. (1998)
Toluene	n-Hexane	U or N	Ali and Tardif (1999)
Methyl chloroform	m-Xylene	C	Tardif and Charest-Tardif (1999)
Toluene	Trichloroethylene	C	Thrall and Poet (2000)
Ethylbenzene	Xylenes	C	Jang et al. (2001)
Simvastatin	Itraconazole	C	Ishigam et al. (2001)
Chloroform	Trichloroethylene	C	Isaacs et al. (2004)

Note: C = competitive, N = noncompetitive, U = uncompetitive.

metabolic inhibition consists of describing the change in V_{max} of the substrate by taking into account the activation of the enzyme in relation to the concentration of the inhibitor. Recently, this type of inhibition has been modeled by interactions-based PBTK models for mixtures of triazolam and erythromycin (Kanamitsu et al. 2000a), 5-fluorouracil and sorovidine (Kanamitsu et al. 2000b), and trichloroethylene and its metabolite dichloroacetate (Keys et al. 2004).

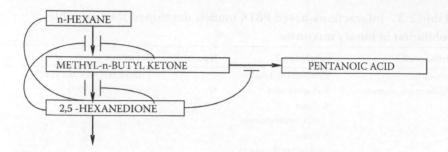

Figure 2.5 Interactions between n-hexane and its metabolites. Arrows (→) represent enzymatic reaction, and lines terminating in '–\' depict a competitive inhibition. (After Andersen and Clewell [1983]).

2.3.3.1.3 Enzymatic Induction

It is well known that exposure to certain chemicals results in increased levels of specific enzymes (e.g., TCDD induces CYP1A1 and ethanol induces CYP2E1), and therefore can lead to changes in metabolic rates of other coexposed chemicals. This phenomenon, called "enzyme induction," results in an increase of V_{max}. Because the enzyme levels are dependent on its synthesis and degradation rates, the rate of change in enzyme levels can be the result of 2 mechanisms: increased enzyme synthesis or decreased enzyme degradation. As mentioned for enzyme inhibition, many examples of inducers are available from the therapeutic drug database for specific CYP enzymes in humans (Dorne et al. 2007b).

Induction by increased enzyme synthesis has been described for induction of CYP2B1/2 by octamethylcyclotetrasiloxane (Sarangapani et al. 2002) and CYP1A1 and CYP1A2 by TCDD via interaction with the Ah-receptor (Andersen 1995; Andersen et al. 1997). Induction by decreased enzyme degradation has been described by Chien et al. (1997) for induction of CYP2E1 by ethanol, acetone, and isoniazid via enzyme stabilization. Exposure to such enzyme inducers and stabilizators would result in increasing the clearance of all other coexposed chemicals that are metabolized by the stabilized enzyme.

2.3.3.1.4 Reversible Protein Binding Interaction

During circulation in the blood stream chemicals may bind to proteins (called "binding proteins"), which makes the chemical unavailable for enzymatic metabolism or for causing toxic effects. Chemicals can thus modify each others "free and available" concentrations through interactions at the level of protein binding where the mechanisms can either be competition for the binding site or induction of binding protein levels (through mechanisms analogous to those described earlier for enzymatic induction). Sugita et al. (1982) developed a PBTK model describing the competition between tolbutamide and sulfonamides (sulfaphenazol, sulfadimethoxin, and sulfomethoxazol) for binding to plasma protein. The model was mathematically similar to that of competitive metabolic inhibition, apart from the fact that here it is not the chemical and the inhibitor chemical competing for the enzyme, but the "inhibitor" is the binding protein that competes with the enzyme for the substrate chemical to lower the rate of metabolism.

2.3.3.2 Modeling Multichemical Mixtures

The interactions-based PBTK modeling examples cited so far dealt with binary mixtures. PBTK modeling of multichemical mixtures is at the embryonic stage, and the modeling strategies that have been put forward to tackle this subject for defined and undefined mixtures are presented below.

Pioneering efforts in the PBPK modeling of more complex chemical mixtures were from a research group led by Professor Kannan Krishnan, Université de Montréal, Canada. Earlier work from this group concentrated on interactions and PBTK modeling between 2 chemicals, for instance, toluene and benzene (Pelekis and Krishnan 1997), toluene and m-xylene (Tardif et al. 1995), and toluene and ethylbenzene (Tardif et al. 1993). As progress was made, these investigators began to build up the mixtures and devoted their effort to PBTK modeling of more complex chemical mixtures. The hypothesis put forward and tested by this group was that the pharmacokinetics of interacting chemical mixture components may be simulated if all binary interactions occuring in the mixtures are described in the model. Briefly, the approach consists of taking the PBTK models for each mixture constituent and connecting them by the description of toxicokinetic interactions at the binary level, which is characterized a priori. By doing so, a network of binary toxicokinetic interactions is created (Figure 2.6) and the kinetics of the mixture constituents becomes interdependent. Any change in the levels of a mixture component therefore necessarily has direct or indirect repercussions on the kinetics of chemicals that are part of the interaction network (Haddad and Krishnan 1998; Krishnan et al. 2002). Conceptually, this approach can be applied to mixtures of any complexity as long as information of all interacting pairs is taken into account in the model.

Using this approach, Tardif et al. (1997) simulated interactions occurring in a ternary mixture of toluene, m-xylene, and ethylbenzene. Then Haddad et al. (1999, 2000b) demonstrated that additional mixture components could be added to the model by simply incorporating mechanistic information on the new binary

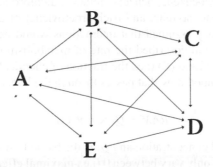

Figure 2.6 A representation of a network of binary toxicokinetic interactions between 5 chemicals. (A, B, C, D and E). All mixture constituents that interact are connected directly or indirectly through this interactions network.

interactions (Figure 2.7). This led to the development and validation of interactions-based PBTK models for mixtures of 4 chemicals (benzene, toluene, m-xylene, and ethylbenzene) and even of 5 chemicals (benzene, toluene, m-xylene, ethylbenzene, and dichloromethane). Applying the same approach, the research group of Professor Raymond S. H. Yang was able to model interactions occurring in a ternary mixture of trichloroethylene (TCE), tetrachloroethylene (PERC), and 1,1,1-trichloroethane (methyl chloroform (MC)) in rats and humans, as well as a complex mixture, gasoline, in rats (Dobrev et al. 2001, 2002; Dennison et al. 2003, 2004a, 2004b; Yang et al. 2004). Although all these studies dealt strictly with interactions that were of a metabolic competitive inhibition nature, there is no reason to believe that the approach using a binary interactions network may not apply to all types of toxicokinetic interactions.

To use this mixture modeling approach, a priori studies must be made on all binary interactions existing in the mixture in order to obtain information on their mechanisms and the values of parameters for their description. This is a major limitation for the construction of mixture modeling since the number of binary interactions (N) increases greatly with the number of mixture components (n) (i.e., $N = n(n - 1)/2$). Accordingly, for mixtures of 5 and 10 chemicals, 10 and 45 binary interactions must be previously studied, respectively. Considering the complexity of some of the mixtures to which humans or other species are exposed to, the characterization of all binary interactions is tedious or even impossible. Some alternative mixture modeling methods have therefore been proposed.

One of the first proposed alternative approaches to binary interactions-based PBPK modeling of mixtures was brought forth by Haddad et al. (2000c). They suggested that when there is insufficient mechanistic information or data on the binary interactions or on the mixture composition, one could get a worst-case estimate by modeling the maximal impact of interactions on the kinetics of the chemicals. Although this approach does not allow the exact prediction of the venous concentrations, it does permit the prediction of the upper- or lower-bound concentrations that could occur in the presence of interacting chemicals. The authors demonstrated the applicability of such an approach for mixtures of volatile organic compounds (VOCs) (m-xylene, ethylbenzene, benzene, toluene, dichloromethane, perchloroethylene, p-xylene, o-xylene, styrene, and trichloroethylene) of up to 10 constituents. For these chemicals, they assumed that interactions would occur only at the level of hepatic metabolism and described the rate of metabolism (RAM) as a function of the hepatic extraction ratio (E) (i.e., the fraction of chemical that is eliminated or metabolized from the blood flow that passes through the liver), as follows:

$$RAM = Ql \times E \times Ca$$

where Ca is the arterial concentration and Ql is the blood flow to the liver. Knowing that the values of E can only vary between 0 (i.e., maximal effect of inhibition) and 1 (maximal effect of induction), they simulated the concentration vs. time profiles for a determined exposure condition of 10 chemicals in various mixture scenarios at $E = 0$, $E = 1$. Their results showed that the limits were well predicted for 9 chemicals out of 10 (Figure 2.8). This approach allows generating conservative but realistic

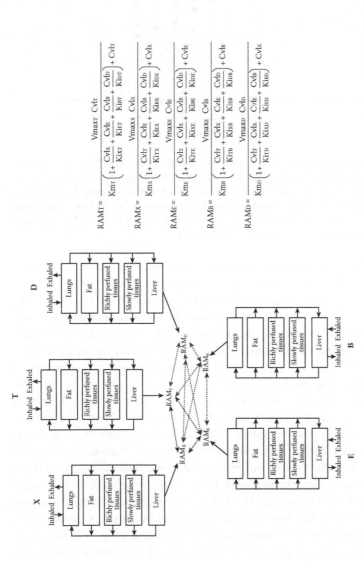

$$RAM_T = \frac{V_{maxT}\, C_{vlT}}{K_{mT}\left(1+\dfrac{C_{vlx}}{K_{ixT}}+\dfrac{C_{vlE}}{K_{iET}}+\dfrac{C_{vlB}}{K_{iBT}}+\dfrac{C_{vlD}}{K_{iDT}}\right)+C_{vlT}}$$

$$RAM_X = \frac{V_{maxx}\, C_{vlx}}{K_{mx}\left(1+\dfrac{C_{vlT}}{K_{iTx}}+\dfrac{C_{vlE}}{K_{iEx}}+\dfrac{C_{vlB}}{K_{iBx}}+\dfrac{C_{vlD}}{K_{iDx}}\right)+C_{vlx}}$$

$$RAM_E = \frac{V_{maxE}\, C_{vlE}}{K_{mE}\left(1+\dfrac{C_{vlT}}{K_{iTE}}+\dfrac{C_{vlx}}{K_{ixE}}+\dfrac{C_{vlB}}{K_{iBE}}+\dfrac{C_{vlD}}{K_{iDE}}\right)+C_{vlE}}$$

$$RAM_B = \frac{V_{maxB}\, C_{vlB}}{K_{mB}\left(1+\dfrac{C_{vlT}}{K_{iTB}}+\dfrac{C_{vlx}}{K_{ixB}}+\dfrac{C_{vlE}}{K_{iEB}}+\dfrac{C_{vlD}}{K_{iDB}}\right)+C_{vlB}}$$

$$RAM_D = \frac{V_{maxD}\, C_{vlD}}{K_{mD}\left(1+\dfrac{C_{vlT}}{K_{iTD}}+\dfrac{C_{vlx}}{K_{ixD}}+\dfrac{C_{vlE}}{K_{iED}}+\dfrac{C_{vlB}}{K_{iBD}}\right)+C_{vlx}}$$

Figure 2.7 Representation of the PBTK model developed for a mixture of 5 VOCs (m-xylene, toluene, ethylbenzene, benzene, and dichloromethane). All binary interactions that occur at the level of the rate of metabolism (RAM) are taken into account between the mixture constituents as shown by the dotted arrows. Because all chemicals interact by competitive inhibition, the K_m of all mixture constituents is modulated by the presence of other chemicals as can be seen in the RAM equations. Cvl refers to venous blood concentrations. V_{max} and K_m refer to the maximal rate of metabolism and Michaelis affinity constant, respectively. K_{iij} is the constant describing competitive inhibition of the metabolism of chemical i by chemical j. (Figure adapted from Krishnan et al. [2002]).

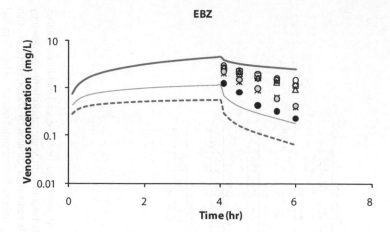

Figure 2.8 Simulations of venous blood concentrations of toluene and ethylbenzene when maximal impact of inhibition (E = 0) (thick line) and maximal impact of induction (E = 1) (dotted line) are considered in rats exposed for a duration of 4 hours at 100 ppm. Experimental data are shown for these chemicals exposed alone and in mixtures of 2 to 10 VOCs, each indicated by a different symbol. The thin line represents the simulation of the single chemical exposure scenario. (Redrawn from Haddad et al. [2000c].)

predictions of internal exposure for risk assessment purposes when insufficient information on the binary interactions in the mixture is available.

Another approach recently proposed for very large or undefined mixtures is the lumping of chemicals. Dennison et al. (2003, 2004a, 2004b) modeled the toxicokinetics of the constituents of 2 mixtures of petrol (i.e., summer blend and winter blend). In these studies, the mixture was treated as a mixture of 6 chemicals: benzene, toluene, ethylbenzene, o-xylene, n-hexane, and a pseudochemical, which is an aggregate or lumping of all the other chemicals in the mixture (e.g., aromatics, isoparaffins, naphthalenes, olefins, paraffins, oxygenates, and 4 to 10 carbon alkanes). They used the binary interactions-based approach to model the toxicokinetics of all 6 components based on estimated parameter values. Competitive inhibition was the principal mechanism of pharmacokinetic interactions among these 6 chemicals. Computer simulation results from the 6-chemical interaction model matched well with gas uptake pharmacokinetic experimental data from single chemicals, a 5-chemical mixture, and the 2 blends of gasoline. The kinetics of the pseudochemical was well predicted, partly due to the fact that interactions had very little impact on the overall venous blood concentration. This lumping approach is interesting for reducing the number of binary interaction studies and should be applicable for lumping chemicals that have similar and particular toxicokinetic behavior and interaction mechanisms.

Price and Krishnan (2005) proposed that for chemicals that are known substrates for the same enzyme, a priori characterization of binary inhibition constants was not necessary for simulating the interactions within a multichemical mixture. The hypothesis tested was that since metabolic interactions for these chemicals are generally competitive inhibition, the inhibition constants (K_i) for cosubstrates should reflect their affinity to the enzyme and hence be equal to their K_m, as stated earlier. They validated this hypothesis with a mixture of VOCs (dichloromethane, toluene, ethylbenzene, m-xylene, o-xylene, p-xylene, trichloroethylene, and styrene) in rats. Furthermore, it is interesting to note that the V_{max} and K_m of the chemicals in this study were estimated by QSAR methods. The resulting simulations yielded predictions of toxicokinetics within a 1.5-fold error compared to experimental values. Using such approaches in mixture PBTK modeling should greatly reduce the number of binary interaction studies for the development of multichemical PBTK models. This approach is suitable if the assumption holds that all chemicals are effectively strict competitors for the same enzyme and do not interact in any other way. This approach could be used as a first step in identifying and validating the default assumption of competitive inhibition as the default mechanism of binary interaction.

Recently, an effort was made by Emond et al. (2005) to model the toxicokinetics of mixtures of 6 PCB congeners in rats. Because these PCBs are highly lipophilic, they modeled their kinetics by simply considering the lipid content of every tissue compartment and lipid flow by blood circulation. Instead of characterizing all 15 binary metabolic interactions in the mixture, they used toxicokinetic data for all these compounds when administered as mixtures to optimize the metabolic rate values of each compound. The values for metabolic rates generally increased as a function of dose, which is consistent with an enzymatic induction mechanism (possibly expected between PCBs). Although this mixture modeling approach allows simulating a posteriori the kinetics of mixture constituents without needing the full set of data for all 15 binary interactions, it is constrained to being predictive only for mixture exposure scenarios (i.e., ratios and concentrations) to which the metabolic rates were optimized. This approach has the merit of facilitating the elucidation of the mechanisms of toxicokinetic interactions between mixture components.

2.3.4 APPLICATIONS OF MIXTURE TK MODELING IN ECOTOXICOLOGY

Since ecotoxicology aims to protect species at the population or community level of organization, it logically follows that regulatory testing for risk assessment focuses on endpoints that are relevant to estimate population or community sensitivities to chemicals (i.e., survival, growth, and reproduction). For ethical and cost reasons, the standard test systems developed focus largely on nonvertebrate testing with as few vertebrate tests as possible being undertaken. The spin-off from all this is that there has not yet been much drive to produce PBTK models in ecotoxicology, and furthermore, the test systems available often do not provide easy ways of obtaining the data needed. In the case of vertebrates, their size makes it possible to subdivide the organism into different compartments and enables exploration of a chemicals distribution within the organism. And while genomic, proteomic, and metabolomic studies on target organs are relatively easy to perform technically, the major constraints in vertebrate testing are ethical considerations, and hence studies are scarce. Conversely, while there are many relevant studies in invertebrate species, the organisms used are physically so small (e.g., springtails, nematodes, isopods, daphnids, algae) that it is difficult, if not impossible, to study distribution and transformation of the chemical in these organisms. Hence, this group is especially difficult to perform TK and TD studies on, and one can only look at the effect triggered by a mixture of chemicals and then try to model what is happening inside the organism. Even if efforts were made in 1 organism, then meaningful extrapolation to others would be very difficult, as invertebrates consist of ca. 50 classes, which differ considerably in physiology and biochemistry. This means that, to be useful, models would need to be developed for at least each of the most common classes. In general, the greatest number of studies was performed with aquatic test species where the routes of exposure and uptake are perhaps not too dissimilar between organisms, which could aid extrapolations. Tests with terrestrial species are more uncommon because the exposure and uptake routes vary considerably between the most common terrestrial test species of earthworms, enchytraeids, isopods, and springtails, and in addition their physiology and biochemistry are rather unknown. The adoption of species such as the nematode *Caenorhabditis elegans* and the *Drosophila* fruit fly as model organisms within the postgenomic research community, as well as research efforts to protect beneficial arthropods such as honey bees and bumble bees against agrochemicals, may provide information on mechanisms of toxicity that can be implemented in ecotoxicology-based TK modeling.

At present, however, most TK studies in ecotoxicology focus on uptake and elimination kinetics and can be modeled only by the DBTK approach. These studies generally only address the uptake and elimination of the chemicals that are detectable and how these are influenced by the other chemicals and still represent black box models. However, for many applications, this may be sufficient.

In the current practice of mixture toxicity studies in ecotoxicology, toxicokinetics is usually not considered explicitly. Sometimes, measured whole body residues are used as an additional dose metric in the dose–response analysis (e.g., van Gestel and Hensbergen 1997). Many TK approaches are available for single compounds, but very few methods have been specifically aimed toward mixtures. Nevertheless, it must be noted that the uptake and elimination kinetics of mixtures are very commonly studied in ecotoxicology. These studies are, however, not considered to be "mixture studies," but rather are designated "bioaccumulation studies," although they provide usable mixture data. For example, many studies follow the time course of uptake and elimination of various compounds from field polluted samples (e.g., for earthworms, Janssen et al. 1997; Matscheko et al. 2002; Jager et al. 2005). Furthermore, in laboratory kinetic studies, organic chemicals are often spiked together into the same medium (e.g., for earthworms, Belfroid et al. 1994; Jager et al. 2003). The advantage of such a setup is that the kinetics of all compounds can be followed after simultaneous exposure, thereby decreasing the number of animals that need to be tested. The (implicit) assumption is that these chemicals do not affect each other's toxicokinetics. As long as we stick to nontoxic levels, this assumption is possibly valid for most organic chemicals, but would not be true for metals. The following text covers the current TK-related thinking in ecotoxicology and gives examples of mixture interactions.

2.3.4.1   Organic Chemicals

Contrary to the general assumption of noninteraction between organic chemicals during uptake and distribution, there are examples where TK interactions have been reported, although mainly relating to interactions on biotransformation. For example, triazine herbicides are known to enhance the toxicity of organophosphorus (OP) insecticides, possibly by increasing biotransformation to the active oxon form (e.g., Lydy and Linck 2003). PAHs seem like a rather homogeneous mixture of substances, and are often considered in terms of the total PAH concentration during risk assessment and quality criteria setting. However, PAHs are extensively metabolized in fish, isopods, and Collembola (but not in, e.g., earthworms), and in mixtures they and their metabolites can interfere in a rather complicated manner, even showing synergistic effects (see, e.g., Willett et al. 2001; Wassenberg and Di Guilio 2004). The petrol additive methyl tert-butyl ether was shown to enhance bioconcentration of fluoranthene in fish (Cho et al. 2003), which may relate to an interaction with biotransformation of the PAH. Piperonyl butoxide is a well-known inhibitor of mixed-function oxidases (cytochrome P450), and used as a synergist for pyrethroid insecticides (Amweg, 2006), but may therefore also be a synergist for PAHs (Lee and Landrum 2006a), and an antagonist for organophosphates that need to be activated through this pathway (Ankley et al. 1991). These interactions are probably not only chemical specific, but also species specific (e.g., variations in P450 system activity), which makes it difficult to find general rules. The potential for metabolic interactions to occur is high, especially for chemicals such as pesticides that are designed and used to have specific biological effects. While some interactions are very foreseeable, like the use of piperonyl butoxide as a pyrethroid synergist, others are not. For example, pesticides may have a difference in the reversibility of binding to AChE; binding of OP pesticides tends to be more irreversible than that of carbamates. As a result, effects of both chemicals may not be additive, although they do have the same mode of action and could in theory be expected to follow CA. Another point is that most OP pesticides have to be metabolized to become active, while carbamates don't require such metabolic activation. So while in some cases carbamates and OP pesticides (both inhibitors of AChE) are found to be noninteractive, as by Scholz et al. (2006) in salmon, there are plenty of cases where interactions have been observed. The interaction between malathion and ergosterol biosynthesis inhibitor (EBI) fungicides in birds is an illustration of synergism due to interactions at the metabolic level (Walker and Johnston 1993; Johnston et al. 1994).

The application of the 1-compartment TK model to mixtures, with the explicit aim to predict toxicity, was first presented by Mackay et al. (1992b) for several sets of organic chemicals. In this simple extension of the 1-compartment model, it is assumed that none of the chemicals is transformed, and that no interactions occur in the toxicokinetics phase. More recently, Baas et al. (2009) proposed a similar approach for complex mixtures of organic chemicals. Some approaches also model metabolites explicitly and are therefore essentially mixture approaches. The damage assessment model (DAM) of Lee and Landrum (2006b) assumes a 1-compartment model for the accumulation of the parent compound, and a Michaelis-Menten biotransformation to a metabolite taking no interactions into account. Both the parent compounds and the metabolite cause "damage" inside the organism. Some of this damage is repaired and the total residual amount of damage in time becomes evident as the observed mortality (see Figure 2.9). A similar DAM extension for mixtures

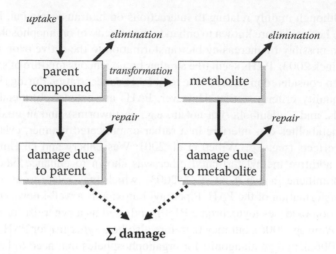

Figure 2.9 Schematic representation of the multicomponent damage assessment model of Lee and Landrum (2006b). The total damage level is linked to the toxic effect.

was presented by Ashauer et al. (2007a), who applied it to sequential pulse exposure of *Gammarus pulex* to 2 AChE inhibitors.

2.3.4.2 Metals

For metals, interactions at the level of uptake and kinetics are more common. The reason lies in the fact that most free metals are charged ions and require transport proteins such as ion channels and specific carrier enzymes to cross biological membranes. A further complication is that several metals are essential for physiological processes within the body (e.g., zinc, copper, and iron) and therefore are actively regulated. Additionally, organisms have evolved in order to live in an environment where it is not uncommon to find huge differences in the availability of free metal ions for uptake driven by different environmental conditions (e.g., due to pH, water hardness, and the presence of ligands). Organisms, therefore, have developed a wide range of physiological mechanisms to take up the metals that are needed by the organism and, on the other hand, to deal with excess levels of such metals (see, e.g., Posthuma and Van Straalen 1993; Rainbow 2002). Consequently, just the uptake side of PBTK models for metals is likely to include more details than the equivalent models for organic compounds. Additionally, due to the similarity between metals, they can replace each other within functional parts of proteins and hijack each other's biochemical metal handling pathways; thus, interactions between excess metals inside the body are likely. For example, work on shore crabs showed that the amount of metallothioneins induced by heavy metals can show either a synergistic or an antagonistic response in binary mixtures of these metals (Martín-Díaz et al. 2005b). These interactions can be species specific, as demonstrated by an antagonistic interaction of zinc on cadmium uptake found in a centipede (Kramarz, 1999a), but not in a ground beetle (Kramarz 1999b). Similarly, Fraysse et al. (2002) exposed Asiatic clam (*Corbicula fluminea*) and zebra mussel (*Dreissena polymorpha*) to Cd and Zn to determine uptake and depuration of ^{57}Co, ^{110}Ag, and ^{134}Cs. Zn and the Cd + Zn mixture increased the ^{110}Ag uptake process in both clams and mussels. The 2 metals also increased the depuration of this radionuclide in mussels, whereas this phenomenon was only observed in clams exposed to cadmium. This demonstrates that great care must be taken when extrapolating between species. Even if the functional biochemical units involved (e.g., metallothionein) may be highly conserved between species, there is a likelihood that their regulation is different among organisms.

The complexity of metal TK is illustrated by the work on biotic ligand models (BLMs) (for an overview see Paquin et al. 2002a). In these models, the target site for the toxic metal is assumed to lie on a biological surface, such as the gill membrane in fish. BLM approaches explicitly consider speciation, interactions with inorganic and organic ligands in the medium, and competition with other ions at the target site (e.g., Ca^{2+} and H^+). However, the BLMs currently consider only the equilibrium situation and are therefore perhaps not strictly TK models, as extensively discussed by Slaveykova and Wilkinson (2005). Furthermore, BLMs do not consider internal redistribution (toxicity is thought to be related to targets on outer membranes, which is widely accepted for acute effects of several metals in freshwater fish (Paquin et al. 2002a)) and have only recently been suggested for multiple toxic metals (Playle 2004). As demonstrated by Playle (2004), the BLM approaches offer options for dealing with mixtures of metals, although it may be quite difficult in practice, especially for metals that have different target ligands (see Paquin et al. 2002a). Additionally, there are possibilities to move toward a more kinetic BLM approach (Paquin et al. 2002b).

In a fashion similar to the discussion presented on organic chemicals, Baas et al. (2007) applied the 1-compartment model without TK interactions for the analysis of-time series survival data for the springtail *Folsomia candida* exposed to binary mixtures of heavy metals. It must be stressed that no internal concentrations were measured in these experiments; instead, the toxicokinetics parameters were solely determined from the survival pattern in time. In this case, the toxicity data were well described without assuming interactions, which stresses that even though we know that interactions on toxicokinetics can occur, this does not mean that they will significantly influence toxicity for every metal mixture in each organism.

2.3.4.3 Mixtures of Organic Compounds and Metals

TK interactions between metals and organic compounds are also possible; phenanthrene appears to enhance the uptake of cadmium from sediment in the amphipod *Hyalella azteca* (Gust and Fleeger 2005). In the same species, chlorpyrifos enhances the accumulation of methyl mercury, but methyl mercury reduces acetylcholinesterase inhibition caused by chlorpyrifos, presumably due to the formation of a chlorpyrifos-MeHg complex (Steevens and Benson 1999).

2.4 TOXICODYNAMICS

2.4.1 GENERAL ASPECTS

The previous section dealt with the toxicokinetic aspect of chemicals and specifically focused on new methods (e.g., physiologically based modeling) used to understand and predict the behavior of mixtures in mammalian toxicology and ecological research. As indicated earlier, it is difficult to distinguish where toxicokinetics ends and toxicodynamics starts (see Figure 2.1). Hence, both aspects should be considered in order to have a full understanding of the toxicology of a chemical or chemical mixture. This section discusses the toxicodynamic aspect of chemicals and illustrates this with examples drawn from both mammalian biology and ecology.

Toxicodynamics is the study of the toxic actions on living systems, including the reactions with and binding to cell constitutents, and the biochemical and physiological consequences of these actions (IUPAC 1997). In many respects, the level of complexity of toxicodynamics in ecotoxicology can be perceived as far higher than in human toxicology. The main argument is the simple fact that ecology is dealing with a large number of species interacting in a dynamic environment and human toxicology is only dealing with humans (as covered in detail in Section 2.2.2). In terms of effects, the interest for humans is usually in very specific endpoints, such as individual birth weight, tumor induction and growth, liver or any target organ necrosis, or even enzyme induction. Hence, in human TD research the endpoints of main interest are usually very specific, and with a mechanistic link to the effect of interest and preferably at the precursor level. Contrastingly, ecotoxicology usually deals with endpoints that are much more integrated at the overall performance of the individual or population, such as total reproductive output, mortality, and population growth, and only in some cases reaching down to the level of alteration of biochemical parameters measured in the whole body, such as small-molecule metabolite profiling (metabolomics), gene expression (genomics), and protein profiles (proteomics). Even where ecotoxicology does reach the levels of genes and metabolites, they have as yet rarely been directly mechanistically linked to relevant endpoints. Used in combination, they may, however, bring us 1 step closer to such mechanistic linkages as demonstrated by Bundy et al. (2008), identifying energy metabolism as the major target of Cu in the earthworm *Lumbricus rubellus*.

In practical terms, these differences mean that for human toxicology, once one knows the maximum chemical concentration in a certain target organ, or the time-weighted average (area under the curve) (TK) and the toxicity endpoint, it is possible to hypothesize the potential health risks. Conversely, in ecotoxicology, a more process-based approach is generally used to go from the internal concentration to the effects of interest. As an illustration, Poet et al. (2004) described a TK/TD model for the organophosphorous pesticide diazinon in rats and humans, but their TD endpoint is the inhibition of several esterases, and the acceptable inhibition linked to feeling slightly ill or long-term neurological problems. In ecotoxicological applications, AChE inhibition is not a toxic endpoint, and would not be measured unless it could be related to life history changes (see, e.g., Callaghan et al. 2001; Fulton and Key 2001), and then used as a biomarker of exposure and correlated to likely effects on parameters like survival and reproduction. Along this line, Jager and coworkers attempted to combine a TK model for organophosphorus pesticides, including interactions with AChE, with TD models that link receptor occupation to survival (Jager and Kooijman 2005) and growth/reproduction (Jager et al. 2007).

2.4.2 Basic Pharmacodynamic, Toxicodynamic, and Dynamic Energy Budget Models

2.4.2.1 Pharmacodynamic Studies in Human and Mammalian Toxicology

The methods and approaches developed for modeling the biochemical and physiological effects chemicals have on organisms are introduced through the early developments in the pharmaceutical industry; they are therefore often designated as "pharmacodynamics" (PD). The consideration of chemical mixtures was largely due to the concerns for potential drug–drug interactions and the interest in combination therapy (polytherapy). The PD endpoints involved in these studies cover a wide range, including effects on the central nervous system, kidney, cardiovascular, and antimicrobial activities. Studies were primarily performed to acquire some form of predictive power, and the results of the pharmacodynamic interactions were analyzed by a variety of mathematical and statistical analyses. Consequently, different approaches of pharmacodynamic modeling were developed. Examples of such PD modeling are quite prevalent in the pharmaceutical literature, and it is not our intention to review this area thoroughly in this book. Interested readers are referred to a summary review in a computational toxicology book (Reisfeld et al. 2007), and to further study the individual papers listed therein. In Section 2.4.3.1, we concentrate on the pharmacodynamics, or as toxicologists would say, toxicodynamics (TD), relevant to biologically based toxicodynamic modeling of toxicological interactions.

In general, the earlier, drug-related models do not include time as an independent variable. Although we label them as PD models in this chapter, it is a potential subject of debate whether or not, in the strictest sense, we may consider such modeling to be pharmacodynamic modeling. The more recent approaches have moved beyond the descriptional- or regression-based mathematical analysis and have incorporated mechanistic or physiological information into the models. These recently developed pharmacodynamic and toxicodynamic models are therefore referred to as mechanism-based pharmacodynamic models and physiological pharmacodynamic models (see Section 2.4.3.1).

To provide a general illustration of the development and relevance of the earlier PD models in pharmaceutical literature, we briefly summarize the work on the sigmoid E_{max} model by Jonkers et al. (1991). These investigators studied the pharmacodynamics of racemic metoprolol, a cardioselective beta-blocker, and the active S-isomer in 2 human subpopulations: extensive metabolizers (EMs) and poor metabolizers (PMs). The drug effect studied was the antagonism by metoprolol of terbutaline-induced hypokalemia (abnormally low potassium concentration in the blood). The time course human plasma concentrations of potassium in relation to the drug concentrations (i.e., pharmacodynamic interaction) were successfully simulated by a sigmoidal function (i.e., pharmacodynamic model), basically a modified Hill equation derived from the earlier work of Holford and Sheiner (1981) for competitive antagonism (i.e., sigmoid E_{max} model). This model is represented by the following equation consisting of effect (E, plasma concentration of potassium), concentration of terbutaline (C, the beta-2 adrenoceptor agonist), and the inhibitor concentration (IC, metoprolol, racemic or enantiomer, beta-blocker):

$$E = E_0 - \frac{\left(E_0 - E_{\max}\right) \cdot C_e^n}{EC_{50}^n + \left[EC_{50}^n \cdot \dfrac{C_{meto}}{IC_{50}}\right] + C_e^n}$$

Here, specifically, E_0 is the potassium concentration in the absence of terbutaline, and $E_{\max}$, EC_{50}, and n (the Hill coefficient) are the "pharmacodynamic parameters" of terbutaline estimated after placebo pretreatment. Thus, EC_{50} was defined as an effect equal to the mean of E_0 and maximum effect ($E_{\max}$), and EC_{50} and $E_{\max}$ are, therefore, related to terbutaline treatment. C_e is the terbutaline concentration in the "effect compartment" (as Jonkers et al. 1991 would define it; it is a hypothetical compartment linked to the plasma or the central compartment by a first-order process), C_{meto} is the metoprolol concentration, and IC_{50} is the metoprolol concentration that corresponds with 50% maximum receptor occupancy. The exponent, n, is "the factor expressing the sigmoidicity of the concentration effect relationship" (Jonkers et al. 1991, p 961). By that, the authors meant that n is an important factor determining the shape of the curve fitting the experimental data. It is important to note that such a modified Hill equation is mainly for the purpose of curve fitting of the experimental data; it does not have too much biological meaning.

Similar approaches and pharmacodynamic analyses, based on a sigmoid $E_{\max}$ model as described above, have been used for quite a number of studies on drug–drug interactions (Mandema et al. 1992a, 1992b; Mould et al. 1995; Lau et al. 1997; Strenkoski-Nix et al. 1998; Tuk et al. 2002; Jonker et al. 2003; Huang et al. 2005). It is, however, important to remember that there are a number of different pharmacodynamic models in the pharmaceutical literature in addition to the sigmoid $E_{\max}$ model.

2.4.2.2   The Ecotoxicological Approaches to Toxicodynamics

For reasons covered earlier, ecotoxicology deals chiefly with endpoints at higher levels of biological organization (e.g., growth rate) than used in human toxicology (e.g., enzyme inhibition). Due to the high diversity in species covered in ecotoxicology, the diversity in mechanisms studied and modeled is much larger in ecotoxicology. Each species has its own specific characteristics and parameters that vary widely between organism classes and species, different life strategies and niches, before reaching the complexity of differences within species, such as between life stage subgroups and sexes. So it is partly from tradition (protection of the population or ecosystem) but also from lack of detailed and transferable knowledge that ecotoxicology applies very different and currently more general concepts than seen in human toxicology. This is done in order to address endpoints such as survival, overall growth, and reproduction, within organisms and systems where the mechanistic links between chemical exposure and measured effect are not known. These concepts are different between survival and subethal endpoints and are described in turn below.

2.4.2.2.1 Approaches for Survival

The exact mechanisms behind mortality are not well understood; an animal can die from all kinds of toxic insults in its body. There are 2 basic competing concepts: the individual tolerance concept and the stochastic death concept. In the individual tolerance concept, it is assumed that an animal dies instantly when its internal concentration exceeds a certain critical level (Kooijman 1981; Mackay et al. 1992b; McCarty and Mackay 1993). Not all exposed animals die at the same time because the value of the threshold differs between individuals. This concept is closely linked to the critical body residue (CBR) approach (for an overview see McCarty and Mackay 1993). It is assumed that an internal level at the CBR leads to a certain effect size in the tested population (usually 50% mortality). This approach is very popular in ecotoxicology, and is generally mentioned as the mechanism behind the typical s-shaped dose–response curves (e.g., a lognormal distribution of sensitivities in the population is assumed in probit analysis). Nevertheless, there are some serious problems with this approach. First, plenty of studies show that CBRs are actually not constant but decrease in time (De Maagd et al. 1997; Lee et al. 2002b; Landrum et al. 2004; Schuler et al. 2006). Another problem is the existence of latent mortality; that is, mortality can continue after exposure has ceased (which is not possible in the individual tolerance concept) (Zhao and Newman 2004). In other words, the kinetics of the total body residue does not match the observed dynamics of mortality. To address this discrepancy, Lee et al. (2002a) introduced the damage assessment model (DAM; see Section 2.3.4.1 and Figure 2.9). This model includes an additional state variable of "damage," and thereby allows for more freedom in the dynamics (at the cost of 1 additional parameter). In line with the individual tolerance concept, the animal dies immediately when the damage exceeds a certain critical level. This model has been successfully applied to several cases (e.g., Landrum et al. 2004; Schuler et al. 2006), but it has only been tested at the single effect level (i.e., describing LC50 vs. time). As with most applications of the individual tolerance concept, no attempt has yet been made to describe the entire dose–response-time surface.

Most of the work on the individual tolerance concept has focused on organic chemicals. One of the reasons may be that for metals a CBR on a whole body basis has only limited applicability. Due to compartmentalization of metals in the body and the presence of regulation and detoxification mechanisms, it is unlikely that the total body residue is simply related to toxicity (see, e.g., Lock and Janssen 2001; Vijver et al. 2004). The biotic ligand models (BLMs) assume a critical level of metal accumulation at the biotic ligand and do not include a time aspect, although more TD-like approaches have been suggested (Paquin et al. 2002b).

The individual tolerance concept has some unrealistic properties (Kooijman 1996; Newman and McCloskey 2000). Most importantly, if there is a distribution in sensitivities, this would imply that the survivors from an experiment are the less sensitive individuals. Experiments with sequential exposure show that this prediction fails (at least as the dominant mechanism) (Newman and McCloskey 2000; Zhao and Newman 2007). There is sufficient reason to conclude that the individual threshold model is not sufficient to explain the observed dose–response relationships, and that mortality is a stochastic process at the level of the individual

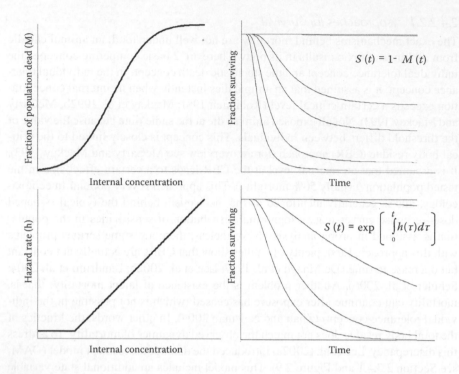

Figure 2.10 Two approaches for dynamic survival analysis. The fraction surviving (S) is shown for a range of concentrations (each line represents a different dose). Top: the individual tolerance concept assumes that an organism dies instantly when its threshold is exceeded (the threshold is normally distributed in this example, yielding an s-shaped relation between internal concentration and fraction dead, M). Bottom: the stochastic approach assumes that the internal concentration increases the probability to die (here with a threshold, and a linear relation between body residue and hazard rate, h).

(Bedaux and Kooijman 1994; Kooijman 1996; Newman and McCloskey 1996, 2000; Zhao and Newman 2007). This is not just an academic discussion; the 2 theories lead to different time courses of mortality at constant exposure (Kooijman 1996) (see Figure 2.10) and have very different consequences for sequential exposure (Newman and McCloskey 2000; Zhao and Newman 2007). In reality, both sensitivity difference and stochasticity are likely to play a role in mortality. Individuals also differ in sensitivity, especially in field populations, but there is clearly a substantial stochastic component involved in mortality that cannot be ignored. The method to deal with stochastic events in time is survival analysis or time-to-event analysis (see Bedaux and Kooijman 1994; Newman and McCloskey 1996). For industrial practices, this method has a long history as "failure time analysis" (see, e.g., Muenchow 1986). Bedaux and Kooijman (1994) link survival analysis to a TK model to describe survival as a function of time (i.e., the hazard rate is taken proportional to the concentration above a threshold value). Newman and McCloskey (1996) take an empirical relationship between external concentration and hazard rate.

2.4.2.2.2 Approaches for Growth and Reproduction

For sublethal endpoints, the individual tolerance concept cannot be applied. Ecotoxicologists are usually not dealing with quantal responses (counting the number of responding individuals), but with continuous responses (reproductive output or grams of body weight). At the LC50, 50% of the test population has died. In contrast, at the EC50, we do not see half the population stopping reproduction, and the other half reproducing at the rate of the control. Instead, all individuals show a decreased reproductive output (on average 50% less than in the control). The s-shaped dose response relationship therefore cannot be the result of differences between individuals (see Lutz et al. 2006 for further discussion). Furthermore, it is also not practical to link a certain effect size to a body residue (as done in the CBR approach for survival), because ECx (and LCx) values can change in time in rather strange ways, even at constant exposure (Alda Álvarez et al. 2006a). There is no such a thing as an "incipient ECx" for continuous responses.

Stressors cause effects through perturbations of certain processes in the normal system, so it is essential to first understand the normal system. For ecotoxicological applications, this means that we have to treat growth and reproduction for what they are—tightly interlinked processes in time—and make assumptions on how toxicants interfere with these processes. Because we are dealing with highly integrated endpoints, it is not enough to attempt to model processes in detail at the molecular level only. Apart from the sheer number of compounds, genes, enzymes that are involved in a process such as reproduction, the resulting model would also be highly species specific. Instead, it is more practical to focus on the level of resource allocation and energy budgets. Results from molecular studies giving a global profile of an organism's response to a stressor can provide good indications of how the energy levels or allocations thereof are affected (e.g., Swain et al. 2010). Molecular response profiling therefore may provide either a starting hypothesis for energy-budget-based analysis or assist in explaining its findings.

All animals use food to grow, develop, and produce offspring. Energy budget approaches apply the conservation laws for mass and energy to understand how the energy in the food is used to fuel all metabolic processes. When a toxicant reduces reproduction, there are energetic consequences; apparently, less energy is devoted to offspring formation, which could mean that less resources are taken up (e.g., a reduction in feeding rate) or that resources are allocated to other metabolic processes (e.g., defense against toxic damage). Another possibility is that the toxicant affects the growth process, which has indirect consequences for reproduction, as body size determines feeding rates and affects the start of reproduction. An understanding of toxic effects on growth and reproduction requires a quantitative understanding of feeding, growth, development, and reproduction as closely linked dynamic processes. Probably the most complete and best-tested energy budget approach in ecotoxicology is the work on dynamic energy budget (DEB) theory (Kooijman 2000, 2001; Nisbet et al. 2000). This theory describes simple rules for metabolic organization, that is, how organisms acquire and use resources for growth, development, and reproduction. Toxic effects can then be viewed as a disruption of normal allocation processes. Such modes of action are based on

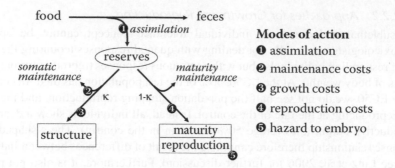

Figure 2.11 Schematic representation of a dynamic energy budget model. Numbers represent some of the positions where toxicants may affect resource allocation. (Adapted from Kooijman and Bedaux 1996.)

physiological allocation processes, and are therefore not directly related to mode of action approaches, describing molecular mechanisms (e.g., AChE inhibition) or overall toxic syndromes (e.g., narcosis). The application of the DEB theory to ecotoxicological test data was first demonstrated by Kooijman and Bedaux (1996), linking TK modeling (a 1-compartment model, accounting for growth) to effects on growth and reproduction of *Daphnia* in time. The internal concentration of the toxicants is assumed to act on 1 or several processes of resource allocation. In most ecotoxicological studies, nothing is known about the toxicokinetics or energetic parameters, and at best we have the time course of effects. Therefore, these parameters generally have to be inferred from the time course of toxicity alone. Also, the physiological mode of action has to be inferred from the time course of effect. However, the different modes of action (see Figure 2.11) have distinct consequences for the growth and reproduction effects in time (see Kooijman and Bedaux 1996; Jager et al. 2004; Álda Álvarez et al. 2006b). Because, in general, no detailed information is available about the toxicants effects on energetics, the physiological modes of action can only represent quite broad allocation patterns (see list in Figure 2.11).

Of course, the acceptable level of model complexity is closely related to the level of detail of the experimental data. Energy budget approaches are generally most informative when data on body size and reproductive output are available over time for a considerable part of the life cycle. Examples of the application of DEB theory to life cycle toxicity data for single compounds can be found in Jager et al. (2004), Alda Álvarez et al. (2005), and Pieters et al. (2006). More complicated TD approaches require other types of experimental data (e.g., at the molecular level), but should never ignore the level of resource allocation when endpoints such as growth and reproduction are concerned.

The DEB theory is also applicable to birds and mammals (Kooijman 2000), but has not been applied extensively to toxic effects here yet (because human toxicology is concerned with more specific effects than disturbance of resource allocation). However, the paper of Van Leeuwen et al. (2003) on tumor growth shows the potential of DEB approaches in this area.

2.4.3 Applications of Mixture TD in Human and Mammalian Toxicology

2.4.3.1 Toxicodynamic Studies of Chemical Mixtures Using Physiologically or Biologically Based Modeling

Because physiologically based toxicodynamic (PBTD) modeling of chemical mixtures is still at its infancy, we choose to only briefly discuss PBTD modeling by concentrating on a number of published and 1 unpublished example on binary mixtures. For more details on the basic principles of PBTD modeling and specific information on the several case studies discussed below, the reader is encouraged to consult the original references.

2.4.3.1.1 Toxicodynamic Modeling of Noncancer Effects of Chemical Mixtures

One of the earliest examples for noncancer effects was the application of computer technology to the PBTD modeling of a toxicological interaction between kepone (also known as chlordecone) and carbon tetrachloride (CCl_4) based on mechanisms of interactive liver toxicity leading to death. Briefly, CCl_4 is a well-known hepatotoxin. Following free radical formation through the cytochrome P450 enzyme system, the toxic effect of CCl_4 can be an accumulation of lipids (steatosis, fatty liver) and degenerative processes leading to liver cell death (necrosis). Kepone is found in the environment as a result of photolytic oxidation of Mirex, a pesticide used for the control of fire ants, or as a pollutant from careless and irresponsible discharge. At relatively low levels (e.g., 10 mg/kg in the diet), repeated dosing of kepone alone in the diet for up to 15 days caused no apparent toxicity to the liver. The toxicological interaction between kepone and CCl_4 was elucidated to be the impairment, by kepone, of the liver's regeneration process. These mechanistic studies were summarized in a number of publications (Mehendale 1984, 1991, 1994). The conceptual model of this toxicodynamic interaction is shown in Figure 2.12.

El-Masri et al. (1996a) constructed a PBTD model based on the mechanism of toxicological interaction between kepone and CCl_4. This PBTD model was verified by literature information, and it was capable of providing time course computer simulations of mitotic, injured, and pyknotic (dead) cells after treatment with CCl_4 alone or with kepone pretreatment. This PBTD model was further linked with Monte Carlo simulation to predict the acute lethality of CCl_4 alone and in combination with kepone. The linkage with Monte Carlo simulation on the computer (i.e., in silico toxicology), in this case, provided a sample size of 1000 animals per dose for acute toxicity studies; such a large sample size, or larger, in a real-world laboratory setting, is highly improbable because of resource limitations and ethical considerations. Thus, the El-Masri et al. (1996a) PBTD study on kepone and CCl_4 provided an illustration of the usefulness of in silico toxicology to develop models to predict the toxicity of mixtures. While these models often remain unvalidated, they do provide options for situations such as, for example, low dose regions, where the generation of good data or validated models would require an ethically unacceptable number of animals.

The second case study involved PBTD modeling of toxicodynamic interactions between trichloroethylene (TCE) and 1,1-dichloroethylene (DCE) regarding their binding and depletion of hepatic glutathione (GSH) in relation to the intrinsic hepatic GSH synthesis (El-Masri et al. 1996b), which is a protective mechanism

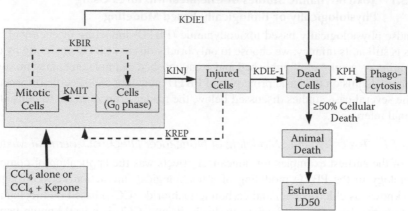

Figure 2.12 A conceptual physiologically based pharmacodynamic (PBTD) model for CCl_4 and kepone interaction. KMIT = rate constant for mitosis; KREP Rate constant of inj. cells repair; KBIR = rate constant for cell birth; KINJ = rate constant for cell injury by toxicants; KDIEI = rate constant for general cell death; KDIE-1 = rate constant for cell death due to injury; KPH = rate constant for phagocytosis. (Redrawn from El-Masri et al. [1996a], with permission.)

toward DCE toxicity. A PBTD model was used to identify the critical time point at which hepatic GSH is at a minimum in response to both chemicals. PBTK models for interactions leading to depletion of hepatic glutathione had been developed by several investigators (D'Souza et al. 1988; Frederick et al. 1992). Model-directed gas uptake (inhalation) experiments with DCE revealed that DCE was the only chemical capable of significantly depleting hepatic GSH. TCE exposure higher than 100 ppm (in the inhaled air) to the rats obstructed the ability of DCE to deplete hepatic GSH, indicating metabolic competitive inhibition of DCE biotransformation to reactive metabolites. TCE exposure lower than 100 ppm was ineffective in inhibiting DCE from significantly depleting hepatic GSH. El-Masri et al. (1996c) further applied these quantitative analyses in establishing an "interaction threshold" between TCE and DCE. This example again illustrates the difficulty in defining what is a TK and what is a TD model. Here TCE inhibits the biotransformation of DCE (TK), but does so as the TCE-DCE interaction directly affects the GSH levels in the liver, making it TD (GSH depletion being considered the effect).

A recent publication on PBTK/TD modeling to determine dosimetry and cholinesterase inhibition for a chemical mixture of 2 organophosphorus insecticides, chlorpyrifos and diazinon (Timchalk and Poet 2008), deserves some special discussion here. Based on the individual PBTK/TD models developed earlier by the same laboratory, Timchalk and Poet (2008) reported their development of a binary interaction PBTK/TD model for chlorpyrifos and diazinon. In their development of the model, Timchalk and Poet (2008) took into consideration a number of important metabolic steps (CYP450 mediated activation and detoxification of a number of esterases—B-

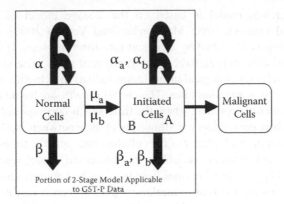

Figure 2.13 A simple 2-stage model of carcinogenesis. Within the framework, the possibility of two initiated populations for adequately describing the kinetics of liver GST-P foci growth was evaluated. The cell division and death rates of normal hepatocytes are denoted by α and β, respectively. The probability of mutation per cell division to A or B initiated cells per day are donoted by μ_a and μ_b. The division and death rates of A, B cells are denoted by α_a, β_a, α_b, β_b, respectively.

esterase or paraoxonase-1 (PON-1)), as well as metabolic interactions. Thus, in their binary PBTK/TD model, the following mechanistic bases were incorporated: 1) each insecticide inhibits the other's metabolism through CYP450, 2) additive B-esterase metabolism of the 2 insecticides, and 3) no PON-1 interactions existed. The resulting model simulations were consistent with their experimental data (Timchalk and Poet 2008), and they concluded that at low environmentally relevant exposure doses of the 2 insecticides, the toxicokinetics of the 2 chemicals are expected to be linear, and their cholinesterase inhibition to be dose additive. Again, here arguments could be had about TK or TD. This is published as TD because the TK interaction is viewed to affect AChE inhibition. The interaction, however, is not at the AChE level but at the metabolic level (TK).

2.4.3.2　Toxicodynamic Modeling of Cancer-Related Effects of Chemicals and a Binary Chemical Mixture

A further example of biologically based toxicodynamic modeling is computer simulation of clonal growth of initiated liver cells in relation to carcinogenesis. The impetus of this research development came from the desire to finding a way to evaluate carcinogenic potentials of chemicals or chemical mixtures without going through the resource-intensive chronic cancer bioassays (Yang 1994, 1997; Yang et al. 2004). The experimental animal model used for this research development involved a modification of the medium-term initiation-promotion bioassay of Ito et al. (1989a, 1989b). It is an initiation-promotion assay of a relatively short duration, and the modification incorporated toxicokinetics into the experimental design. This protocol allows the evaluation of carcinogenic potential within 8 weeks by identification of hepatic preneoplastic foci expressing glutathione S-transferase placental form (GST-P) as an endpoint marker.

The clonal growth model is based on the 2-stage model of carcinogenesis (Moolgavkar and Luebeck 1990; Moolgavkar and Venzon 2000), where the carcinogenesis process is described by 2 critical rate-limiting steps: 1) from normal to initiated cells and 2) from initiated cells to malignant states. The model allows the incorporation of relevant biological information such as the kinetics of tissue growth and differentiation and mutation rates. The clonal growth stochastic model adopts a discrete time numerical approach, where the time axis is decomposed into a series of time intervals, where parameters are allowed to change between, but not within, segments. To represent the multiplicity of the cellular states and the time-varying nature of the numerous cell behavior variables, the numerical model resorts to a recursive simulation. The growth of normal liver is described deterministically, whereas other cellular events use stochastic simulation. This approach facilitates description of complex biological processes with time-dependent values, allowing data gained on the rate-limiting steps in short-term experiments to be extrapolated to statements about carcinogenicity under chronic exposure.

A further important concept in the clonal growth stochastic model (Thomas et al. 2000; Ou et al. 2001, 2003; Lu et al. 2008) is the incorporation of the hypothesis of 2 initiated cell populations (referred to as A and B cells) for the GST-P foci development. Figure 2.13 provides a conceptual schematic for this hypothesis. The B cells are initiated cells that display selective growth advantage under conditions that inhibit the growth of initiated A cells and normal hepatocytes. The use of the 2-cell hypothesis for the description of diethylnitrosamine (DEN) control data (with partial hepatectomy) also indicated the presence of multiple phenotypes of initiated clones following DEN treatment, with resistant phenotypes arising during early carcinogenesis.

2.4.3.3 Biochemical Reaction Network Modeling and Gene–Protein Network Modeling

The ultimate goal in studying chemical mixtures is to be able to predict the toxicities of chemical mixtures irrespective of their complexity. Given the astronomically large number of possible combinations of available chemicals in commerce, it is impossible to rely on conducting bench-level experimentation alone to address the problems of chemical mixtures. Therefore, we believe that the utilization of computer modeling, in conjunction with experimental work, is essential in the studies of toxicology of chemicals, chemical mixtures, and multiple stressors. Biology is well served by the application of computer technology as an alternative research method to conserve resources and minimize the use of laboratory animals. In the past few years, tools such as "reaction network modeling" have become available to support computer simulation at the molecular interaction level (Klein et al. 2002; Liao et al. 2002; Liao 2004; Reisfeld and Yang 2004; Yang et al. 2004, 2006; Mayeno et al. 2005). While such a tool is still at the developmental stage and no application has been implemented yet, an approach has emerged for the potential predictive capability on the toxicology of chemical mixtures and their subsequent risk assessment (Yang et al. 2010). Given below is an outline of such an approach. It should be noted that biochemical reaction network modeling, though fundamentally a toxicodynamic modeling tool, because it is intimately associated with reaction mechanisms, is nevertheless an integration of toxicokinetics and toxicodynamics. Detailed discussion on biochemical reaction network modeling is beyond the scope of this chapter. Readers are referred to Yang et al. (2010) for more explanation:

1) For a given class of chemicals (e.g., PCBs), a "training set" of chemicals or congeners is selected and the enzyme-based reaction rules involved are derived from the known biochemical or chemical mechanisms of the reactions. This information is entered into a "transformation" database. Generation of qualitative biochemical reaction networks using appropriate software for the chemicals in the training set is possible at this stage. Here, "qualitative" indicates that the inventory of metabolites and their interconnections has been predicted, but the "quantitative" amounts of these chemicals have not.

2) Critical enzyme kinetic studies are conducted on the chemicals in the training set using human enzymes (e.g., cytochrome P450 isoforms, epoxide hydrolase, phase II enzymes). The studies include individual chemicals, as well as chemical mixtures, to discern kinetic rate constants (e.g., K_m, V_{max}) under independent or interaction conditions. The rate constants are entered into a "kinetics" database. Using appropriate software, the generation of qualitative and quantitative biochemical reaction networks for the chemicals in the training set is possible at this stage.

3) Derivation of kinetic rate constants for chemicals or congeners outside of the training set, but within the chemical class, is possible through QSAR and molecular modeling calculations based on appropriate descriptors (e.g., physicochemical and quantum chemical properties of the chemicals, as well as enzyme active site conformation).

4) Prediction of both qualitative and quantitative biochemical reaction networks, for chemicals outside of the training set but within the chemical class, including chemical mixtures, is possible at this stage. The confidence level for such predictions will increase as more and more validations are made.

5) By coupling the biochemical reaction network model with a general PBTK model, the toxicokinetics of critical metabolites in the whole organism can be discerned. Because toxic or reactive species in a given organ or tissue can be identified and quantified, investigators with toxicology knowledge of the given class of chemicals can assess the resulting toxic effects in that organ or tissue. In this manner, the overall toxicity of the chemical mixtures within that class of chemicals in the whole organism can be deduced.

6) As more and more classes of chemicals (and corresponding biotransformation and kinetic data) are introduced into the modeling framework, the scope of its predictive capability expands accordingly. At some point, the toxicology of chemical mixtures of any composition and component numbers can be assessed and predicted through such an approach.

2.4.4 Applications of Mixture TD in Ecotoxicology

The constant CBR approach (see Section 2.4.2.2) has been applied to mixture toxicity of narcotic chemicals at a single time point (e.g., Van Wezel et al. 1996; Leslie et al. 2004). However, there are only a few published examples in ecotoxicology (that we know of) where mixture effects are modeled in time. For effects on survival, the CBR concept has been extended to mixtures by McCarty et al. (1992), describing the LC50 of mixtures in time by linking a 1-compartment TK model to a fixed CBR. The work of Lee and Landrum (2006a, 2006b) links a more elaborate TK model to the individual tolerance concept to describe effects of a mixture of the parent compound and the metabolite (see Figure 2.9). They follow the individual tolerance concept, using a fixed critical level of damage. Their model is equivalent to assuming an exponential distribution for this threshold in the test population. A limitation is that both approaches have only been tested at the 50% effect level. More dedicated experimental work is needed to test the power of these approaches. A comparable mixture extension of the DAM approach was presented by Ashauer et al. (2007a), and tested with sequential pulses of AChE inhibitors.

Recently, the stochastic approach of Bedaux and Kooijman (1994) has been extended to mixtures, and tested on the full dose–response surface of binary mixtures in time. This extension to mixtures was undertaken by Baas et al. (2007), who analyzed survival data for 6 binary mixtures of heavy metals in the springtail *Folsomia candida*. The crucial aspect was that survival was scored daily over a period of 21 days. Because the approach used is a combined TK/TD approach, it is used to fit the survival data for all time steps simultaneously, using a set of 8 parameters. Note that in this approach the raw survival data were described, not just the

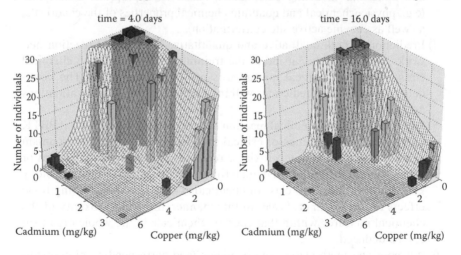

Figure 2.14 Example of the fit of a stochastic model to the effects of a mixture of copper and cadmium on the survival of the springtail *Folsomia candida*. Only two time observations are shown, but the model is fit to all data for all time steps simultaneously. (After Baas et al. 2007.)

LC50 in time (see Figure 2.14). Classic dose–response surface analysis (e.g., Haas et al. 1996) generally does not work with data in time, so results from each time point have to be analyzed separately. Employing the descriptive response surface analysis framework of Jonker et al. (2005) (see Section 4.5.2 for details) requires 5 to 7 parameters per time point (which implies using over 100 parameters to fit the whole data set), and indicates a range of interactions in the data set, that change in time. Interestingly, the stochastic model could fit most of the mixtures without any additional interaction parameter. This could indicate that the identification of interactions may either occur following from differences in toxicokinetics between the compounds (see Figure 2.2) or be an artifact of random errors at a particular time point, which may also explain the lack of reproducibility of specific interactions (Cedergreen et al. 2007).

For sublethal endpoints, there are very few studies that follow endpoints in time. Examples are the studies by Van Gestel and Hensbergen (1997) and Jonker (2003). However, these authors did not apply TK/TD models to analyze their data, but simply treated the data at each time step as a new data set. Interestingly, they found that the apparent mixture interactions changed over time, emphasizing the need for more mechanistic models to analyze and understand this behavior.

The DEB approach for sublethal endpoints has been successfully applied to a range of chemicals and organisms, but has not been applied to mixtures, although Kooijman (2000) speculates on the possible implementation (a first demonstration using a simple mixture of 2 PAHs is currently in press; Jager et al. 2010). However, some promising combinations of chemicals with another stress (especially food limitation) have been presented (e.g., Heugens 2003; Pieters et al. 2006) and simulated in Jager et al. (2004) and Alda Álvarez et al. (2005). Interestingly, if 2 compounds affect different energy allocation processes (e.g., maintenance and assimilation), a physiological interaction between these processes has in fact become inevitable because of the allocation rules in the DEB model. Furthermore, the DEB approach also implies interactions on toxicokinetics, as toxicokinetics is influenced by body size and growth rate. Therefore, it is to be expected that a model analysis based on energy budgets shows deviations from classical CA and IA even without imposing statistical interactions between the compounds. This approach has a lot of potential, but still awaits rigorous testing. An optimal data set would comprise data on growth and reproduction in time over a considerable part of the life cycle, which for a mixture can be quite resource demanding (especially if observations can only be obtained by destructive sampling). It is quite likely that an energy budget approach will not provide a perfect fit for all but the simplest mixtures on the first try. However, the nature of the deviations from model expectations provides more information on the underlying mechanisms than purely descriptive statistical interaction terms, and thus delivers directions for further mechanistic research.

2.5   SUMMARY AND CONCLUSIONS

2.5.1   Toxicokinetics

2.5.1.1   Human and Mammalian Toxicology

As described in Section 2.3, there have been considerable efforts in developing models for simulating and predicting the toxicokinetics of mixture constituents, but the scientific community is still clearly at the embryonic stage. So far, the binary to multichemical extrapolation approach for modeling toxicokinetic interactions within mixtures has proven to be successful (see Section 2.3.3.2), but remains to be challenged for interactions that are not of a metabolic nature. This approach is scientifically the most appealing one because it allows the use of the wealth of information that has already been generated on binary interactions. However, its major drawback is the amount of experimental studies and resources needed to obtain missing information on multichemical mixtures. Chemical lumping is a potential way to reduce the information needed, and the use of in vitro studies has proven promising for obtaining mechanistic information and parameter values on binary interactions more efficiently.

Other emerging technologies surely enhance our ability to determine mechanisms of binary interaction, and hence facilitate the use of binary-interactions-based PBTK models of mixtures. These include the different in silico technologies that are being developed to predict ligand–enzyme interactions such as QSAR or 3D modeling of the different enzymes implicated in the biotransformation of xenobiotics (De Graaf et al. 2005). Additionally, biochemical reaction networks are also very promising tools that could help predict the rate of reactions and inhibition constants (Mayeno et al. 2005).

2.5.1.2   Ecotoxicology

There are only a few published examples in ecotoxicology for which a TK model has been used in the analysis of toxicity data for mixtures. McCarty et al. (1992) and Baas et al. (2007, 2009) used a 1-compartment model, without kinetic interactions. Lee and Landrum (2006a, 2006b) proposed and tested a more elaborate model, accounting for biotransformation and an additional state of damage (also without interaction) (Section 2.3.4.1 and Figure 2.9). For single compounds, a broad range of models has been developed and successfully applied in ecotoxicology. In the near future, most of these TK models can probably be easily adapted to accommodate mixtures of compounds, although some further work is needed on interactions in uptake and excretion kinetics, metabolic transformation interactions, and binding interactions at the target site.

At present for many ecotoxicological applications, regarding the organisms as a 1-compartment model is a good initial choice for whole body residues, and probably also for mixture TK. We may not have to use more complex models (depending on the problem at hand). There are, of course, several situations were such simplicity does not apply and where more elaborate PBTK modeling may be required. These include larger organisms where internal redistribution is not fast enough to make a 1-compartment model reasonable, species with known tissues for handling and detoxification of chemicals where these are likely to interact, and also situations where the kinetics of receptor interactions or the accumulation of internal damage may need to be modeled explicitly as additional state variables (see Section 2.3.4.1). However, it is generally advisable to start with the simple 1-compartment model, and build more complex models when needed (in view of the limited data available). For single metals, there is growing evidence in the aquatic field that advocates the concept of biodynamic modeling for linking internal metal concentrations to toxicity. This has been achieved by taking into account what proportion of the total metal is in free form and able to be biologically active (Rainbow 2002; Luoma and Rainbow 2005). To get better estimates for the uptake and effects of organic chemicals, data based on membrane–water partitioning (e.g., Escher and Sigg 2004) and chemical activity (e.g., Reichenberg and Mayer 2006) should be incorporated, and this would provide information to perform cross-species extrapolation. For mixtures of metals, approaches for addressing joint toxicology have to account for speciation and competition between metals (in different forms) at the target organ level. The combination of multiple BLM approaches offers great possibilities for the interpretation of toxicity data for metal mixtures (Playle 2004), but has not yet been applied to actual data. Nevertheless, this complexity may not be necessary for all mixture studies, as demonstrated by Baas et al. (2007).

For wildlife ecotoxicology, PBTK modeling may be a viable choice, applying the knowledge available from human toxicology combined with data from the most relevant test species available. Examples of mixture TK models in this area are currently unavailable.

2.5.2 Toxicodynamics

2.5.2.1 Human and Mammalian Toxicology

Although PBTD modeling of chemical mixtures may still be in its infancy, it holds great promise in various fields, as shown by the examples given in Section 2.4.3.

For noncancer effects the use of PBTD models has elucidated the fundamental mechanisms of toxicological interactions. Such mechanistic knowledge linked with Monte Carlo simulations has initially been employed in in silico toxicology to develop models that predict the toxicity of mixtures in time. The combination of PBTK/TD models for individual compounds with binary PBTK/TD models can be achieved by incorporating key mechanistic knowledge on metabolism inhibitions and interactions through shared enzyme pathways. Simulations of such models can then be compared to experimental data and allow conclusions to be reached about their pharmacokinetics and the likelihood of effects being dose additive.

In cancer research the resource-intensive chronic cancer bioassays originally needed to evaluate carcinogenic potentials of chemicals or chemical mixtures has led to the development of computer simulation of clonal growth of initiated liver cells in relation to carcinogenesis. This model describes the process of carcinogenesis based on the 2-stage model with 2 critical rate-limiting steps: 1) from normal to initiated cells and 2) from initiated cells to malignant states. Because this approach can incorporate relevant biological and kinetic information available, it usefully facilitates description of the carcinogenesis process with time-dependent values without the need for chronic exposures.

To stretch toward the goal of having the ability to predict the toxicity of the infinite number of possible mixture exposure scenarios, it is clear that computer modeling must be employed in conjunction with experimental work. The latest development toward such abilities is initial biochemical reaction network modeling (see Section 2.4.3.3). While a seemingly insurmountable task, such network models provide a system for collating the ever-building TK and TD information in a way where the toxicology of chemical mixtures of increasing component numbers can be assessed and predicted.

2.5.2.2 Ecotoxicology

The problem of the application of any TD model in ecotoxicology to mixtures is that there is an extreme lack of data on toxic effects of mixtures measured as a function of time. Almost all studies focus on the effects of the mixture after a fixed exposure period only, which is of limited use for the application of dynamic approaches.

For mortality, the individual tolerance concept (using a fixed CBR, or the more elaborate DAM) has been applied to mixtures at the 50% effect level, but more work needs to be done to validate its applicability, and to test it against the stochastic approach. For the stochastic approach, several mixture toxicity studies have been performed where the full dose–response surface could be explained, which is promising.

For sublethal responses, the level of resource allocation is essential. The DEB approach offers great promise as a TD model in ecotoxicology. It has been applied to the combination of a toxicant with another stressor (food limitation), but its application to mixtures of toxicants requires further work, and a comparison to dedicated experimental data. The EU sixth framework project NoMiracle has delivered such data, which in time will help develop the DEB mixtures approach to also cover sublethal endpoints.

2.6 RECOMMENDATIONS

TK modeling within human toxicology is well established, and there are good data and models at the level of binary mixtures. In terms of modeling toxicokinetic interactions within multiple mixtures, the approach of binary to multichemical extrapolation is the most promising, but is often hampered by the need for data on unknown binary mixtures. Chemical lumping is an innovative approach for reducing the amount of information needed on binary interactions for modeling the TK interactions in mixtures. There are other avenues that are worth investigating for reducing costs and experiments to elucidate binary interaction mechanisms, such as extrapolation from in vitro data.

  1) Explore ways of reducing the amount of binary mixture information needed for binary to multichemical extrapolation, or for collecting it more effectively.
     a) There should be more efforts invested in this chemical lumping approach in order to better define the criteria (e.g., metabolic rates, partitioning, volatility, interactions with different metabolic enzymes).
     b) Increased efforts to obtain mechanistic information and parameter values on binary interactions from in vitro studies (e.g., incubations in microsomal preparations, hepatocyte suspensions).

c) Improve and validate extrapolation approaches for in vitro data. Several have been suggested and can be used as starting points, but so far the success rate often depends on the physicochemical properties of the chemicals studied (Suzuki et al. 1995; Carlile et al. 1997; Obach 1997; Schmider et al. 1999; Witherow and Houston 1999; Houston and Kenworthy 2000; Venkatakrishnan et al. 2000, 2001; Bachmann and Ghosh 2001; Zhou et al. 2002; Houston and Galetin 2003; Ito and Houston 2004, 2005; Egnell et al. 2005; Hakooz et al. 2006; Miners et al. 2006; Obach et al. 2006).

d) Further efforts should also be directed toward developing a better mechanistic hepatic clearance model by more accurately describing the complexity of its physiology (i.e., acinus, enzyme location, transport processes) (Theil et al. 2003). The development of successful in vitro–in vivo extrapolations of binary interactions would be quite important advancements for the PBTK modeling of mixtures.

Within the ecotoxicology field TK modeling is basically still at the single chemical stage, but most of these TK models could be adapted for work with mixtures of compounds. This would, however, require data and knowledge building in some key areas. For all these areas, considerable experience can be gained from the data generated in the mammalian toxicology area, as outlined in Section 2.3.4.

2) Prioritize research that provides more detail of chemical interaction potential at key levels (ADME) within nonmammals.

a) Interactions in uptake and excretion kinetics, for example, competition between metals (and with other ions) for binding to ligands and uptake through ion channels. For nonpolar organic chemicals, current ecotoxicological data suggest this factor may be of minor importance. However, the building knowledge on transporters mentioned in Section 2.3.1.1 should be born in mind when defining assumptions.

b) Metabolic transformations, for example, limitations in the capacity of enzymes to transform 2 chemicals, chemicals inhibiting the induction of metabolizing enzymes, or metals competing for binding to metallothioneins or displacing essential metals from functional proteins. It should be noted that the potential for biotransformation differs between species. Clearly, there is a need for predictive estimations, and in terms of metabolic inhibition, genomic and proteomic information may aid in judging enzyme homology. Enzyme induction would also have to be investigated using transcriptomic data.

c) Interactions at the target site. For narcotic chemicals, this may be a minor factor, but it would be a major factor for chemicals targeting specific receptors. For example, 2 OP esters may compete for binding with AChE. Currently, such interactions are rarely considered.

While the 1-compartment model has great applicability within ecotoxicology and can be used for mixtures, there are situations where its application would be an unjustifiable simplification. These areas need further work in order to extend the 1-compartment model to a more complete TK consideration.

3) Further work on extending the 1-compartment model to improve its applicability in specific situations.

   a)   Examine at what size internal redistribution in organisms (e.g., fish species such as trout) can become so slow that explicit consideration of TK may be required. For small organisms (e.g., 2 cm fathead minnows in acute toxicity tests), internal redistribution may be rapid enough to justify using the 1-compartment model, but we need to know the size to which such data can be extrapolated.

   b)   Development of more elaborate models in species where interactions between compounds or with receptors take place in a specific target tissue or organ (e.g., the hepatopancreas in woodlice, or the chloragogenous tissue in earthworms).

   c)   For metals combining multiple BLM approaches, the concept of biodynamic modeling holds great promise for addressing mixtures.

   d)   For organic chemicals the use of octanol–water partitioning coefficients (K_{ow}) should be extended with more studies using concepts such as liposome-water distribution for ionizable chemicals (Escher and Sigg 2004) and the more general application of chemical activity (Mayer and Holmstrup 2008).

   e)   The explicit inclusion of additional state variables where the kinetics of receptor interactions or the accumulation of internal damage may need to be modeled.

Toxicodynamic studies involving chemical mixtures are relatively scarce in the human and ecological arenas. Of the few available, a greater portion of such TD studies are relatively simple, without much mechanistic insight. In that sense, TD, in comparison with TK, is a much less mature scientific discipline. Within the field of ecotoxicology, promising approaches for explaining and predicting effects as a function of exposure time using biologically based models such as DEB show a lot of promise, but need developing and testing for mixtures.

4) TD studies on chemical mixtures is an important area for development and promises to result in significant contributions.

   a)   Increase efforts to progress the application of TD approaches to mixture studies.

   b)   Develop DEB modeling to include mixture effects on sublethal parameters and test these as widely as possible.

There are moves for the disciplines of human and ecological risk assessment to start closing the gap between considering effects at the individual and the population level. Human risk assessment has undergone fundamental changes in recent years, in order to consider the population-level variation. This has particularly involved PBTK modeling, which has gone on to population PBTK modeling using Bayesian statistics and Markov chain Monte Carlo simulations. Nevertheless, while including population-level variability, human risk assessment still aims to protect the individual. Within ecological risk assessment there are developments to include receptor characteristics, which, when perfected, would allow the latter to better employ the approaches of human risk assessment, but with the challenge of accounting for 1 extra level of biological organization (i.e., interspecies differences).

5) Actively encourage exploration of the scope for increased crossover and better inclusion of the real-world issues of population variability and uncertainty.

a) Further advancement and integration of Bayesian statistics and Markov chain Monte Carlo simulations into the PBTK/TD modeling of chemical mixtures would be a significant scientific contribution. Examples of such work already exist in the literature and can be used as guiding principles (Gelman et al. 1996; Bernillon and Bois 2000; Jonsson and Johanson 2003; David et al. 2006; Hack 2006; Marino et al. 2006; Covington et al. 2007; Lyons et al. 2008; Redding et al. 2008).

b) Continue the recent advancement of -omics and computational technologies to converge the approaches of toxicodynamics between ecotoxicology and mammalian toxicology. Such work may be achievable at the level of computer integration of knowledge at the subcellular levels. This could help overcome the limitations owing to the small size of many of the organisms utilized in ecotoxicity studies that make sampling and analyses at the organ or tissue levels impossible.

3 Toxicity from Combined Exposure to Chemicals

Andreas Kortenkamp and Rolf Altenburger

CONTENTS

3.1 INTRODUCTION

Toxicologists and risk assessment experts are confronted with demands to respond to this situation by weighing whether exposures to multiple chemicals are associated with risks to human health, wildlife, and ecosystems. However, established risk assessment procedures, with their focus on dealing with environmental and human health risks on a chemical-by-chemical basis, are ill-equipped to deal with these challenges (Office of Emergency and Remedial Response 1991). Nevertheless, awareness is growing among experts that this is a genuine problem worthy of serious attention. In approaching the topic from a scientific viewpoint, a number of issues can be distinguished:

- Can the effects of mixtures be predicted from the toxicity of individual components?
- Are risks to be expected from exposure to multiple chemicals at low doses?
- How likely is it that chemicals interact with each other, leading to synergistic effects, and which factors determine the potential for synergisms?

In this chapter, we outline the issues and principles that are relevant to toxicity assessments of combined exposures. The scope of this overview is limited to combinations of chemicals, but excludes the topic of nonchemical stressors acting in concert with chemicals. Because the issues are of a generic nature, we draw on examples from human, environmental, and ecological toxicology. Section 3.2 briefly outlines approaches to mixture effects assessment (Chapter 4 elaborates these approaches in more detail), Section 3.3 discusses mixture effects in relation to modes and mechanisms of action, and Section 3.4 addresses the problems and possibilities of predicting mixture effects. In Sections 3.5 and 3.6, emphasis is on the predictability of synergism and on effects at low concentration or dose levels of chemicals in mixtures. Section 3.7 provides an overview of scarcely available data on mixture effects in real-world exposure scenarios. This chapter ends with an outlook to the future.

3.2 APPROACHES TO MIXTURE EFFECT ASSESSMENT

In studying mixtures, many researchers have followed what can be called a "whole mixture approach" (USEPA 1986), where a combination of many chemicals is investigated as if it were 1 single agent, without assessing the individual effects of all the components. Refer to Chapters 1 and 4 for a discussion about this approach from an exposure and a test design perspective, respectively. This type of experiment is useful for studying unresolved (complex) mixtures or specific combinations on a case-by-case basis (for an example from wastewater pollution, see Thorpe et al. 2006, and from epidemiology, see Ibarluzea et al. 2004). However, with whole mixture approaches it is not possible to identify which chemicals make a contribution to the overall mixture effect or how the constituents work together in producing a joint effect. Furthermore, the composition of mixtures subjected to whole mixture approaches is often undefined and may change in ways difficult to recognize (see Chapter 1), with potentially significant consequences for the resulting effects. For all these reasons it is necessary to understand how chemicals act together in producing combination effects, and this requires more analytical methods, so-called "component-based" analyses. The aim of component-based mixture analyses is to explain the effect of a mixture in terms of the responses of its individual constituents. Thus, an attempt is made to anticipate joint effects quantitatively from knowledge of the effects of the chemicals that make up the mixture. These methods are particularly suited to combinations where all the components are able to induce the effect of interest. However, they are of little use with mixtures where some constituents might influence the effects of others, without producing a response on their own.

3.2.1   ADDITIVE EFFECTS OF TOXICANTS IN MIXTURES

When chemicals in a mixture act together to produce an effect, but do not enhance or diminish each other's actions, the resulting mixture effect is commonly considered additive. It is important to realize that this particular use of the term "additivity" is specific to mixture toxicology and must not be confused with additivity in the mathematical sense. Sometimes the term "noninteraction" is used synonymously with "additivity."

Methods are required that allow more reliable calculations of additive mixture effects from information about the responses of an organism to individual mixture components. For this purpose, 2 concepts have emerged: concentration addition (often also called "dose addition" or "Loewe additivity") and independent action (also referred to as "response addition" or "Bliss additivity") (Greco et al. 1992). These concepts are based on 2 entirely different ideas about how the joint action of chemicals can be perceived. See Chapter 4 for a more detailed elaboration of these concepts; here only a brief description is given.

3.2.2   CONCENTRATION ADDITION

The concept of concentration addition (CA) looks at mixture effects of chemicals in terms of a "dilution" principle. It assumes that 1 chemical can be replaced totally or in part by an equal fraction of an equi-effective concentration of another (hence the term "similar action"), without diminishing the overall combined effect (Loewe and Muischneck 1926). If the assumption of CA holds true, these fractions of equi-effective concentrations, which are also called "toxic units," sum up to a value of 1—thus the name "dose" or "concentration addition." CA implies that every chemical in any concentration contributes to the overall toxicity of a mixture. Whether the individual doses are also effective alone does not matter. Thus, combination effects should also result from chemicals at or below effect thresholds, provided sufficiently large numbers of components sum up to a sufficiently high total effect dose.

As already introduced in Chapter 1, a widely used application of CA is the "toxic equivalence factor" (TEF) concept for the assessment of mixtures of polychlorinated dioxins and furans (PCDDs/Fs) (Van den Berg et al. 2006). Under the additional assumption of parallel dose–response curves, doses of specific PCDD/F isomers are all expressed in terms of the dose of a reference chemical, 2,3,7,8-

TCDD, needed to induce the same effect ("equivalent" or "equi-effective" dose), and assessment of the resulting combined effect is obtained simply by adding up all equivalent TCDD doses.

3.2.3 INDEPENDENT ACTION

Independent action (IA) assumes that the joint effect of a combination of agents can be calculated from the responses of individual mixture components by adopting the statistical concept of independent events (Bliss 1939). This means that agents present at doses or concentrations below effect thresholds (i.e., 0 effect levels) will not contribute to the joint effect of the mixture, and if this condition is fulfilled for all components, there will be no combination effect. This central tenet of the concept of IA is commonly taken to mean that exposed subjects are protected from mixture effects as long as the doses or concentrations of all agents in the combination do not exceed their no-observed-effect levels (NOELs) (COT 2002).

3.2.4 SYNERGISMS AND ANTAGONISMS

It is possible that the combined effect of a mixture of chemicals is different from the effect expected to occur on the basis of the assumption that neither mixture component influences another's action (the additivity or "noninteraction" assumption). Such deviations from expected additivity are usually assessed in terms of synergisms or antagonisms. If the mixture effect is larger than expected, there is said to be synergism; if the mixture effect is smaller than expected, there is antagonism.

Thus, synergisms and antagonisms are defined in quantitative terms, in relation to a specific additivity expectation. The additive effects of a combination of chemicals that can be calculated by using CA or IA serve as a point of reference for identifying a synergism or antagonism.

The occurrence of synergisms or antagonisms shows that the basic assumption underlying additivity is not fulfilled, namely, that chemicals in a mixture produce joint effects without influencing each other's action. Instead, the effects of some chemicals have a bearing on those of others, such that the overall combination effect is either smaller or larger than expected. In other words, mixture components are thought of as interacting with each other. In the specialist language of mixture toxicology, synergisms and antagonisms are therefore often termed "interactions," to make a distinction from the case of noninteraction or additivity.

Interactions may occur when, for example, 1 mixture component interferes with the uptake or the metabolism of a second one, such that increased (or decreased) amounts of an active toxicant are present compared to the situation when the second chemical acts on its own (see Chapter 2). In the first situation, synergisms are likely to occur, while the latter case may lead to antagonisms.

3.3 MECHANISMS AND MIXTURE EFFECTS

Although the fundamental assumptions that lie at the heart of CA and IA are incompatible, the ways in which both concepts think of combination effects represent 2 equally valid approaches to dealing with the issue. However, when faced with the task of evaluating specific mixtures, the issue arises as to which of the 2 concepts is appropriate for the mixture in question, and therefore should be chosen for assessment. This issue becomes all the more virulent in cases where the 2 assessment concepts produce different predictions of mixture effects. Similarly, regulators are faced with the problem of which criteria to use when grouping chemicals together for purposes of joint (cumulative) risk assessment and regulation (see Chapter 5).

In an approach to deal with these decision problems, the assumptions that underpin CA and IA have been allied to a broad mechanism of combination toxicity, as follows: CA is thought to be applicable to mixtures composed of chemicals that act through a similar or common mode of action (COT 2002; USEPA 1986, 2000b, 2002b). Although the original paper by Loewe and Muischneck (1926) contains little that roots CA in mechanistic considerations, the idea of similar action probably derives from the "dilution" principle, which forms the basis of this concept. Because chemicals are viewed as dilutions of each other, it is implicitly assumed that they must act via common or similar mechanisms. Conversely, IA is widely held to be appropriate for mixtures of agents with diverse or "dissimilar" modes of action. Although rarely stated explicitly, this presumably stems from the stochastic principles that underpin this concept (see Chapter 4). The idea that chemicals act independently is equated with the notion of action through different mechanisms. By activating differing effector chains, so goes the argument, every component of a mixture of dissimilarly acting chemicals provokes effects independent of all other agents that might also be present, and this feature appears to lend itself to statistical concepts of independent events. Often, IA is held to be the default assessment concept when the similarity criteria of CA appear to be violated (COT 2002). Implicitly taking "dissimilar action" as the simple negation of "similar action," it is then assumed that IA must hold, even without further proof that the underlying mechanisms indeed satisfy the dissimilarity criterion.

Although these ideas are plausible, their application to specific combinations of chemicals is far from clear. The difficulty lies in defining reliable criteria for "similar modes of action." Often, the induction of the same phenomenological effect is deemed sufficient for accepting similar action. However, this could be inappropriate for certain combinations of chemicals that operate by distinct molecular mechanisms. At the other extreme of the spectrum of opinions, an identical molecular mechanism, involving the same active intermediate, is required to fulfill the similarity assumption. This position, with its very strict similarity criterion, may mean that only very few chemicals qualify for inclusion into mixture effects assessments, leaving out a large number of others that also provoke the same response. In effect, this would provide an unrealistically narrow perspective on real existing mixtures. A middle position is occupied by the view that interactions with the same site, tissue, or target organ should qualify for similarity (Mileson et al. 1998).

How reliable are assumptions about mechanisms or modes of action in guiding choices between CA and IA as the appropriate assessment concepts? Figure 3.1 shows an example with a mixture of anticancer drugs composed of agents with widely differing sites and modes of action. This diversity suggested that IA might provide a

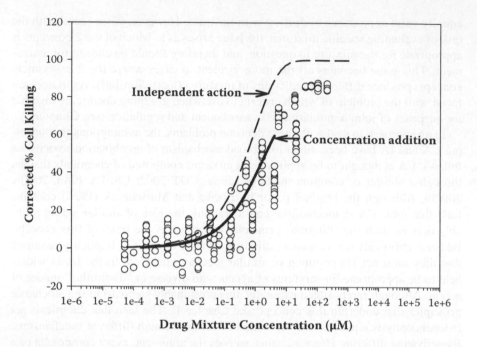

Figure 3.1 Combined effects of a combination of 7 anticancer drugs with different sites of action. Shown are the predicted (solid line: concentration addition, broken line: independent action), and the observed (circles) cytotoxic effects on DU 145 prostate cancer cells. Doxorubicin, daunorubicin, vincristin, cis-platin, etoposide, melphalan and 5-fluorouracil were combined at equitoxic concentrations. The mixture effect predictions were derived from the concentration–response curves of the individual anticancer drugs

way of correctly anticipating the effects of the combination. On the other hand, the fact that all drugs were able to induce cell death might equally well have justified use of CA. The question as to which of the 2 concepts was valid for the evaluation of this particular mixture is of practical relevance because both concepts produced quantitative predictions of the resulting mixture effects that differed substantially. The issue could only be decided through experimental validation. In this case, CA proved to yield the prediction that agreed best with the experimental observations. This example shows that assumptions about similarity or dissimilarity in the mode of action of mixture components are problematic in guiding decisions about the choice of IA or CA as assessment concepts. In this example, the question could only be resolved pragmatically by using the 2 concepts side by side for comparisons with the experimentally observed effects.

Figure 3.2 shows the outcome of a mixture experiment with toxicants that disrupt algal reproduction by diverse mechanisms (Faust et al. 2003). The mixture was carefully composed to contain chemicals with widely differing modes of action. In this case, IA was the concept that produced the prediction that best reflected the observed effects of the mixture. CA led to an overestimation of the experimentally observed responses.

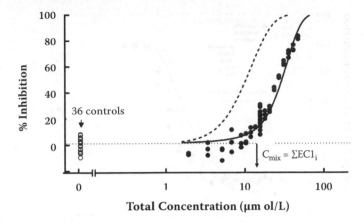

Figure 3.2   Observed and predicted algal toxicity of a mixture of 16 dissimilar acting substances. Shown are the predicted (solid line: concentration addition, broken line: independent action), and the observed (circles) cytotoxic effects on cell reproduction in the unicellular green alga *Scenedesmus vacuolatus*. All components are provided at equitoxic concentrations in the ratio of their individual EC_{01}. The mixture effect predictions were derived from the concentration–response curves of the individual compounds. (Modified after Faust et al. [2003], with permission.)

An assessment dilemma can arise when neither the additivity expectation of CA nor that of IA is fulfilled such that the observed mixture effects fall in the middle between the predictions derived from either CA or IA. An example is shown in Figure 3.3 for the inhibition of algal reproduction by a multicomponent mixture of nitrobenzenes (Altenburger et al. 2005). The joint effects of this particular mixture could be described as antagonistic in terms of CA or as synergistic with respect to IA. For multicomponent mixtures where the constituents do not meet the requirements of one or the other concept exclusively, Walter et al. (2002) have proposed a "prediction window" bracketed by the 2 concepts. Declarations of synergistic or antagonistic effects would thus be confined to combined effects larger or smaller than predicted by either concept.

These examples show that complications may arise when mechanistic considerations form the starting point for discussions about choosing CA or IA as assessment concepts (Borgert 2007), or when criteria for the grouping of chemicals into classes for cumulative risk assessment and regulation are derived from ideas about modes of action. Especially the strict similarity criteria for CA, where identical molecular targets and mechanisms are required, seem overly limiting and narrow. Apart from the case shown in Figure 3.1, another example has recently been published in the literature where CA yielded the valid mixture effect prediction, and IA led to underestimations of effect, although the combinations in question did not fulfill strict similarity criteria. Rider et al. (2008) presented an investigation of the joint effects of various antiandrogens in disrupting male sexual differentiation in rats. Some chemicals in this mixture led to feminizations of male offspring by antagonizing the effects of testosterone, and others by

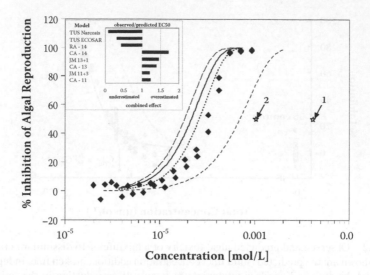

Figure 3.3 Observed and predicted algal toxicity of a mixture of 14 nitrobenzenes. Shown are the predicted (long dashed line: concentration addition, short dashed line: independent action), and the observed (diamonds) cytotoxic effects on cell reproduction in the unicellular green alga *Scenedesmus vacuolatus*. All components are provided at equitoxic concentrations in the ratio of their individual EC$_{50}$. The mixture effect predictions were derived from the concentration–response curves of the individual compounds. The solid line and the dotted line represent predictions where part of the components is regarded as similar acting and the subgroups are subsequently modeled by independent action. The arrows signify combined effect predictions based on quantitative structure activity relationships. The inset summarizes the ratio of observed to predict median effects for the different models. (Modified after Altenburger et al. [2005], with permission.)

suppressing steroid synthesis through interfering with uptake processes of steroid precursors, or by inhibiting enzymes important in synthesis. Considering these diverse mechanisms, IA was expected to be the valid prediction concept for joint effects. However, the experimental data agreed better with the additivity expectation according to CA.

There is currently no convincing alternative approach that would solve these difficulties. For this reason, it is important to take a pragmatic stance and to briefly review the literature for evidence of the usefulness of CA and IA as mixture assessment concepts.

3.4 THE PREDICTABILITY OF MIXTURE EFFECTS—A BRIEF REVIEW

Since its beginnings as a recognized scientific discipline in the 1980s, ecotoxicology has provided important stimuli for the study of the effects of exposures to mixtures. One of the original goals was to predict the toxicities of untested mixtures on the basis of knowledge of the effects of their components. Notable are the pioneering studies by Könemann (1980, 1981) and Hermens et al. (1984, 1985b) of the effects of multiple mixtures of industrial organic compounds on fish and daphnids. Since then, considerable progress has been made in mixtures toxicology. Today, the outcomes of assessments of a wide variety of chemical mixtures in a multitude of biological assays are available.

The following section provides a brief overview of experimental studies where additivity expectations based on CA or IA were explicitly formulated and used to evaluate data in terms of synergisms or antagonisms. Unfortunately, many studies with relevance to mammalian toxicology have been conducted without reference to an explicitly formulated additivity expectation. As a consequence, it is impossible to assess which of the 2 concepts, CA or IA, produced accurate predictions of combination effects (COT 2002). The section is structured according to mixtures made up of certain groups of chemicals or compound classes (such as pesticides, mycotoxins, and endocrine disruptors), various endpoints representative of different levels of biological complexity (from the molecular to the community level), the mode of action thought to be at play with certain combinations (similar or dissimilar action), and exposure typologies (exposure to many chemicals simultaneously or in sequence). Focus is on organic chemicals.

3.4.1 CHEMICALS INDUCING UNSPECIFIC CELL MEMBRANE DISTURBANCE (NARCOSIS)

A large variety of organic chemicals have the ability to induce unspecific cell membrane disturbances. This may lead to a plethora of effects in aquatic organisms, from narcosis in fish to inhibition of growth in unicellular organisms. In ecotoxicology, this mode of action is therefore often referred to as narcosis. Könemann (1980, 1981) and Hermens et al. (1984, 1985a) have shown that multicomponent mixtures of such unspecifically acting organic substances induce combined effects in aquatic organisms. With median effect concentrations of all individual mixture components as the input values, median effect concentrations for the combinations could be predicted fairly well by using CA, even for multicomponent mixtures of chemicals.

Common unspecific mode of action of all organic compounds has been taken up in quantitative structure–activity relationships (QSARs; see Chapter 5) as the concept of baseline toxicity and in toxicokinetics as the body burden concept (see Chapter 2). Baseline toxicity refers to the idea that a minimum toxicity expectation may be formulated for any given organic compound based on considerations of a compound's partition properties between hydrophilic and lipophilic chemicals (e.g., between water and octanol). Commonly, this is expressed in terms of the octanol–water partition coefficient (K_{ow}) of a chemical. The partition coefficient allows estimations of a local concentration or body burden for each individual chemical in the mixture. Assuming that this produces the same toxic effect (disturbances of cell membranes), it is then possible to anticipate joint narcotic action by adding together the respective local concentrations or body burdens for each individual mixture component.

3.4.2 Mixtures of Pesticides

Deneer (2000) reviewed the usefulness of CA for describing combination effects of pesticides on aquatic organisms. Pesticides are extremely well characterized in terms of mode of action, and are therefore ideal as reference cases to study the relevance of mechanistic information for anticipating joint toxicity. The review was based on experimental investigations of 202 mostly binary pesticide mixtures from 26 different studies, dating from 1972 to 1998. Results were reported for toxicity assays using fish, crustaceans, insects, mollusks, and algae. For more than 90% of the studies, CA was found to predict mixture effect concentrations correctly within a factor of 2, despite the fact that the assumption of a similar mode of action was violated by 85 of the mixtures under investigation.

In a more recent review, Belden et al. (2007) evaluated 45 studies dealing with 303 pesticide mixture experiments. The authors quantified the difference between predicted and observed mixture effect concentrations. In 88% of the studies that could be evaluated using CA, the predicted mixture effect concentrations differed by no more than a factor of 2 from the observed effect concentrations, again irrespective of the involved mode of action of the mixture components.

3.4.3 Mixtures of Mycotoxins

Speijers and Speijers (2004) reported on a series of studies that attempted to identify and assess the combined effects of mycotoxins with endpoints relevant to animal health. Using considerations of the mode and mechanisms of action of the various known mycotoxins as a starting point, they reviewed in vivo and in vitro studies. Unfortunately, the reviewed studies often lacked a clear point of reference for the formulation of an additivity expectation. For the few investigations where explicit modeling and design of the mixture study was available, the results showed that combined exposure to several classes of mycotoxins generally results in an additive effect with a few minor exceptions indicating synergistic interaction.

3.4.4 Combinations of Endocrine Disruptors

Endocrine disruptors are a diverse group of chemicals with the ability to interfere with the normal action of hormones. A diversity of mechanisms is at play, ranging from mimickry of endogenous hormones to disruption of normal hormone action by antagonism of receptor binding or interference with hormone metabolism. In the last 10 years, good evidence (reviewed by Kortenkamp 2007) has become available to show that the combined effects of endocrine disruptors belonging to the same category (e.g., estrogenic, antiandrogenic, or thyroid-disrupting agents) can be predicted by using CA. Eighteen studies have been published that were designed to examine the validity of the concept of CA for the predictive hazard assessment of mixtures of endocrine disruptors on the basis of knowledge about dose–response relationships of individual chemicals. These studies investigated 28 different mixtures of estrogenic, antiandrogenic, or thyroid-hormone-disrupting chemicals with 2 to 18 components. Twelve different types of endpoints in 13 different assays, organisms, and exposure regimes were studied, covering various levels of biological complexity from in vitro receptor activation to in vivo effects in laboratory animals. All in all, these efforts resulted in 34 different test cases for the validity of CA. In 25 of these cases no significant departures of observed responses from responses expected under the assumption of CA were detected. In 5 cases deviations from expected concentration additivity were seen for parts of the dose ranges or concentration ratios studied. In only 4 of the 34 test cases, a general departure of observed dose–response curves from predicted ones was found. In most of the cases where a significant departure from CA was detected, this resulted in an overestimation of the anticipated joint endocrine-disrupting effect. More than dose-additive joint actions were reported for only 3 of the test cases, and only for parts of the dose levels examined. As yet, no endocrine disruptor mixture has been shown to follow IA, although comparatively little is known about mixtures composed of chemicals from different classes of endocrine disruptors that might be expected to exhibit a diverse range of mechanisms.

Recently, evaluations of mixtures of endocrine disruptors have appeared where combination effects were anticipated on the basis of the responses observed with all individual mixture components. Crofton et al. (2005) analyzed 18 thyroid-disrupting chemicals in terms of their ability to induce changes in thyroxin levels in rodents and observed clear mixture effects that were slightly stronger than anticipated by CA. Hass et al. (2007) studied combinations of 3 androgen receptor antagonists in an extended rat developmental toxicity model. The male offspring of female rats, which were dosed over the entire duration of pregnancy, showed significant signs of feminization (reduced anogenital index, retained nipples) with a mixture of anti-androgens. With respect to changes in anogenital distance, the joint effects of the 3 chemicals were predicted well by CA, but with retained nipples the effects were slightly stronger than those anticipated by CA. Similar results were obtained when additional endpoints indicative of disruption of male sexual differentiation were analyzed (Metzdorff et al. 2007).

3.4.5 Mixture Effects and Level of Biological Complexity

The majority of the studies described in the preceding sections have been carried out using organism-based assays with endpoints representative of physiological effects in individual organisms. The question arises to what extent the likelihood of agreement of observed mixture effects with additivity expectations varies when moving up or down the level of biological complexity, from the molecular to the population level or even the community level.

Experimental evaluations of combination effects at the cellular or subcellular level from suspected endocrine disruptors or mycotoxins have demonstrated the usefulness of CA (Speijers and Speijers 2004; Kortenkamp 2007). However, on the grounds of theoretical considerations, this may not be surprising. For biological responses at the molecular level, for example, enzyme activities or receptor interactions, the dilution principle that is at the heart of CA can be readily interpreted in terms of molecular interactions. For this reason, other than concentration-additive effects may be difficult to envisage, and there is little scope for IA or substantial deviations from additivity suggesting synergism or antagonism. The likelihood of observing mixture effects accurately described by IA might increase as we move from the molecular to the cellular level. Cellular responses may be the result of interacting signaling pathways and diverse mechanisms, and these effects might follow the IA principle. Similarly, the potential for synergisms or antagonisms increases from the cellular level onward, because interferences with uptake processes and metabolic steps may come into play (see Chapter 2). Such phenomena are not accessible at the level of isolated enzymes or with biological responses very close to receptor activation.

In one of the few examples of this kind, Metzdorff et al. (2007) followed effects typical of antiandrogen action through different levels of biological complexity from receptor-mediated effects to physiological responses, in one and the same organism, the rat. Changes in reproductive organ weights and of androgen-regulated gene expression in the prostates from male rat pups were chosen as endpoints for extensive dose–response studies. With all these endpoints, the joint effects of 3 antiandrogens were dose additive. It appears that the effector chains triggered by antiandrogens feed through to higher levels of biological complexity and organization, without violating the principles of CA. This insight may be of relevance for risk assessment and regulation.

With endpoints representative of higher biological complexity the concern is that mixture effect assessments might be complicated by increased biological variability, which is always more likely to occur at higher levels of biological organization. In studies with natural algal communities, Arrhenius et al. (2004) and Backhaus et al. (2004) were able to show that the concepts of CA and IA were suitable to describe combined effects for specifically acting environmental contaminants.

3.4.6 EVIDENCE FOR THE VALIDITY OF INDEPENDENT ACTION

Comparatively few examples exist where the validity of IA has been assessed. Hermens et al. (1985b) combined 33 chemicals, which can be grouped into 3 classes with presumably differing modes of action. The observed combination effects were slightly lower than predicted using CA, but IA was not specifically evaluated. In a study utilizing a cell proliferation assay with human breast cancer MCF-7 cells, Payne et al. (2001) tested a mixture of 2 estrogen receptor agonists (o,p'-DDT, p,p'-DDT), 1 antiandrogenic agent (p,p'-DDE), and a chemical that induces cell division by as yet poorly defined mechanisms (β-HCH). IA and CA predicted the observed effects equally well.

Walter et al. (2002) assessed the effect of a mixture of 11 aquatic priority pollutants on algal reproduction. The chemicals were selected for structural diversity by using chemometric methods. In this study, statistical estimates of effect concentrations down to effect levels of 1% were derived by regression analysis of concentration–response data. Based on these estimates of low effects, IA yielded quite accurate predictions of mixture toxicity.

All the above studies used groups of similarly acting chemicals, where each group had a different presumed mode of action. Often, dissimilarity was inferred on the basis of diverse chemical structures, but proof of dissimilar action could not be provided because the actual mechanisms involved were unclear. There is the possibility that many of these experiments in fact utilized chemicals that at least partly acted in similar ways. Thus, there is a need to consider studies that have employed very strict criteria for dissimilar action.

Backhaus et al. (2000a) studied a multicomponent mixture of chemicals that all interfered specifically with differing sites in bacteria. The observed effects agreed excellently with IA. A diverse mixture of 16 chemicals, all known to specifically interact with different target sites in algae, was assessed for inhibition of reproduction in algae by Faust et al. (2003). Similar to the approach taken by Walter et al. (2002), estimates of low effects, down to 1%, were produced by regression analysis of concentration–response data of individual chemicals. These estimates were utilized to calculate mixture effect predictions according to IA. This produced fairly accurate predictions of the observed combination effects, while CA fell well short of observations.

3.4.7 MIXTURE EFFECTS AND EXPOSURE TYPOLOGY: SEQUENTIAL EXPOSURES

The majority of experimental mixture studies have analyzed the effects that arise from simultaneous exposure to chemicals. Very few studies exist where sequential exposure to several chemicals was analyzed. Only a concept founded on an understanding of the relationship between dose or concentration and exposure duration, time to effect, and recovery can hope to deal with the effect of sequential exposures. Conceptual frameworks for descriptions of time-dependent toxicity from a mechanistic perspective are available (e.g., Rozman and Doull 2000; Ashauer et al. 2006). However, the link between existing dose-time response models and a framework for mixture effect analysis from sequential exposure has yet to be made. A recent example of an interesting study that looked at sequential exposures is from Ashauer et al. (2007b), who base their analysis on a 1-compartment model for substance uptake, plus additional parameters for effect propagation and recovery. Generalizations are not yet in sight.

3.5 DEFINING DETERMINANTS OF SYNERGISTIC MIXTURE EFFECTS

It is clear that scientific approaches to assessing mixture toxicity require adequate hypotheses about the expected combined effects for the case of noninteraction (additivity). However, the success of CA and IA in anticipating additive mixture effects does not diminish the importance of the question: How likely is it that chemicals interact with each other, leading to exacerbations of effects (synergisms)?

It is important to realize that a straightforward answer to this question is difficult to give because deviations from expected additivity are impossible to predict quantitatively. No mathematical algorithm exists that would allow the anticipation of synergisms or antagonisms, not even in qualitative terms. On the basis of data about dose–response relationships of individual mixture components, synergisms and antagonisms are essentially unpredictable. Nevertheless, what can be learned from documented cases of synergisms? Are there general principles that might help to foresee whether synergisms (or antagonisms) are likely?

Lichtenstein et al. (1973) provided an early example of synergistic combination effects between insecticides and herbicides, with insect lethality as the endpoint. They found that photosystem II–inhibiting herbicides, such as atrazine, simazine, or monuron, increased the lethal effects of the insecticides parathion and DDT in various insects. Conversely, some insecticides seem to enhance the plant-damaging effects of certain herbicides (Kwon and Penner 1995). These interactions have been explained in terms of competition for cytochrome P450 mixed-function oxygenases, which resulted in inhibitions of detoxification reactions leading to elevated levels of the herbicides and therefore prolonged activity. Similar types of interaction were described for the effect of the fungicide prochloraz on the adverse effects of lambda-cyhalothrin in the honey bee *Apis mellifera* (Pilling et al. 1995). Bocquene et al. (1995) studied acetylcholinesterase inhibition with mixtures of organophosphates and carbamates in 4 different marine species. Though acting on the same physiological process (reduction in acetylcholinesterase activity in motor neuron synapses), various mixtures consistently showed higher than concentration-additive effects. Belden and Lydy (2006) have shown that simultaneous exposure to the pyrethroid esfenvalerate and chlorpyrifos produced enhanced effects on the mobility of *Chironomus tentans* larvae and fathead minnows compared to the effects expected for CA or IA.

In all these cases, certain mixture components altered the internal concentrations of other mixture constituents by enhancing or reducing metabolic conversion reactions. This kind of toxicokinetic interaction cannot be anticipated by the mathematical algorithms of CA or IA. If such interactions occur, substantial deviations from expected additivity are the result. See also Chapter 2.

A second type of nonadditive mixture response is likely to occur when some components undergo chemical reactions with each other before biological responses are initiated. Examples for this are rare. Escher et al. (2001) found slightly higher than additive toxicities with mixtures of phenolic uncouplers in bacterial assays. They explained these synergisms by invoking ion pair formation in the affected membranes, which leads to greater transport rates of the effective compounds. Other examples discussed in the literature (e.g., Ferreira et al. 1995; Mu and LeBlanc 2004) deal with observations of alterations of doses available at the target level, alterations of time to effects, and interacting signaling pathways. Modeling approaches, such as physiologically based pharmacokinetic modeling (PBPK; Yang et al. 2004) have been successfully employed to describe specific cases (see Chapter 2).

In summary, deviations from expected additive effects are well documented and can be assessed in terms of synergism or antagonism. In some cases, the mechanisms underlying such deviations are well understood. A typical case of interaction occurs when 1 substance induces toxifying (or detoxifying) steps effective for another mixture component, which in turn alters profoundly the efficacy of the second chemical.

3.6 MIXTURE EFFECTS AT LOW LEVELS

A key issue that drives chemical risk assessment as well as evaluations of the health impacts of existing pollution is whether chemicals produce combination effects when they are present at levels that individually do not induce observable effects. It is often

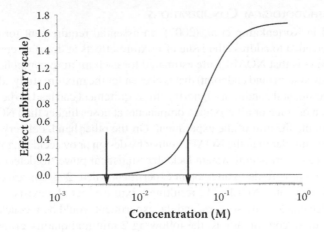

Figure 3.4 Illustration of a "sham" mixture experiment with chemicals that all exhibit the same dose–response curve. At the low dose to the left (arrow, 4×10^{-3} M), the effect is hardly observable. A combination of ten agents, at this dose (total dose: 4×10^{-2} M) produces a significant combination effect, in line with expectations following dose addition.

argued that this question cannot be decided without considering the mode of action of the chemicals that occur together at low levels, and that a distinction should be made between agents that act through a common mechanism ("similar action") and those that show diverse or "dissimilar" mechanisms of action (COT 2002).

This distinction is regarded as one of considerable practical relevance: because similarly acting chemicals can replace one another, without loss of effectiveness, combination effects can be expected even at doses well below no-observed-adverse-effect levels (NOAELs). The idea can be illustrated by considering a dose fractionation experiment (Figure 3.4) where, for example, a dose equivalent to 4×10^{-2} M produces an effect of measurable magnitude. The same effect is reached when instead this dose is administered in 10 portions of 4×10^{-3} M, or when 10 portions of 10 different chemicals of equal potency are used, even though the response to one of those dose fractions is not measurable. This means that combination effects occur at doses well below NOAELs for single chemicals provided a sufficiently large number of chemicals are present. In such situations, traditional risk assessment approaches with their focus on single chemicals are problematic because considerations of individual NOAELs do not reveal much about possible risks without information about exposure to other simultaneously occurring agents.

The situation is thought to be completely different when exposure is to chemicals with diverse modes of action. Because these chemicals interact independently with different subsystems of the affected organisms, so the assumption, mixtures pose no health concerns as long as the levels of each component stay below their NOAELs (Feron et al. 1995; COT 2002).

In this section we discuss experimental evidence for combination effects at low doses. A comprehensive review of the topic can be found in Kortenkamp et al. (2007).

3.6.1 Methodological Considerations

As reviewed in Kortenkamp et al. (2007), an essential requirement for experimental studies intended to address the issue of mixture effects at doses or concentrations below NOAELs is that NOAELs are estimated for each mixture component by using the same assay system (and endpoint) that is chosen for the mixture study, ideally under identical experimental conditions. Ignoring this requirement can lead to the inadvertent administration of some or all mixture components at doses higher than NOAELs, and would undermine the aim of the experiment. On the other hand, delivery of doses or concentrations smaller than the NOAEL, either by design or by accident, might present problems if the experimental system lacks the statistical power to detect effects. For example, it would be futile to attempt an experiment where 2 agents are combined at 1/100 of their individual NOEL. The resulting mixture effect, if it exists, would be too small to be detectable in most cases, and the experiment would be inconclusive.

Based on these considerations, the following 2 minimal quality criteria for low-dose mixture experiments suggest themselves for assessments of studies published in the literature: 1) The effects of individual mixture components should have been determined under conditions similar to those of the mixture. 2) NOAELs (or NOELs and NOECs when a neutral effect concept is adopted) should have been estimated for each mixture component, and the absence of observable effects demonstrated directly. In addition to these 2 minimal requirements, it would be desirable to calculate quantitative additivity expectations. This would allow evaluations of combination effects in terms of synergism, antagonism, or additivity.

3.6.2 Early Studies with Unspecifically Acting Organic Chemicals

In the 1980s, a series of studies of the effects of multicomponent mixtures of unspecifically acting organic chemicals on fish and other aquatic organisms was published. Könemann (1980) combined 50 agents at concentrations of 2% of their LC50 for fish and observed a joint mortality of 50%. Evaluating a broader range of endpoints, Hermens et al. (1984, 1985a, 1985b) and Broderius and Kahl (1985) found strong mixture effects in experiments with 21 to 50 chemicals on daphnids, fish, and marine bacteria. In all these studies, a joint effect of 50% was observed when the mixture components were administered at concentrations equivalent to 2.4 to 9.6% of their individual EC50. Considering the steepness of the concentration–response relationships of unspecifically acting organics in acute aquatic toxicity assays, it is reasonable to assume that these concentrations were below the no-observed-effect concentration (NOEC) of each chemical. However, the validity of this assumption was confirmed by actual determinations of NOEC values in just 1 of these studies and for only 5 of the mixture components (Hermens et al. 1984). For all the other substances and studies this ultimate proof is missing. It is therefore necessary to consider mixture studies where NOEL/NOEC estimates for every mixture component were provided explicitly.

3.6.3 Experiments with Chemicals Exhibiting a Specific Similar Mode of Action

Jonker et al. (1996) administered female rats with a mixture of 4 nephrotoxicants: tetrachloroethylene, trichloroethylene, hexachloro-1,3-butadiene, and 1,1,2-trichloro-

3,3,3-trifluoropropene. All 4 chemicals produced kidney toxicity through a pathway involving conjugation to glutathione. Increased kidney and liver weights were observed in rats that received the agents at 25% of their individual lowest observed nephrotoxic effect level, which the authors presumed to be equivalent to NOELs. This study is suggestive of combination effects at doses around NOELs, but it suffers from a lack of proof that the chosen doses were indeed NOELs.

Backhaus et al. (2000b), Faust et al. (2001), and Arrhenius et al. (2004) have presented mixture studies on marine bacteria, algae, and algal communities where combinations of chemicals were selected according to very strict similarity criteria. The mixtures included 10 quinolone antibiotics (inhibitors of bacterial DNA gyrase), 18 s-triazines, and 12 phenylurea herbicides (inhibitors of photosynthetic electron transport). All agents were tested at concentrations equal to or below their individual NOECs. In all cases, significant mixture effects ranging from 28% to 99% of a maximum possible effect were observed, and these effects could be predicted quite accurately by application of the CA concept.

3.6.4 Evidence for Low-Dose Mixture Effects from Studies with Estrogenic Chemicals

A number of experiments where groups of estrogenic, thyroid-disrupting, and antiandrogenic chemicals were combined at low doses have been published in the literature. In all of these studies, well-founded statistical criteria were used to derive low-dose estimates for single compounds (often NOELs), and the experimental power of the chosen assays was sufficient to demonstrate mixture effects of the combinations at their respective NOELs.

Silva et al. (2002) combined 8 xenoestrogens at levels equivalent to 50% of their individual NOECs in the yeast estrogen screen. Responses of up to 40% of a maximum estrogenic effect were observed. Using the same assay, Rajapakse et al. (2002) investigated whether low levels of weak xenoestrogens would be able to modulate the effects of estradiol. A combination of 11 xenoestrogens, all present at levels around their individual NOECs, led to a doubling of the effects of estradiol. The overall mixture effect was concentration additive.

In subsequent studies, the analysis of low-dose mixture effects has been extended to in vivo endpoints. Tinwell and Ashby (2004) combined 8 estrogenic chemicals at doses that gave no statistically significant uterotrophic responses when tested on their own. When administered together, quite strong uterotrophic effects were observed in the rat. In this study, no attempts were made to anticipate combination effects quantitatively and to examine their agreement with an additivity expectation. Brian et al. (2005) examined a mixture of 5 estrogenic chemicals in fish (*Pimephales promelas*) with vitellogenin induction as the endpoint. Marked effects well in agreement with CA where found when all 5 chemicals were combined at concentrations that individually did not induce vitellogenin synthesis.

Crofton et al. (2005) analyzed 18 thyroid-disrupting chemicals in terms of their ability to induce changes in thyroxin levels and observed clear mixture effects (slightly stronger than anticipated by CA) when all chemicals were combined at doses equivalent to their individual NOELs, or even below.

Hass et al. (2007) studied combinations of 3 androgen receptor antagonists in an extended rat developmental toxicity model. The male offspring of female rats that were dosed over the entire duration of pregnancy showed significant signs of feminization (reduced anogenital index, retained nipples) with a mixture of antiandrogens at their individual NOELs for these endpoints. Quantitatively, these effects agreed well with the responses anticipated by CA.

3.6.5 EVIDENCE FROM STUDIES WITH DISSIMILARLY ACTING CHEMICALS

In view of the widely held notion that mixtures composed of chemicals with a dissimilar mode of action pose no health concerns as long as the levels of each component stay below their NOAELs, it is of particular interest to assess whether this is borne out by experimental evidence.

Hermens et al. (1985b) combined 33 chemicals that can be grouped into 2 classes with presumably differing modes of action. The mixture produced 50% mortality in fish when all components were present at 4% of their individual EC50. It was assumed that these concentrations were below NOECs, although NOECs were not estimated in this study. It is therefore conceivable that some chemicals may have been present at levels above their NOECs, and this point may be particularly relevant with compounds that exhibit shallow dose–response curves. Utilizing a cell proliferation assay with human breast cancer MCF-7 cells, Payne et al. (2001) tested a mixture of mitogenic agents with differing modes of action (estrogen receptor agonists, an antiandrogenic agent, and others). A significant proliferative effect was observed when these chemicals were present at concentrations equivalent to 25% to 100% of their individual NOECs.

Walter et al. (2002) assessed the effect of a mixture of 11 aquatic priority pollutants on algal reproduction, which were selected for structural diversity by using chemometric methods. Combined at their NOECs, the pollutants produced a joint effect of 64%.

A diverse mixture of 16 chemicals, all known to specifically interact with different target sites in algae, was assessed for inhibition of reproduction in algae by Faust et al. (2003). When these chemicals were combined at concentrations equivalent to 6.6 to 66% of their NOECs, a combined effect of 18% was observed.

In demonstrating that dissimilarly acting chemicals too have the propensity to produce significant mixture effects when combined at levels below NOECs, these studies contradict received expert opinion that mixtures of dissimilarly acting chemicals are safe at doses below NOAELs. However, before definitive conclusions can be drawn, it is important to review some papers that have produced equivocal evidence of low-dose mixture effects.

3.6.6 LACKING OR EQUIVOCAL EVIDENCE OF LOW-DOSE MIXTURE EFFECTS

Jonker et al. (1990) prepared mixtures of 8 arbitrarily chosen chemicals that were fed to rats. Each chemical affected a different target organ, by differing modes of action. In 1 mixture, the agents were combined at doses equivalent to their NOAEL, and 2 further mixtures representing 1/3 and 1/10 NOAEL were investigated. Rats exposed to the NOAEL mixture for 4 weeks showed darkened livers, decreased hemoglobin levels, and increased kidney weights. The experiment with the 1/3 NOAEL mixture yielded increased kidney weights, which the authors interpreted as "chance finding." No effects became apparent with the 1/10 NOAEL mixture. Although the authors concluded that there was "some, but no convincing evidence for an increased risk from exposure to a combination of chemicals when each chemical is administered at its own individual NOAEL (p 630)," it is debatable whether the NOAEL and 1/3 NOAEL mixtures were entirely devoid of effects. In fairness to the authors, however, it is important to point out that the chosen endpoints are quite difficult to quantify.

Jonker et al. (1993) also examined a mixture of toxicants that act by different mechanisms but affect the same target organ. This mixture included 4 different kidney toxicants. The chemicals were combined at doses presumed to be NOAELs on the basis of range-finding tests, and at 1/4 of NOAELs. Rats exposed to the NOAEL combination experienced slight growth retardations, increased relative kidney weights, and elevated numbers of epithelial cells in their urine. However, rats given 1 of the individual chemicals at doses equal to the presumed NOAEL showed similar effects. Thus, at least 1 dose higher than its actual NOAEL was used in the mixture experiment. The combination of 1/4 of NOAEL did not provoke significant observable effects.

The effects of mixtures of 18 endocrine active organochlorine pesticides and environmental contaminants, including 2,3,7,8-TCDD on rats, were investigated by Wade et al. (2002). The animals were treated for 70 days with a combination of all chemicals at their respective minimal residue levels or acceptable daily intake (ADI) values. This so-called "ADI mixture" did not produce observable effects. The experiment is difficult to interpret, because minimal requirements for low-dose mixture studies were not met: none of the agents were tested individually, and information about their NOAELs in relation to the endpoints examined in this study was not available. The ADI values for most of the chemicals were derived based on endpoints unrelated to those measured in this study. This raises serious doubts as to whether the statistical power of this experiment (10 animals per dose group) was sufficient to demonstrate effects at these low doses. A combination equivalent to doses 10 times higher than those administered in the "ADI mixture" study led to decreases in epididymus weights. However, TCDD alone, at the dose present in the "10 times ADI mixture," also produced this effect, which indicates that the observed responses were attributable solely to TCDD and that the contribution of the remaining agents to this effect was negligible. It is likely that a true mixture effect had not occurred.

Lacking or equivocal evidence in some studies (like the one discussed above and that of Van Meeuwen et al. (2007), not discussed here) can be explained in terms of problematic selections of dose levels, and these examples illustrate the difficulties with experiments aimed at investigating mixture effects at low doses and highlight the need to consider experimental power and choice of chemicals.

Taken together, there is good evidence that combinations of chemicals are able to cause significant mixture effects at doses below NOAELs for individual chemicals, irrespective of perceived similarity or dissimilarity of the underlying modes of action. This issue is therefore of considerable relevance for chemicals risk assessment.

In chemicals regulation, NOAELs are combined with so-called "safety factors" to derive acceptable or tolerable daily intakes for humans (ADI, TDI), which denote exposures that can be tolerated for a lifetime without any effect. The question is whether this claim is tenable for chemicals when exposure is to a large number of substances, all at levels around their ADI. On the basis of the available experimental evidence as well as theoretical considerations, the possibility of combination effects cannot easily be ruled out. On the other hand, this possibility cannot readily be confirmed either. To decide the issue on a sound scientific basis, knowledge about relevant exposures, in terms of the nature of active chemicals, their number, potency, and levels, is essential.

3.7 WHAT ABOUT THE REAL WORLD?

Mixture toxicology, although an interesting issue, is only of theoretical relevance when the levels of chemicals that occur together in the "real world" are too low to be of concern. Therefore, what is the evidence that large numbers of chemicals occur together, and are there indications that chemicals might have a combined effect?

Because pesticides are inherently toxic and are deliberately released into the environment, their distribution and fate are comparatively well monitored. Multiple pesticide residues occur regularly in food. US Department of Agriculture food residue monitoring programs have shown that 24% of all food items contained residues of more than 1 pesticide (Mattsson 2007). In Germany, the situation is similar. According to estimates of the Federal Institute for Risk Assessment (BFR 2005), about 1/3 of food samples analyzed in routine monitoring programs showed more than 1 pesticide residue.

Considering that there are approximately 350 approved food additives and 300 natural flavoring complexes, the potential for combined exposure through the diet is generally acknowledged (Feron and Groten 2002). Furthermore, xenobiotic residues in the human body can combine with food toxins of natural origin such as mycotoxins (Mattsson 2007). In terms of their levels and their known toxic potencies, these mycotoxins may be of unexpected concern.

Medication with drugs is another area where humans are purposefully exposed to combinations of compounds with known biological activity. During the treatment of heart diseases 3/4 of patients take more than 4 different drugs at the same time, and 1/3 of patients even have voluntary exposure to more than 8 different drugs. The serious biological consequences of exposure to drug combinations recently came into sharp focus when the lipid-regulating drug Lipobay was introduced into the US market. Other lipid-regulating drugs were already in widespread use. The active ingredient of Lipobay, cervistatin, functions by reducing cholesterol synthesis. However, interactions with the widely used gemfibrozil-based drugs occurred, whereby gemfibrozil inhibited the metabolic conversion of cervistatin. As a result, the internal concentrations of cervistatin built up to levels far higher than in patients who took cervistatin on its own (Psaty et al. 2004), and this led to sometimes severe toxic effects in the affected individuals. Ultimately, this episode precipitated the withdrawal of Lipobay from the market. The dangers of adverse drug reactions from drug combinations are a long-standing concern in drug regulation, which is dealt with by detailed instructions and warning notes in patient information sheets.

Unintentional exposure to mixtures of pollutants occurs regularly not only in several occupational settings, such as welding or chemicals production, but also through environmental pollution. Combined exposures of concern include fine particles and gases in ambient air (Feron and Groten 2002), or volatile organic compounds from building materials (Feron and Groten 2002). Several nonpolar, lipophilic organic compounds are of high environmental persistence. Through partitioning processes they may distribute into all environmental media until they are ubiquitous and will accumulate in adipose tissue. Many of these persistent chemicals occur together in human tissues (Fernandez et al. 2007a).

As these few examples show, there can be no doubt that exposure situations where chemical agents occur together are the norm rather than the exception. Refer to Chapter 1 for more examples.

Human impact assessments of the effects of chemical exposures through food, personal care products, drinking water, or air have traditionally focused on single substances. Yet, detrimental human health effects due to individual environmental pollutants have normally not been demonstrated at exposure levels encountered in industrialized societies. While it may be artificial to reduce human impact assessments to single chemicals, there are indications that combinations of chemicals may play a cumulative role. Recent findings from the area of endocrine disruptors may illustrate the point.

Ibarluzea et al. (2004) measured 16 persistent organochlorine pollutants in blood serum of breast cancer patients and in women not suffering from breast cancer. While none of the 16 individual chemicals were elevated in the serum of breast cancer cases, the total estrogenic load found in patients was higher than that in controls. Damgaard et al. (2006) observed an association between congenital cryptorchidism[1] and a summative parameter of the levels of certain organochlorine pesticides in mothers' milk. Swan et al. (2005) found that decreases in anogenital distance[2] among male infants are associated with prenatal phthalate exposure. Earlier, Pierik et al. (2004) identified paternal exposures to pesticides and smoking as factors associated with these congenital malformations. Very recently, Main et al. (2007) found associations between the sum of polybrominated diphenyl ethers in breast milk and cryptorchidisms in newborn boys. Fernandez et al. (2007b) reported associations between cryptorchidisms in boys and the estrogenic load from nonsteroidal estrogens in mothers' placentas.

In most of these publications, risks were not associated with individual chemicals, yet indications of effects were apparent when health outcomes where correlated with indicators of cumulative or summative exposures. In this, there are echoes with the results of experimental low-dose mixture studies, where chemicals, although present at levels that individually did not produce observable responses, worked together to

---

[1] The absence of 1 or both testes from the scrotum.

[2] Measure of male feminization measuring the distance between the anus and the base of the penis. Measuring the anogenital distance in neonatal humans has been suggested as a noninvasive method to predict neonatal and adult reproductive disorders.

yield effects. However, before firm conclusions can be drawn, epidemiology needs to embrace the reality of mixture effects at low doses by developing better tools for the investigation of cumulative exposures. The application of biomarkers able to capture cumulative internal exposures, such as in the studies by Fernandez et al. (2007b) and Ibarluzea et al. (2004), holds great promise in this respect.

Combined exposures have also been documented for surface waters. An example includes the Swiss river Aa, where mixtures of 5 herbicides comprising triazine and phenylureas were found. Seasonal variations in pollution levels were expected to lead to considerable combination effects in aquatic organisms (Chèvre et al. 2006).

Environmental samples often contain swathes of different chemicals in mixtures. An important question for risk assessment, regulation, and remediation is to establish whether the majority of chemicals contribute to the overall mixture effect, or whether joint toxicity can be traced back to a few substances. This issue has been the topic of considerable research efforts in the field of ecotoxicology. Its resolution has required whole mixture approaches, where environmental samples were subjected to extraction procedures, followed by fractionation and chemical analysis (toxicity identification evaluation (TIE), bioassay-directed fractionations). There are interesting examples in the literature where such approaches were combined with component-based mixture assessments with the aim of identifying chemicals that contribute to mixture effects (see Chapter 4).

In some cases, the effects of complex environmental mixtures could be accounted for in terms of concentration-additive effects of a few chemicals. In sediments of the German river Spittelwasser, which were contaminated by chemical industries in its vicinity, around 10 chemicals of a cocktail of several hundred compounds were found to explain the toxicity of the complex mixture to different aquatic organisms (Brack et al. 1999). The complex mixture of chemicals contained in motorway runoff proved toxic to a crustacean species (*Gammarus pulex*). Boxall and Maltby (1997) identified 3 polycyclic aromatic hydrocarbons (PAHs) as the cause of this toxicity. Subsequent laboratory experiments with reconstituted mixtures revealed that the toxicity of motorway runoff could indeed be traced to the combined concentration-additive effects of the 3 PAHs. Svenson et al. (2000) identified 4 fatty acids and 2 monoterpenes to be responsible for the inhibitory effects on the nitrification activity of the bacteria *Nitrobacter* in wastewater from a plant for drying wood-derived fuel. The toxicity of the synthetic mixture composed of 6 dominant toxicants agreed well with the toxicity of the original sample.

There are examples where the application of CA led to predictions of combination effects for the complex mixture that were larger than the experimentally observed values. Burkhard and Durhan (1991) identified 3 insecticides in effluents to cause toxicity to *Ceriodaphnia dubia*. At the concentrations measured in the mixture, they found the combined effects of the 3 identified toxicants to be higher than the original toxicity of the effluent. Kosian et al. (1998) evaluated the toxicity of sediment pore water samples to the oligochaete *Lumbriculus variegates*. The samples were fractionated and 6 fractions subjected to toxicity testing. The toxicity observed in the 6 fractions was higher than the effects observed in the original sample. Thorpe et al. (2006) investigated estrogenic chemicals in UK sewage treatment works effluents. By using a toxicity equivalency factor approach (an application of CA), the estrogenicity of the samples in a yeast-based screening assay was predicted on the basis of the measured levels of various steroidal estrogens and other estrogenic chemicals. In many cases, the observed estrogenicity of the effluent samples fell short of the predicted effects. Thus, it appears that CA may overestimate the toxicity of mixtures of compounds identified in studies of bioassay-directed analyses.

However, the converse has also been observed: for multiple samples of contaminated groundwater from the Bitterfeld area in Germany, the joint toxicity was anticipated on the basis of the 6 most prevalent chlorinated hydrocarbons, assuming concentration additivity, but this could not fully explain the ecotoxicological effects of the original samples (Küster et al. 2004).

When aiming to single out major contributors to overall effect in mixtures occurring in environmental samples, a major challenge is the confirmation of suspected toxicants. Grote et al. (2005) and Altenburger et al. (2004) undertook an analysis of confirmation procedures for mixtures effects in effect-directed identification procedures of contaminated sediment samples. They demonstrated that there are several difficulties in unequivocally attributing individual effects to an overall joint effect when basing the analysis on sums of toxic units at median effects only, as is often done. Methodologically, the findings point to the more general issue of sensitivity consideration in combined effect assessment, that is, reflection of the ability to detect real but quantitatively small changes in effects within a given bioassay system. Moreover, a further complication lies in the fact that mixture ratios in environmental samples are unlikely to occur at equitoxic ratios, so it is hard to detect more than 3 to 6 components as significantly contributing to an observable overall effect, though this might be due to methodological reasons only (see also Chapter 4).

Taken together, there are strong indications from human epidemiology that environmental pollutants may act together at existing exposure levels to produce health effects. However, more evidence is needed to substantiate the case. In ecotoxicology, the evidence is much stronger. In both fields, however, considerable advances are needed to better quantify combination effects.

3.8 SUMMARY AND CONCLUSIONS

The concepts of concentration addition and independent action allow valid calculations of expected effects, when the toxic potencies of the individual mixture components are known.

In the majority of cases, concentration addition yielded accurate predictions of combination effects, even with mixtures composed of agents that operate by diverse modes of action. In ecotoxicology, concentration addition usually produces more conservative predictions than independent action. There are indications that this is true also for mammalian toxicology, but more data are needed to come to more definitive conclusions.

Disregard of mixture effects may lead to considerable underestimations of hazards from chemicals, and application of either concept of CA or IA has to be seen as superior to ignoring combined effects.

Various methods, models, and tools based on either concept are available and allow calculation of predictions and assessment (Solomon et al. 2008; Chapters 4 and 5).

Deviations from expected additive effects can be assessed in terms of synergism or antagonism. In only a few cases are the mechanisms underlying such deviations well understood. Interaction typically occurs when 1 substance induces toxifying (or detoxifying) steps effective for another mixture component, which in turn alters profoundly the efficacy of the second chemical.

There is good evidence that combinations of chemicals are able to cause significant mixture effects at doses well below NOAELs, irrespective of perceived similarity or dissimilarity of the underlying modes of action. On the basis of the available experimental evidence as well as theoretical considerations, the possibility of combination effects therefore cannot easily be ruled out. On the other hand, this possibility cannot readily be confirmed either. Knowledge about relevant exposures, in terms of the nature of active chemicals, their number, potency, and levels is essential for a proper decision on this issue.

Human epidemiology strongly indicates that environmental pollutants may act together at existing exposure levels to produce health effects. However, more evidence is needed to substantiate the case. In ecotoxicology, the evidence is much stronger. In both fields, however, considerable advances are needed to better quantify combination effects.

3.9   RECOMMENDATIONS

Mixture toxicology has made enormous progress in the last decades, and ecotoxicology in particular was instrumental in refining and validating methods for the anticipation of combination effects on the basis of information about the potency of individual mixture components. Mammalian toxicology is catching up fast. Despite these achievements, considerable challenges lie ahead if the insights of mixture toxicology are to find entry into chemical risk assessment and regulation.

Chief among the knowledge gaps that currently impede progress is a lack of information about cumulative exposure scenarios. Comparatively few studies have measured multiple chemicals in one and the same sample, and consequently, information about how many pollutants co-occur, and at what levels, is patchy. Viable concepts for cumulative exposure assessment strategies need to be developed, and the experiences that have occurred in the areas of bioassay-directed fractionations and with the toxicity identification evaluation (TIE) concepts no doubt provide valuable stimuli.

The advances made with assessing the effects of multiple chemicals at low doses in laboratory experiments have yet to be fully realized in human epidemiology. Although there are interesting recent epidemiological studies that deal explicitly with the mixtures issue (Hauser et al. 2003a, 2003b; Swan et al. 2003, 2005; Damgaard et al. 2006; Main et al. 2007), too much of human epidemiology is still focused on individual chemicals. Epidemiology needs to embrace the reality of mixture effects at low doses by developing better tools for the investigation of cumulative exposures. The application of biomarkers able to capture cumulative internal exposures, such as in the studies by Fernandez et al. (2007b) and Ibarluzea et al. (2004), holds great promise in this respect. Only an approach that fully integrates epidemiology with laboratory science can hope to achieve this task.

An extension of concepts of mixture toxicology to dealing with sequential exposures is timely.

Last, concepts need to be developed that allow translation of the insights of experimental mixture toxicology into viable approaches to chemical regulation. Progress in this field depends not least on whether a consensus can be achieved as to which assessment concept to adopt as a default. Another issue is the question of how knowledge gaps, particularly in exposure assessment, can best be dealt with, and how decisions can be made on the basis of incomplete scientific data (see also Chapter 5).

ACKNOWLEDGMENT

Thanks are due to Jean Lou Dorne, John Groten, Jan Kammenga, and Ryszard Laskowski, for their valuable contribution to the discussions at the International SETAC/NoMiracle Workshop on Mixture Toxicity in Krakow that led to this chapter.

4 Test Design, Mixture Characterization, and Data Evaluation

Martijs J. Jonker, Almut Gerhardt,
Thomas Backhaus, and Cornelis A. M. van Gestel

CONTENTS

4.1   INTRODUCTION

This chapter discusses test design, analysis, and assessment of results for mixture concentration–response analysis. This chapter does not cover test designs for other types of experiments in (mixture) toxicity research, such as designs for physiologically based pharmacokinetic/pharmacodynamic (PBPK/PD) modeling, or uptake elimination kinetics measurements, or capture-recapture studies for assessing mixed chemical exposure to wildlife. Mixture concentration–response analyses have been performed and discussed since the early days of mixture toxicity research (Bliss 1939) and encompass a large body of literature. In this chapter we do not aim to present new statistical designs and analysis methods. The goals of this chapter are rather to give an overview of the approaches that do exist in the literature, to specify which test design and statistical method can be used for which research aim, and to indicate potential pitfalls when results are assessed. In this way we hope to provide general guidelines for setting up mixture toxicity concentration–response experiments. The researcher can then, after deciding on the design and statistical model, make more detailed decisions, such as on number of replicates and concentrations or doses to test after performing a power analysis. Yet, specific subjects, such as power analysis, are beyond the scope of this chapter.

The term "toxic unit" (TU) plays an important role in mixture concentration–response analysis. It is defined as the actual concentration of a chemical in the mixture divided by its effect concentration (e.g., c/EC50; Sprague 1970). The toxic unit is equivalent to the "hazard quotient" (HQ), which is used for calculating the hazard index (HI; Hertzberg and Teuschler 2002). The term "hazard quotient" is generally used more in the context of risk assessment (see Chapter 5 on risk assessment), and the term "toxic unit" is used more in the context of concentration–response analysis, and therefore the latter term is used here. Toxic units are important for 2 reasons. First, toxic units are the core of the concept of concentration addition: concentration addition occurs if the toxic units of the chemicals in a mixture that causes 50% effect sum up to 1. Second, toxic units can help to determine which concentrations of the chemicals to test when a mixture experiment needs to be designed.

Efficient planning of a mixture concentration–response analysis very much depends on the aim of the study. The aims that are frequently encountered in the literature are therefore discussed and organized first in Section 4.2. The concepts of concentration addition (CA) and independent action (IA) can then be used to make quantitative statements about the toxicity of a mixture in these experiments, as discussed in Section 4.3. Section 4.4 focuses on practical experimental issues and the endpoint dependency of the statistical analyses. In Sections 4.5 and 4.6 methods to analyze component-based and whole mixture experiments are discussed. Each method is subjected to an overview of the pros and cons, and an assessment of inherent possibilities and difficulties. More specifically, we indicate which assumptions are made, whether the data can be compared with CA or IA, and which deviations from these reference concepts can be tested for. In Section 4.7 some case studies are discussed, where research goal, experimental design, and analysis are well connected. After summarizing the main findings of this chapter in Section 4.8, we discuss future developments, unresolved issues, and important topics for a robust effect assessment in Section 4.9.

4.2 TESTING AIMS

At least 3 major aims for mixture toxicity studies can be identified:

1) mechanistic research,
2) product optimization and chemical hazard, and
3) risk assessment.

In each of these disciplines several more specific goals can be identified (see text box). Because of this, a diverse set of approaches has been developed for analyzing and assessing the toxicities of chemical mixtures, which can be grouped into 3 major classes: 1) mixture experiments in which the toxicity of the mixture is characterized without making any effort to connect it to the toxicities of the components; 2) whole mixture approaches, that is, inferring from mixture effects the toxicity contributions of the individual components; and 3) component-based approaches, that is, inferring from the mixture components their joint toxicity.

Whole mixture approaches may also be indicated as diagnosis instruments, and are often used for site-specific, retrospective investigations, and hence often deal with complex mixtures, that is, those that have at least partly an unknown chemical composition. This may, for example, concern industrial or field samples containing a mixture of chemicals that is only partly or incompletely characterized. Whole mixture approaches may include bioassays, effect-directed analysis (EDA), and toxicity identification evaluation (TIE). Bioassays may be used to determine actual toxicity of an environmental sample, and do not necessarily bother about composition of the mixture or toxicity of the components. EDA and TIE approaches may be used to identify the (groups of) chemicals that are the main cause of toxicity. Mixture toxicity concepts may be useful to explain how chemicals present in the sample could have interacted to cause its toxicity.

Component-based approaches are used as prognosis instruments, to predict the effects of chemical mixtures, or to unravel interactions between mixture components. They are restricted to mixtures with a defined chemical composition. As component-based approaches allow predicting the toxicity of mixtures that are not yet found in the environment, they are the fundamental option for prospective assessments, such as the setting of environmental quality standards.

THREE PRINCIPAL AREAS OF MIXTURE STUDIES

1) Mode or mechanism of action research
  • Do the components share the same primary site of action?
2) Product optimization
  • Improving robustness, for example, against fluctuations in the quality of the components of a chemical product
  • Identifying main interactions
  • Increasing the spectrum of activity, for example, in pesticidal preparations
  • Maximizing wanted responses, for example, pharmaceuticals
  • Reducing the risk of resistance development, for example, for pesticides
  • Improving physical or chemical properties such as solubility or stability
3) Hazard and risk assessment
  • Assessment whether there is a risk to be expected for human health or the environment from a current or anticipated exposure situation
  • Identification of the most sensitive components of the exposed system (human body, particular ecosystem)
  • Identification of the most important chemicals in terms of their contribution to overall toxicity
  • Comparison and ranking of pollution sites
  • Definition of cleanup goals
  • Monitoring of remediation processes
  • Analysis of the relationship between modeled or measured exposure concentrations and observed (eco)toxic effects
  • Setting of environmental quality and health standards
  • (Comparative) hazard assessment of chemical products

4.3 CONCENTRATION ADDITION AND INDEPENDENT ACTION: EMPIRICAL MODELING

Numerous methods and approaches for component-based approaches have been described in the literature. They can be roughly grouped into 2 classes: specific mechanistic models that are tailored toward precisely describing a predefined situation (chemical mixture, biological system) and simpler, more general statistically based concepts. Here, focus is on the second, more general concepts, which can be traced back to 2 different fundamental concepts, commonly called concentration addition (CA) and independent action (IA), or response addition (RA) (Boedeker et al. 1992). Both concepts can also be found under various other names (Faust et al. 2001) and are implemented in a diverse set of models for predicting or assessing mixture toxicities; see compilations in Berenbaum (1989), Boedeker et al. (1990), Kodell and Pounds (1991), and Grimme et al. (1994).

Before introducing the concepts of CA and IA, it is important to note that basic to these concepts is the term "interaction." It was defined by Plackett and Hewlett

(1952, p 145) as follows: "A and B are said to interact if the presence of A influences the amount of B reaching B's site of action, or changes produced by B at B's site of action; and/or reversely with A and B interchanged." Interaction thus concerns the processes affecting the final amount of a chemical reaching a target site or its activity expressed at that target site, both possibly resulting in a changed response of the test organism or the endpoint measured. As an example, Walker and Johnston (1993) found that malathion in a mixture with prochloraz was more toxic to birds than expected from the toxicity of the single compound. They were able to explain this from an induction by prochloraz of the activity of the cytochrome P450 system that transforms malathion in its more toxic metabolite malaoxon. Activity of malathion was increased by prochloraz, leading to a higher toxicity.

The outcome of the interaction between chemicals in a mixture may be a toxic effect that is greater or smaller than expected from the toxicity of the single compounds. When a chemical is not toxic by itself, but causes another chemical to be more toxic, this is called "potentiation." When a mixture is more toxic than expected from the toxicity of its individual compounds, this is termed "synergism." "Antagonism" is used to indicate that a mixture is less toxic than expected from the toxicities of the individual components. In all cases, the interpretation of the mixture effects depends on the choice of the reference model. For that reason, it is important to discuss here in more detail the most commonly used reference concepts of CA and IA.

4.3.1 Concentration Addition

CA was first formulated in a publication by the German pharmacologist Loewe in 1926 (Loewe and Muischneck 1926). For a mixture of n components, the concept can be mathematically expressed as

$$\sum_{i=1}^{n} \frac{c_i}{ECx_i} = 1$$

(4.1)

where c_i gives the concentration (or dose) of the ith component in an n-compound mixture that elicits $x\%$ total effect and ECx_i denotes the concentration of that substance that provokes $x\%$ effect if applied singly. The ratio c_i/ECx_i—the "toxic unit"—gives the concentration of a compound in the mixture scaled for its relative potency. If the sum of the toxic units of the mixture components equals 1 at a mixture concentration provoking $x\%$ effect, CA holds. Consequently, CA assumes that an arbitrary mixture component can be exchanged by another chemical with the same mechanism of action, without changing the overall mixture toxicity, as long as its concentration expressed in toxic units remains the same.

Equation 4.1 is equivalent to

$$ECx_{Mix} = \left(\sum_{i=1}^{n} \frac{p_i}{ECx_i} \right)^{-1}$$

(4.2)

Here p_i denotes the relative fraction of chemical i in the mixture, that is,

$$\sum_{i=1}^{n} p_i = 1$$

Equation 4.2 allows the direct calculation (prediction) of an ECx value for the mixture. In general, no explicit formulation of the CA-expected mixture effect $E(c_{Mix})$ is possible; direct calculations are restricted to the level of effect concentrations (ECx values) (Faust et al. 2001). Only in the so-called "simple similar action" cases can CA-expected mixture effects be directly calculated. Since the Saaresilkä agreement (consensus agreement on mixture toxicology terminology; Greco et al. 1992) simple similar action may be considered a special case of CA, which assumes that the individual curves of the components are concentration parallel; that is, there is an effect-level independent constant potency factor between the individual concentration–response curves. On this condition, the CA-expected mixture effect can be explicitly formulated as

$$E(c_{mix}) = f\left(\alpha, c_i + \sum_{i=2}^{n} g_i c_i \right) \qquad (4.3)$$

where f is an appropriate concentration–response model, α a vector of model parameters, c_i the concentrations of the $i = 1, \ldots, n$ chemicals in the mixture, and g_i the potency factor mentioned before. The effect of a mixture that contains n components at concentrations $c_1, c_2, \ldots, c_n$ is assumed to be identical to the effect of, for example, the first compound at a concentration

$$c_i + \sum_{i=2}^{n} g_i c_i$$

All components behave as if they were simple dilutions by a factor g of this first chemical; hence, all concentrations of component 2 ... n can be rescaled to the first chemical, independent of the considered effect level. A widely used application of this approach is the toxic equivalence factor (TEF) concept for the assessment of mixtures of polychlorinated dioxins and furans (PCDDs/Fs). Here, concentrations (or doses) of specific PCDD/F isomers are all expressed in terms of the concentration of a reference chemical, 2,3,7,8-tetrachlorodibenzo-p-dioxin (TCDD), needed to induce the same effect ("equivalent" or "equi-effective" concentration). The assessment of the resulting combined effect is obtained simply by adding up all TCDD-equivalent concentrations.

Parallelism of dose- or concentration–response curves has often been used as a decision criterion on whether to apply CA to a mixture. But it should be pointed out here that the general formulation of CA in Equation 4.1 assumes neither a specific shape of each concentration–response curve of the components nor a specific relationship between the curves. Even if all chemicals in a mixture share an identical receptor binding site, differences, for example, in the toxicokinetic behavior of the substances might lead to concentration–response curves that are not concentration parallel, if the responses of the exposed animals are observed on a higher, integrating level (e.g., reproduction). Also, the biometrical description of the individual concentration–response data might exert an influence on the parallelism of the concentration–response curves. If for all components curves are described by only one, inflexible model (such as the classic probit model), the resulting curves might be more concentration parallel than with a biometrical analysis that uses more flexible models or even different models for different components (Scholze et al. 2001).

From a mathematical perspective (see Equation 4.1), CA simply represents the weighted harmonic mean of the individual ECx values, with the weights just being the fractions p_i of the components in the mixture. This has important consequences for the statistical uncertainty of the CA-predicted joint toxicity. As the statistical uncertainty of the CA-predicted ECx is a result of averaging the uncertainties of the single substance ECx values, the stochastic uncertainty of the CA prediction is always smaller than the highest uncertainty found in all individual ECx values. Perhaps contrary to intuition, the consideration of mixtures actually reduces the overall stochastic uncertainty, which is a result of the increased number of input data.

4.3.2 INDEPENDENT ACTION

In contrast to CA, the alternative concept of IA assumes that the mixture components act dissimilarly (Bliss 1939). This concept, which is also known as "response addition," can be mathematically formulated as

$$E(c_{Mix}) = 1 - \prod_{i=1}^{n} \left[1 - E(c_i) \right] \tag{4.4}$$

where $E(c_{Mix})$ denotes the proportional effect that the total mixture at a concentration

$$c_{Mix} = \sum_{i=1}^{n} c_i$$

provokes and $E(c_i)$ are the proportional effects that the individual components would cause if applied singly at that concentration at which they are present in the mixture. Equation 4.4 shows that IA simply follows the statistical concept of independent random events (Bliss 1939). Due to this probabilistic background, IA assumes strictly monotonic concentration–response curves of the individual mixture components and a Euclidian-type effect parameter scaled to an effect range of 0 to 1 (0% to 100%).

Details about the application of these concepts for actually calculating a mixture toxicity expectation can be found in Backhaus et al. (2000a, 2000b) and Faust et al. (2001, 2003).

IA has been further developed by including a so-called "coefficient of correlation" in order to account for possible correlations between the susceptibilities of the individuals within a population (tolerance correlation) (Plackett and Hewlett 1963a, 1963b). This coefficient is typically denoted as r, which varies from -1 (completely negative correlation) to $+1$ (completely positive correlation). These extensions of the basic IA equation (Equation 4.4) have been formulated for binary mixtures and quantal endpoints only. An application to multicomponent mixtures is a difficult and yet unresolved problem, because the univariate r results in a high-dimensional correlation matrix, which is not only problematic to determine but also extremely difficult to interpret. Furthermore, for scientifically sound correlation, coefficients that are tailored to the mixture of concern have to be available for each assessment. However, this is hardly ever the case. As a consequence, to our knowledge, no multicomponent mixture study has been published that demonstrates the applicability of IA with $r \neq 0$.

4.3.3 CONCENTRATION ADDITION VERSUS INDEPENDENT ACTION

Although both concepts allow the calculation of mixture toxicity on the basis of the toxicity of the individual compounds, there are fundamental differences between CA and IA when it comes to assessing mixtures of low concentrations of toxicants. These differences originate in the opposite conceptual assumptions of both concepts. CA operates on the level of effect concentrations (Equation 4.1), while IA uses the effects of the single toxicants for the calculation of an effect of the mixture (Equation 4.4). Hence, according to CA, every toxicant that is present in the mixture contributes to the overall toxicity (in direct proportion to its toxic unit). In contrast, IA implies that only those components contribute to an overall toxicity that are present in the mixture at a concentration, whose effect—if that concentration would have been applied singly—is greater than 0.

Such threshold values are often estimated using no-observed-effect concentrations or levels (NOECs or NOELs). It might be tempting to substitute the individual ECx values in the CA equation (Equation 4.2) with NOELs in order to calculate a mixture NOEL. But this would imply that all NOELs provoke the same, statistically insignificant effect; that is, all of them must have been determined in an identical experimental setup (in terms of number of replicates, spacing of test concentrations, variance structure), which is hardly ever the case. Nevertheless, a range of methods, such as TEFs or TEQs (see Chapters 1 and 5), makes use of a CA-like approach and sums up NOEL-based hazard quotients. This introduces an additional source of uncertainty in the risk assessment, which is fundamentally different from the question of whether CA is an appropriate concept for the mixture of interest.

NOELs might correspond to effects as high as 5% on average (Allen et al. 1994). IA-compliant mixture effects are therefore to be expected, even if all components are present only at their individual NOELs. If only a certain fraction of the individual NOELs are present, it depends on the number of mixture components, the precision of the experimental data and the steepness of the individual concentration–response curves whether those mixture effects might become visible. It might therefore be too far fetched to generally assume that mixtures composed of low concentrations (below their individual NOEC) of dissimilarly acting substances do not pose a risk (see also Chapter 3 and the recent review by Kortenkamp et al. 2007). Especially keeping in mind that IA is only a theoretical borderline case, its fundamental assumptions—chemicals that are completely independently acting and do not influence each other's toxicity whatsoever—are never completely fulfilled in real biological systems.

Neither CA nor IA makes any assumption on the targeted biological system, nor do they consider any specific properties of mixture components beyond their pharmacological (dis)similarity. This is both the strength and the weakness of the concepts. On the one hand, this simplicity allows the establishment of general rules for mixture toxicity assessment, which is essential for considering the joint action of chemicals in regulatory guidelines. On the other hand, it cannot be assumed that these concepts actually describe biological reality, except perhaps in very simple systems. Even if all components of a mixture are similarly or dissimilarly acting, additional (unspecific) binding sites or differences in toxicokinetics or biotransformation pathways, respectively, interfere. Hence, assuming an appropriate experimental power in terms of accuracy and precision, differences between CA and IA expectations and the actually observed mixture toxicity always become apparent. The crucial question therefore might not be whether deviations between simple concepts and complex biological realities can be observed, but whether CA and IA are oversimplistic, that is, whether their empirical predictive power is sufficient for a certain purpose (e.g., safe risk assessment).

4.4 TECHNICAL ISSUES AND PITFALLS

4.4.1 Use of Existing Knowledge

Both CA and IA concepts only allow the prediction (extrapolation) from single substances to mixtures. Information on dose– or concentration–response relationships for the individual chemicals is required when designing a mixture toxicity study. Such information also is needed when analyzing mixture toxicity data. In case of the CA concept, proper effect endpoints like EC50 or LC50 (or generally ECx and LCx) values must be available for each individual chemical to allow for a calculation of the toxic strength of the mixture and to determine whether deviations from the reference concept occur. Calculation of such endpoints is only possible when concentration–response relationships are available. The IA concept requires a full concentration–response relationship to enable calculation of the contribution of the individual chemical to the toxicity of the mixture at its concentration in the mixture and for analyzing the mixture interactions.

4.4.2 Endpoints in Relation to Test Design and Dose–Response Modeling

Toxicity endpoints need to be chosen according to the following criteria in mixture toxicity:

1) Type of chemical substances and their mode or site of action to be studied. Behavioral parameters are, for example, often linked to acetylcholinesterase (AChE) inhibitors, such as chlorpyrifos, or chemicals reacting with the acetylcholine receptor, like imidacloprid and thiacloprid. That means for a certain mixture, relevant endpoints, that is, which can best reflect the specific toxicity of one or more components of the studied mixture, should be selected. If irrelevant endpoints are studied (which do not show any specific cause–effect relationship from the single substances), mixture effects might be misinterpreted. For example, if a study on chlorpyrifos reveals behavioral effects (plausible for a neurotoxin) and Ni (in many cases no apparent effect on behavior) is the second substance in the mixture, the behavioral endpoint is good only for 1 substance. As Ni by principle might affect other systems (e.g., reproduction) without effect on behavior, the mixture effects might be misinterpreted because "apples" are compared with "oranges." However, if both substances are neurotoxic, then behavior might be the right choice.

In summary, the chosen biological endpoints must be sensitive to each of the individual components. Hence, at least for mixtures of dissimilarly acting substance, so-called "integrating endpoints" (usually mortality, growth, and reproduction) are needed. Mixtures of compounds that share the same mode and mechanism of action can be analyzed using more specific endpoints.

2) Duration of the experiment. Survival and avoidance behavior are, for example, suitable for short-term tests with either only high concentrations (survival) or both high and low concentrations (behavioral parameters). Some endpoints can only be recorded after long-term exposure, for example, reproduction or full-life-cycle effects, like the emergence of chironomids after 21 to 28 days. Long-term exposures are subject to more confounding errors and factors than short-term exposures, just by the factor of time, for example, changes in temperature, growth of fungi over time, but also by the interaction of multiple responses and effects; for example, when a toxin affects survival the lower number of survivors is exposed to more food (less competition) and might grow better. Stark (2006) showed that there is little relation between the acute LC50 and the sensitivity of longer-term population growth parameters. This means that several endpoints have different inherent "uncertainty or noise," and hence might affect the model outcome, for example, a less good fit. Reasons for such an uncertainty might be exposure time (as above) or other confounding factors or errors (see next point). One consequence of this is that acute toxicity tests with survival as the endpoint often tend to show steeper concentration–response curves than long-term tests focusing on sublethal endpoints like growth or reproduction. Steepness of the concentration–response curves of the individual chemicals in the mixture is an important aspect that may affect interpretation of the mixture interactions.

3) Confounding factors. Some endpoints (e.g., egg hatching, molting, and behavior) are more susceptible to confounding factors than others (survival),

often being also more sensitive toward toxic stressors and often showing nonlinear concentration–response profiles. See, for example, Heugens et al. (2001) for an overview of the effects of different confounding factors, like temperature, salinity, and food availability, on the susceptibility of aquatic organisms to toxicant stress. Concepts for the analysis of responses to both single chemicals and mixtures are easy to be used with survival data, but much less with data from chronic sublethal tests with other continuous test parameters. Confounding factors therefore may affect the interpretation of mixture interactions.

4) Characteristics of endpoints. Some endpoints have a higher inherent variability, hence providing "noise," making detection of significant changes due to toxic stress more difficult. This fact also holds for the test species; for example, laboratory-generated clones of a species generally show lower response variability than field-collected individuals of the same species. All these types of noise may affect the shape of the concentration–response relationship and therefore lead to wrong or deviating conclusions about mixture interactions.

5) Dichotomous responses. Some endpoints (e.g., behavioral parameters) can lead to nonmonotonous concentration–response curves, which are more difficult to model than the monotonous concentration–response curves that are observed, for example, in survival studies. Also in case of hormesis, nonmonotonous concentration–response curves are found. Nonmonotonous curves pose inherent problems for the application of both mixture toxicity concepts (Backhaus et al. 2004; Belz et al. 2008).

6) Practical aspects. Some endpoints need more elaborate experimental setups than others, for example, microcosm or mesocosm studies on community responses compared to simple beaker tests for determining effects on the survival of a single species. As a consequence, possibilities for the replication or repeating of tests may be limited. This may affect the possibility of meeting requirements for a proper analysis of the resulting toxicity data against the 2 reference models of CA and IA.

7) Quantal data. The type of data generated by certain endpoints is not always applicable to the concepts of CA and IA, for example, quantal endpoints (providing yes or no answers instead of continuous data). Histological parameters or immunological responses may not be appropriate for concentration–response curve analysis due to serious deviations from the monotonous curve pattern. Such data, therefore, are also less suited for the determination of mixture interactions.

8) Type of endpoint. The type of endpoint recording is essential for the application of different types of mixture toxicity models. Endpoints measured at only 1 point in time may only be used to derive concentration–response-related parameters, such as ECx or LCx or NOECs. Continuous recording or at least repeated recording of responses may allow for time series analysis. Time-related responses may, for example, be used for the derivation of kinetic parameters by applying pharmacokinetic/dynamic models (like the PBPK models in human toxicology; see, e.g., Krishnan et al. 1994) or

the dynamic energy budget (DEB) model (in ecotoxicology; see Kooijman 2000). See Chapter 2 for further details.

9) Resources and manpower available. Resources and manpower effort determine the choice of a particular test design, for example, the trade-off between a fully factorial designed experiment and a single ray experiment. Also these considerations determine the possibilities of adequately establishing mixture interactions.

4.5 COMPONENT-BASED APPROACHES, TEST DESIGNS, AND METHODS

One of the major aims of component-based strategies, that is, when inferring joint action from the toxicity of single chemicals, is to analyze whether the mixture follows the expectations set by CA and IA—or whether interactions lead to higher toxicities than predicted. When setting up such an experiment, choices have to be made on the number of replicates, the number of exposures to the mixtures, and the number of exposures to the single mixture components. Limitations are set by the number of experimental units that can be handled, by ethical considerations concerning the use of test organisms (especially the case for vertebrates) and the number of mixture components.

There are typically 2 specific questions to be answered:

1) Do you want to compare the observed toxicity with both the IA and the CA concepts?
2) Which type of deviation from this reference do you want to detect?

Additional questions relevant in this situation are these:

3) How many experimental units can be handled?
4) Which endpoint is being measured?

4.5.1 FIXED-RATIO DESIGN

Using the so-called "fixed-ratio" or "ray" design, the mixture of interest is analyzed at a constant concentration ratio while the total concentration of the mixture is systematically varied. Hence, a concentration–response curve (a "ray" in the mixture response surface; see Section 4.5.2) of the mixture is recorded, which can then be analyzed just as the concentration–response curve of a single chemical. A comparison with the concentration–response curves of the individual components allows a comparison with both CA and IA predictions. For this purpose $k(n + 1)$ test groups are needed, where n = number of mixture components and k = number of concentrations per concentration–response curve. For the fixed-ratio design the use of D-optimal designs has been proposed as an efficient approach (Coffey et al. 2005). For a fixed sample size, the D-optimal design provides a criterion to select the experimental concentrations and number of subjects per concentration level that result in minimum variance of the model parameters, and thus increased power to detect departures from additivity.

Fixed-ratio designs may be analyzed with mixture concentration–response curve methods. For CA it may be important to validate the slopes of the concentration–response curves of the individual mixture components and the mixture (see Equation 4.3 and Van Wijk et al. 1994). Concentration–response curves have also been used for a methodology that can be described as stepwise fitting, used by van Gestel and Hensbergen (1997) and Posthuma et al. (1997). This method exploits the fact that most equations used for concentration–response analysis are parameterized in such a way that the EC50 is 1 of the parameters, which enables an estimation of the 95% confidence interval around the median effect concentration level. Stepwise fitting starts with fitting the individual concentration–response curves for every individual mixture component, to determine whether the toxicity in the experiment differs from the range finding. This yields updated EC50 values, and these are used to recalculate the toxic unit values of the mixture concentrations. The response to the mixture can now be quantified by fitting a concentration–response curve as a function of the new toxic unit values. If the 95% confidence interval of the mixture EC50 estimation excludes the value of 1.0 TU, the mixture effect may deviate from CA at the 50% effect level. If one wants to take into account different mixture ratios in this analysis, the concentration–response fitting has to be performed on fixed-ratio mixture concentrations only, according to the ray design. The advantage of this approach is that it takes into account the uncertainty in the prediction of the response to the mixture through the use of the 95% confidence interval of the median effect level. Estimations of deviations from CA at other concentration levels can be performed using other parameterizations of the concentration–response function that include the EC10, EC25, or EC90 (Van Brummelen et al. 1996a; van Gestel and Hensbergen 1997; Van der Geest et al. 2000). Stepwise fitting can therefore be performed by people having good training in concentration–response analysis of single toxicants, which most (eco)toxicologists have. The statistical inference does not take into account that the values of the parameters of the individual curves are actually predictors for the complete mixture data set. In addition, when the toxic units are recalculated, the uncertainty in the estimation of the EC50 values cannot be taken into account.

Fixed-ratio designs especially allow a convenient visualization and interpretation of experimental results, even for mixtures with many compounds. If a single ratio is tested, an obvious drawback of this design is that no statement on mixture-ratio-dependent deviations from CA or IA can be made. The mixture concentration–response curve methods have been extended with methods that enable the quantification of effect-level-dependent deviations from CA along mixtures of increasing toxic strength in a fixed ratio of concentrations, moles, TUs (equitoxic), or increasing concentrations in any combination (Van der Geest et al. 2000; Gennings et al. 2002; Crofton et al. 2005). These methods are usually based on constructing a 95% confidence interval around the fitted effect of the mixture, and analyzing whether the effect predicted by CA is captured by this confidence interval. If the real effect is underestimated at low concentrations and overestimated at high concentrations, then synergism at low concentrations and antagonism at high concentrations may be detected (Gennings et al. 2002). The method published by Van der Geest et al. (2000) has, for instance, been used by Banks et al. (2003) to analyze the concentration-level-dependent effect of diazinon and copper on the water flea *Ceriodaphnia dubia*.

4.5.2 ASSESSMENT OF THE COMPLETE CONCENTRATION–RESPONSE SURFACE

4.5.2.1 Research Aim and Experimental Design

If the toxic effect of a chemical combination is tested and compared with the effect of the individual chemicals, it may happen that the effect of the tested mixture deviates from the effect predicted by CA or IA. This mixture can be considered as 1 combination of the endless number of other possible combinations in which these chemicals can be mixed. If more combinations of this specific set of chemicals are tested, it can happen that effects of a number of different combinations at low concentrations differ from CA or IA, but that the effects of high-concentration combinations are well predicted. Such a systematic deviation pattern may be relevant for risk assessment, or may provide insight into the modes of action. Three types of systematic deviations from CA or IA can be defined as biologically relevant, based on studies published in the literature:

1) global synergism or antagonism,
2) concentration-ratio-dependent synergism or antagonism, and
3) concentration-level-dependent synergism or antagonism.

If the aim is to screen for such systematic deviations because a priori knowledge is lacking or interactions are expected, the complete concentration–response surface should be tested. Yet, the number of possible test combinations increases exponentially with the number of chemicals in the mixture. Full concentration–response surface analysis is therefore seldomly performed for testing more than 4 or 5 chemicals simultaneously.

Even if mixtures of a limited number of toxicants are investigated, a robust statistical design needs to be adopted to select the concentration combinations to test. The full factorial design enables full coverage of the complete concentration–response surface. This design is generally applicable and therefore frequently discussed in standard statistical textbooks (see, e.g., Sokal and Rohlf 1995), but it can have disadvantages for toxicity studies that are described below. Apart from the statistical design, the researcher also has to decide on the number of concentration combinations to test. A toxicity test with a single chemical is usually performed with 5 or 6 concentrations, including a control, to estimate the slope and functional form of the single concentration–response relationship in a reliable manner. For a full factorial design with 2 chemicals, this would mean testing 5^2 to 6^2, that is, 25 to 36 concentration combinations. The concentration range and distribution of concentrations have to be considered as well. It is possible to select the concentration combinations on a logarithmic scale rather than the normal scale, to take into account the multiplicative characteristics upon which concentration–response relationships are usually based.

In addition, one could "scale" the complete experimental design using a toxic unit approach. This ensures that relevant concentrations are tested.

For scaling the experiment, the EC50 is usually taken as the basis for the toxic unit, and a reasonable concentration range for the individual toxicants could be 0, 0.25, 0.5, 1, 2, and 4 toxic units (obviously representing a log2 scale). In this way, all the concentrations of the individual components in the mixture are scaled such that differences in "toxic strength" between chemicals are taken into account. This is experimentally elegant and also avoids the problem that the tested concentrations are too low or too high to measure the effect on the endpoint effectively (although 4 TU may be too high when the mixture contains many chemicals). Using the toxic unit concept for the experimental design requires knowledge about the toxicity of the individual chemicals, and range-finding experiments may be necessary. The design of a mixture concentration–response experiment can therefore be broken into 3 steps:

1) Perform range-finding experiments with the individual mixture components or explore existing knowledge, to determine the toxicity of each component by finding the median effect concentration (EC50) for the endpoint of interest.
2) Determine which toxic unit levels need testing for both the individual mixture components and the mixtures.
3) Calculate the required amounts of each chemical for each mixture, considering that 1 TU = c / EC50. This scaling procedure is not strictly necessary for mixture concentration–response analysis, but it is recommended.

Given the usual steepness of concentration–response curves, concentrations with a toxic strength of 4 TU usually provoke quite high toxic effects of >90%. However, they might need to be tested in order to quantify the absolute maximum response for estimating the parameters in the concentration–response function (its asymptote). A major disadvantage of a full factorial design is that, in the given example, 9 of the mixture concentrations would have a combined toxic strength even higher than 4 TU (Figure 4.1). These concentrations are likely to be a waste of experimental effort, assuming that the maximum response already occurs at 4 TU. Hence, unless the underlying concentration–response curves are unusually flat or antagonism at high concentrations is expected, the full factorial design may be an inefficient design for mixture toxicity studies. More efficient and cost-effective for covering the concentration–response surface is to use mixture rays (Gennings et al. 2004) that are based on toxic unit scaling (van Gestel and Hensbergen 1997; see Section 4.5.1).

The procedure to set up such an experimental design is as follows. Once the EC50s of the individual toxicants are established, the chemical concentrations can be expressed in terms of these EC50 as toxic units (c / EC50). Choose the toxic unit levels that need to be tested, for instance, 0, 0.25, 0.5, 1, 2, and 4 toxic units. Choose the ratios to be tested, for instance, 1:0, 2/3:1/3, 1/2:1/2, 1/3:2/3, 0:1.[1] Calculate the

---

[1] Note that the ratio design also includes testing (again) of the single chemicals, simultaneously with the mixtures. This is considered necessary since it is generally accepted that EC50 may differ in time (see Section 4.1).

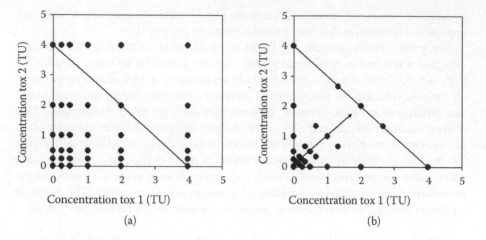

Figure 4.1 Examples of possible designs for determining the toxicity of binary mixtures, including the single chemicals as well as covering the entire concentration–response surface. The left-hand (a) graph shows a full factorial design where all concentrations of the single chemicals are combined to obtain mixtures. The right-hand (b) graph shows the ray design (arrow: one ray), with chemicals in the mixture tested at fixed concentration ratios (e.g. 3:1, 1:1, and 1:3). Both approaches include the testing of the single chemicals and the mixtures in one experimental design.

test concentrations by multiplying the toxic unit levels with the EC50s for each ratio. The result is shown in Figure 4.1, demonstrating that the full response surface is covered. The combined concentrations on the rays now represent the same predicted toxic strength as the individual concentrations, and a maximum of 4 toxic units is tested for the individual chemicals as well as the mixtures. The 1-to-1 ray is called equitoxic, because both chemicals are present in the same toxic strength. Cotter et al. (2000) combined the factorial and ray design in 1 concentration–response study.

4.5.2.2 Data Analysis: Judging Deviations from CA and IA

After setting up and running the experiment, the data have to be analyzed by assessing the deviation of the mixture responses from the responses predicted by CA or IA. This assessment can be performed in many ways. In general, assessment of the complete concentration–response surface has been performed using 3 types of methods:

1) mixture concentration–response curve methods,
2) multiple regression, and
3) nonlinear response surface models.

As indicated, response surface analyses are particularly useful to screen for synergism or antagonism, concentration-level-, and concentration-ratio-dependent deviations, and the data analysis method should accommodate this. Hence, we discuss if and how such a screening can be performed for each of these methods.

4.5.2.2.1   Mixture Concentration–Response Curves

Strictly speaking, fitting mixture concentration–response curves is not really a method suitable for analyzing the response surface of a mixture. Only a part of the response surface is analyzed, or alternatively, the multidimensional response surface is condensed to a single curve. It is mentioned here because mixture concentration–response curves have traditionally been used to assess the complete concentration–response surface since the dawn of mixture toxicity research. Bliss (1939) proposed a method with probit concentration–response curves to quantify synergism, which was later improved by Finney (1942). In the decades since, mixture concentration–response curve methods have been developed further (Chou and Talalay 1983; Barton 1993) and frequently used (Posthuma et al. 1997; Van Gestel and Hensbergen 1997). Concentration–response curves for mixtures are obtained if the total combined concentration is increased and a curve is fitted to the measured response. This approach has been used in 4 ways for mixtures of increasing toxic strength:

1) in a fixed molar ratio,
2) in a fixed concentration ratio,
3) in a fixed ratio of toxic units (equitoxic mixtures), or
4) in any combination.

There is no consensus on the best or most optimal approach, but it obviously influences the interpretation of the outcome of the analysis. See also Section 4.5.1 on fixed-ratio designs.

4.5.2.2.2   Multiple Regression

It has been shown that the multiple linear regression model with a link function is equivalent to CA (Gennings 1995). This concurrence between the 2 approaches is also intuitively reasonable, because both CA and multiple regression models describe straight isoeffective lines (isoboles). It means that the multiple regression model can be used to analyze the mixture toxicity data in order to identify deviations from CA. Deviations from CA can be tested through the interaction terms in the regression model. The likelihood function to be optimized depends on the endpoint measured. This requires detailed knowledge about multiple regression analysis, such as

1) how to choose a proper link function,
2) how to choose a suitable likelihood function,
3) how to judge the model fit,
4) how to detect multicolinearity,
5) how to interpret the model parameters, and particularly,
6) how to interpret the (higher-level) interaction parameters.

The higher-level parameters in the multiple regression model enable quantification of how chemicals influence each other in relation to the measured response. Suppose that $\beta_{1,2}$ (the estimated function parameter for the first-level interaction term between chemical 1 and chemical 2) in a regression model is negative and significant, and $\beta_{1,2,3}$ (parameter for the second-level interaction with chemical 3) is positive and significant. It can then be concluded that chemicals 1 and 2 have an antagonistic relationship, which is decreased by the presence of chemical 3.

Multiple regression is not designed for studying the effects of different concentration ratios. In addition, it does not enable the detection of concentration-level-dependent deviations. But the multiple regression approach has been used to develop a methodology to test if specific mixtures deviate from a CA reference response surface (Gennings and Carter 1995), which could be used to compare, for instance, one specific mixture ratio with another. This procedure uses the single chemical concentration–response data to construct the concentration–response surface under the assumption that deviations from CA are not occurring among the chemicals in the mixture. The effect of a mixture can then be compared to this model prediction using a constructed prediction interval to determine if the joint effect of the chemicals can be described with CA. The advantage of this approach is that the data requirements are only the single chemical concentration–response curves for each mixture component and the mixtures of interest (Teuschler et al. 2000).

The advantage of multiple regression is that methods are established, well described, and available in almost all statistical sofware packages, and that the fitting procedures have been well developed (Neter et al. 1996). Furthermore, the complete $n + 1$ dimensional concentration–response surface is fitted to the complete data set, taking into account that the parameters of the concentration–response relationships of the individual mixture components are actually predictors for the complete mixture data set. The model allows individual concentration–response curves to have their unique slopes.

A disadvantage is that multiple regression, by definition, only allows application of the CA concept; there is no possibility to compare the response with the IA concept. In addition, the researcher is limited to using 1 type of concentration–response curve for the complete data set through the choice of the link function. It may, however, be more appropriate to use different types of concentration–response curves for the different mixture components. Finally, deviations from CA can be properly tested for through the interaction parameters, but concentration-ratio- or concentration-level-dependent deviations from CA cannot be detected.

Multiple linear regression has been used quite extensively to detect deviations from CA. For instance, De March (1987) used it to quantify effects of 5 binary mixtures of metals on the survival of *Gammarus lacustris*. Narotsky et al. (1995) used it to analyze the effect of 5 toxic agents on the development of rats in a full factorial design. Nesnow et al. (1998) used multiple regression to analyze the tumorigenic effect of 5 polycyclic aromatic hydrocarbons (PAHs) in lung tissue in a full factorial design. If multiple regression is the preferred method, it should be noted that this framework enables the development of efficient experimental designs to assess the concentration–response surface in the multiple regression context (Gennings 1995, 1996; see Section 4.5.3 on fractionated factorial designs). Due to such adjustments, other methods, for instance, to assess concentration-level-dependent deviations, cannot be used.

4.5.2.2.3 Nonlinear Response Surface Models

Nonlinear response surface models have been introduced by Hewlett and Plackett (1959), when they formulated simple similar action for mixture components with dissimilar concentration–response curves (note: this is later defined as CA). Since then several response surface modeling methods specifically designed for mixtures have emerged in the literature. Although the various formulations in the literature may look different, their rationale is the same and can be described as follows.

As indicated earlier, it is generally accepted that CA occurs if Equation 4.1 holds (Berenbaum 1985), where ECx_i is the concentration of chemical i that results in the same effect ($x\%$) as the mixture. In case of a 50% mixture effect $ECx_i = EC50_i$, and in case of a 6% mixture effect $ECx_i = EC6_i$. Thus, the goal is to calculate this specific concentration of chemical i solely, that is, associated with a certain specific mixture response. To calculate a response from a concentration (or dose), a concentration–response function can be used, given by

$$y = f(c_i)$$

where y denotes response and $f(c_i)$ is the concentration–response function (e.g., log-logistic). So, to calculate a specific concentration from a response we need to inverse this relationship:

$$c_i = f^{-1}(y)$$

where f^{-1} symbolizes the inversed function. How does this look for a specific concentration–response function? For example, the log-logistic function can be written as

$$y = max/(1 + (c/EC50)^\beta)$$

where max denotes the control response at concentration zero, EC50 is the median effect concentration, and β is a slope parameter. This function can also be written as

$$c = EC50 \times ((max - y)/y)^{(1/\beta)}$$

This expression can be used to explicitly calculate the concentration, c, associated with mixture response, y. One can therefore write

$$ECx = EC50 \times ((max - y)/y)^{(1/\beta)}$$

and substitute it into the CA equation for each toxicant i. The resulting CA mixture model is difficult to apply, because it is an implicit equation and iterative procedures have to be used to find the predicted response for each mixture combination of interest. This model can then be fitted to basically any type of mixture toxicity data set, if enough data points are measured to support the model parameters.

In all formulations that have appeared in the literature thus far, a generalization of the CA reference concept was performed to statistically test for deviations from CA. This means that a function describing interaction is incorporated in the CA model such that if the interaction parameter is 0, the interaction function disappears from the function. This nested structure allows testing whether its appearance in the model improves the description of the data significantly by applying the likelihood ratio test. The various nonlinear response surface approaches do differ in the way this deviation function is formulated.

In general, the advantage of these response surface models is that they enable the description of nonlinear concentration–response relationships, and that differences in slopes and functional form of the individual concentration–response curves can be accounted for. The complete $n + 1$ dimensional concentration–response surface is fitted to the complete data set, which takes into account that the parameters of the individual concentration–response relationships are actually predictors for the complete mixture data set. Different likelihood functions can be used to adjust the analysis for different types of endpoints. Each approach has its own specific advantages, and response surface models for IA have also been developed (Haas et al. 1997; Jonker et al. 2005). The user needs to have some programming skills and statistical knowledge to judge the result. Specifically, the user needs to know how to

1) choose a proper likelihood function,
2) judge the model fit,
3) judge the effect of multicolinearity, and
4) interpret parameter values.

Disadvantages are that these response surface models are not available in standard software packages. Like all nonlinear statistical methods, the methodology is still subject to research, which has 2 important consequences. First, correlation structure of the parameters in these nonlinear models is usually not addressed. Second, the assessment of the test statistic is based on approximate statistical procedures. The statistical analyses can probably be improved through bootstrap analysis or permutation tests.

Greco et al. (1990, 1995) were among the first to introduce such a response surface CA model, specifically designed for taking into account the sigmoid nonlinear characteristics of many concentration–response toxicity data. Their formulation is heavily based on the Hill concentration–response model (Hill 1910), which is equivalent to the commonly used log-logistic model (Haanstra et al. 1985). Deviations from CA were tested using an interaction function in the model, which was also based on the Hill model. The suitability of the model therefore depends on whether the response data can adequately be described with this log-logistic model, because other response functions cannot be used. Other limitations are that the model is only developed for binary mixtures, and that this model does not enable the detection of concentration-level-dependent deviations or concentration-ratio-dependent deviations.

Haas et al. (1996) generalized the response surface modeling approach and showed that it is possible to substitute different concentration–response functions in the CA model, such as exponential, multistage, log-logistic, and the log-Weibull models. They further generalized the CA model with an excess function to describe deviations from CA. This deviation function enabled the description of concentration-ratio- and concentration-level-dependent deviations from CA. Different likelihood equations are used to fit the model, and the significance of additional parameters in the model is assessed through the likelihood ratio test. Obvious advantages are that the data can be screened for synergistic or antagonistic, concentration-ratio-, and concentration-level-dependent effects. A limitation is that the excess function for describing deviations from CA is formulated such that it can only be used if the concentration ranges of the mixture components are the same or very similar. For instance, this model cannot be used for mixtures of 2 compounds if the EC50 for 1 compound is 1 mg/L, and 100 mg/L for the other. Haas et al. (1997) also developed a response surface model for the IA concept. If the IA model is generalized with an interaction term, this interaction term can cause biologically impossible responses, such as survival below 0 or above 0. Haas et al. (1997) therefore used a transformation procedure to make sure that the predicted response was restrained to a biologically relevant range. The disadvantage of the approach described by Haas et al. (1997) is that it only enables the description of synergism or antagonism in comparison with the IA concept. More complicated deviation patterns, such as concentration-ratio- or concentration-level-dependent deviations, cannot be described. So far, both approaches have only been developed for the analysis of binary mixtures.

The response surface approach was further developed by Jonker et al. (2005). In the deviation function they incorporated the characteristic that a small amount of a very toxic chemical in the mixture can have a much larger effect on the biological response than a large amount of a slightly toxic chemical. The deviation function in the CA or IA concept depended on each chemical's relative contribution to toxicity, calculated from the toxic units. Both the CA and the IA concept were generalized to describe synergistic or antagonistic, concentration-ratio-dependent, and concentration-level-dependent deviations from either reference model. The advantage of the methodology is that the models can be very generally used. Different likelihood functions can be incorporated, and the approach can take into account differences in individual nonlinear concentration–response curves (slopes and functional form) and differences in relative toxicities of the individual chemicals. Synergism or antagonism, concentration-ratio-dependent deviations, and concentration-level-dependent deviations, compared to both CA and IA, can be described, and the approach has been shown to be useful for analyzing mixtures of more than 2 chemicals. In order to make optimal use of this flexibility, the user needs to have statistical knowledge and experience with model fitting, and learn to interpret the parameter values.

Response surface models can be generally applied to various experimental designs, but the best possible analysis opportunities exist where the experimental design covers all ratios and concentration levels equally, such as described above in Section 4.5.2.1 (Figure 4.1B). It is possible to apply the analyses to simpler experimental designs of single ratio (e.g., equitoxic) mixtures or combinations at a specific concentration level (e.g., EC50), but this limits the types of deviation for which one can test. Response surface models are therefore very useful as a screening tool for systematic deviations from CA or IA. Replication of concentrations is not essential, as the analysis is regression based, and variance calculations for statistical testing are made from the deviations between data and model values. If the number of experimental units is limited, emphasis should be placed on covering the response surface as best as possible to support the model parameters. Because nonlinear response surface models are not implemented in standard software packages, they have less frequently been used than multiple regressions. Gaumont et al. (1992) used a response surface model to analyze the effect of folic acid on synergistic cytotoxic interactions between different antifolates. In addition, Jonker et al. (2004) used a response surface model to address the toxicity of various mixtures to nematode populations in relation to soil chemistry, and Jonker et al. (2005) used it to assess the effect of 2 simple mixtures on various life cycle parameters of the nematode *Caenorhabditis elegans*. Faessel et al. (1999) used a response surface model to analyze the combined effect of various cytotoxic drugs on sensitive and resistant human tumor cell lines.

4.5.3 FRACTIONATED FACTORIAL DESIGN

The fractionated factorial design is a robust way to reduce the size of experiments that involve many experimental factors. It is therefore particularly suitable for screening studies, exploratory experiments with unknown chemicals, or experiments focused on more complex mixtures. The assumption underlying the use of fractionated factorial designs is that the measured response is driven largely by a limited number of main effects and lower-order interactions, and that higher-order interactions are relatively unimportant. If this assumption holds, then the full factorial design is obviously wasteful and inefficient. A fractionated factorial design achieves the efficiency of providing full information about main effects and low-order interactions with fewer experimental units by confounding these effects with the unimportant higher-order interactions. The data can be analyzed with linear models. Designing such an experiment results in a confounding scheme, which indicates which effects can be estimated (Neter et al. 1996).

The advantage of using a fractionated factorial design is that the method is well developed. Established linear models (with link functions) are used for data analysis, so all advantages and disadvantages described above apply here as well. Implementation of the fractionated factorial design is, however, not trivial, and the user needs to have a fair amount of statistical knowledge and be familiar with design matrices.

Groten et al. (1997) used a fractional 2-level factorial design to examine the toxicity (clinical chemistry, hematology, biochemistry, and pathology) of combinations of 9 compounds to male Wistar rats through a 4-week mixed oral and inhalatory study. They subsequently analyzed the data with multiple linear regression. It was concluded that despite all restrictions and pitfalls that are associated with the use of fractionated factorial designs, this type of factorial design is useful to study the joint adverse effects of defined chemical mixtures.

4.5.4 ISOBOLES

As indicated above, assessing the complete concentration–response surface of a mixture can be costly in terms of labor and resources. If the research question does not demand the assessment of a full concentration–response surface, then several methods can be used to cut down on the experimental design. One possibility is to select concentration combinations on the bases of isoboles. Isoboles are isoeffective lines through the mixture-response surface, defined by all combinations of c_1, c_2, ..., c_n that provoke an identical mixture effect. As indicated, the CA-predicted isoboles are linear. Classical isobole designs aim at experimentally describing 1 or several points on an isobole and comparing them with CA expectations (Sühnel 1992; Kortenkamp and Altenburger 1998). For this experimental design it is therefore required to have a priori knowledge about the toxicity of the individual mixture components. Depending on the number of points on the isobole that are investigated, isobole-oriented approaches can still be rather laborious. A fairly complete mixture concentration–response experiment is necessary for each investigated point on the isobole; that is, $k \times (n + j)$ test groups are needed (j = number of points that are to be investigated on the isobole, n = number of mixture components, k = number of concentrations per concentration–response curve). And such a large isobole design is more or less equivalent to covering the complete concentration–response surface. If only 1 point on the isobole is investigated, the design boils down to a fixed-ratio design, as described in Section 4.5.1. The major advantage of isobole designs is their ability to detect mixture-ratio-dependent deviations (interactions) from predictions and observations. In order to minimize k, isobole-related experiments and subsequent data evaluations are often focused on 1 particular effect level, typically 50%. The possibilities to determine effect-level-dependent interactions are then limited. Designs that overcome this limitation and make use of multiple complete fixed-ratio experiments have been put forward, for example, by Casey et al. (2005).

4.5.5 A IN THE PRESENCE OF B

An approach that is restricted to binary situations is to analyze the shift of the concentration–response curve of the first agent that is caused by a fixed "background" concentration of a second chemical (Pöch 1993). This design requires at least $k \times 2 + 1$ test groups (k = number of test concentrations per concentration–response curve). Under these circumstances it can be assessed whether the increase in toxicity of the first chemical that is caused by the background concentration is in compliance with IA expectations. For a comparison with CA, the concentration–response curve of the second chemical also needs to be recorded. In that case, the extended design requires at least $k \times 3$ test groups.

4.5.6 POINT DESIGN

In a frequently used approach, which might be called a "point design," only 1 mixture concentration is actually tested, and its effects are compared to the effects that the individual components provoke if applied singly at that concentration at which they are present in the mixture. Several examples for this design can be found in the scientific literature. This design, in principle, only requires $n + 1$ test groups, not counting any controls. Nevertheless, visible deviations between observed and predicted effects are not necessarily of relevance, as the experimental variability of effect data is sometimes considerable. Especially the steepness of the concentration–response curves might have a considerable influence. In the case of steep concentration–response curves, small, experimentally unavoidable shifts in the applied concentrations might lead to comparatively huge shifts in experimentally observed effects. An extension of the point design is therefore to record the concentration–response curves of all components and the mixture and use effects data that are the result of a complete concentration–response analysis. One particular application of the point design is to analyze a situation in which all the components are present in a concentration that is presumedly below a predefined threshold, and to see whether the mixture still provokes clear effects (Backhaus et al. 2000a; Faust et al. 2001; Silva et al. 2002). If the concentration–response curves of the mixture components are not available, this design does not allow for a comparison of the observed mixture effect with the CA prediction, as it does not allow estimation of the necessary ECx values. But as it provides the $E(c_i)$ values for all components, this design type at least principally allows assessment of whether the observed mixture effect is in compliance with IA.

4.6 WHOLE MIXTURE APPROACHES, TEST DESIGNS, AND METHODS

"Undefined mixtures" may be defined as mixtures with the chemicals being unavailable for entire testing (Groten et al. 2001). Most often this is the case for "complex" mixtures, mixtures with more than 10 chemicals (McCarty and Borgert 2006).

In "whole mixture" test designs and methods, the mixture is at least partly undefined (Groten et al. 2001), and most often a complex mixture (McCarty and Borgert 2006). In its simplest setting, a whole mixture experiment would be an exposure study using an undefined environmental chemical sample. The response of the test organisms or system is compared with a reference, in order to determine whether the environmental sample induces some effect. A fractionation step has to be included to get some idea of the (groups of) chemicals most responsible for the toxic effect caused by the mixture. When the identity of the mixture is revealed, whole mixture experimental designs may enable analyzing whether the response to the mixture differs from either CA or IA. Whole mixture test designs are usually used for the identification of the most important chemicals in a mixture in terms of their contribution to overall toxicity. Furthermore, they are used in tiered environmental monitoring, aiming at classification, comparison, and ranking of sites (according to chemical stressors and toxicity).

Although approaches to complex mixtures in human toxicology and ecotoxicology may seem different, they also show similarities (see also Chapter 5). In practice, the main questions asked are

- How do we evaluate complex mixtures containing large fractions of unidentified chemicals?
- How do we characterize the toxicity for complex mixtures that degrade or vary in composition from one site to another?
- What statistical, chemical, or toxicological evidence is needed to show that 2 complex mixtures are sufficiently similar to use toxicity data on one mixture to evaluate the other?

4.6.1 BIOASSAYS

To obtain insight into the toxicity of the complex mixture and the main contributing components, first testing of the mixture may be performed. Toxicity testing can be done applying the same procedures applied for single chemicals. By comparing toxicity of the whole mixture with effects of the single constituents at comparable concentrations and duration of exposure, $n + 1$ experimental groups, that is, the number of compounds in a mixture plus the mixture itself, are required (point design; see above). This is an economic design recommended for a first screening of not well-characterized mixtures; however, it does not allow for the evaluation of synergism, potentiation, or antagonism (Cassee et al. 1998).

Direct toxicity assessment allows the evaluation of the effect of exposure to complex mixtures of contaminants and has been considered an important tool when data are lacking on the composition of the mixture present. In practice, it may be rather difficult to find a reference sample that has exactly the same properties as the polluted sample except for the pollution present. This especially seems to be the case for soils and sediments, which by definition are heterogeneous and show rather large variations in physical–chemical properties, like pH, clay, and organic matter contents. For that reason, often sampling is done in a gradient. Another approach is to include a series of reference samples. Toxicity of highly contaminated samples usually is assessed by preparing dilution series, and in that case the substrate used for diluting the contaminated samples is used as the reference or control.

In the practice of soil and sediment analysis, bioassays often are used in conjunction with chemical analysis and ecological field observations. This approach, named the TRIAD approach, was first introduced for sediment analysis by Chapman (1986) and is becoming more common for contaminated land assessment (Jensen and Mesman 2006). Such multimetric methods allow for reduction of uncertainties in risk assessment as evaluation is based on several independent lines of evidence (Chapman et al. 2002).

Comparison and ranking of sites according to chemical composition or toxicity is done by multivariate nonparametric or parametric statistical methods; however, only descriptive methods, such as multidimensional scaling (MDS), principal component analysis (PCA), and factor analysis (FA), show similarities and distances between different sites. Toxicity can be evaluated by testing the environmental sample (as an undefined complex mixture) against a reference sample and analyzing by inference statistics, for example, t-test or analysis of variance (ANOVA).

Apart from the "1-sample case," natural pollution gradients, for example, in a stream below a point pollution source or in whole effluent toxicity (WET) testing, allow for the construction of concentration–response curves of a mixture. Depending on the resources, either ANOVA designs or regression designs can be applied. The outcome is a descriptive evaluation of toxicity, for example, leading to the establishment of a NOEL in case the mixture remains undefined or only partly defined. Even if all components in the mixture from a natural pollution gradient are analyzed chemically, the concentrations of the compounds in the mixture might not be present in the same ratio along the gradient. This is considered a confounding factor when dealing with natural gradients, for example, as studied in an acid mine drainage (AMD) gradient (Gerhardt et al. 2004, 2005; Janssens de Bisthoven et al. 2004), hence, not allowing for CA and IA to be applied.

Toxic potency evaluation quantifies toxicity of undefined mixtures by a combined field and laboratory approach. The environmental sample or extract of the sample is first concentrated, and subsequently a defined geometric or logarithmic dilution series is tested for toxicity. This allows for defining the range of toxicity, for example, deriving NOEL values and concentration–response curves, hence providing more information than the simple 1-sample case. The advantage of this "artificial gradient" over a natural gradient is that the components remain in the same ratio over the different dilutions. Evaluation of the toxicity of the original environmental sample by its position on the generated concentration–response curve is possible (see, e.g., Houtman et al. 2004).

The major drawback of bioassay approaches is that they do not provide much information on the identity of the chemicals responsible for the toxicity of the sample tested.

4.6.2 BIOSENSORS

Nowadays, biosensors can also be a tool to assess bioavailability and toxicity of complex chemical mixtures. Biosensors are usually analytical devices that incorporate biological material or a material mimicking biological membranes associated with or integrated within a physical–chemical transducer or transducing system. Biosensors are based on biochemical biomarker responses at the suborganism level. They are able to contribute to the monitoring of environmental quality, for example, in cases of contaminated land remediation processes or wastewater treatment; however, the link to ecological measures often still has to be established. Biosensors are considered a valuable tool for measuring complex matrices and also for providing information in cases where no other technology is available (Ciucu 2002). Biosensors may also be applied in human toxicology.

Biosensors differ from bioassays mainly by the fact that in bioassays the transducer is not an integral part of the analytical system and biosensors can extract quantitative analytical information of single compounds in complex mixtures. One example is the determination of concentrations of dioxin-like compounds in the blood and environmental samples using the Calux assay, where within a complex matrix its levels are determined with great accuracy (see, e.g., Murk et al. 1997). Additionally, compounds that are difficult to detect (e.g., surfactants, chlorinated hydrocarbons, sulfophenyl carboxylates, dioxins, pesticide metabolites) can more easily be evaluated using biosensors.

4.6.3 Fractionation Methods, TIE, and EDA

To enable identification of the (groups or classes of) chemicals responsible for toxicity of the sample, a TIE approach can be followed. The TIE approach was first developed for the characterization of effluent toxicity (USEPA 1991; Norberg-King et al. 1992; Durhan et al. 1993; Mount and Norberg-King 1993). The first line in the TIE approach is to determine toxicity of the effluent sample using bioassays. The second line includes identification of priority pollutants by chemical analysis and determination of their toxicity either by additional testing or by collecting literature data. The final step is to try to explain the toxicity of the effluent sample from the knowledge on the toxicity of the priority pollutants. The second line usually requires a more sophisticated approach, including, for instance, fractionation schemes and associated chemical techniques to unravel the identity of the toxicants in the complex mixture. Often a stepwise selective removal of certain fractions is applied to identify the chemicals contributing most to the toxicity of the mixture. Such stepwise removal may, for instance, include complexation of metals by adding ethylene diamine tetra acetic acid (EDTA), complexation of certain classes of organic chemicals by C18 solid phase extraction, and separation of fractions upon acidification. Each fraction is separated and the remaining sample can be tested for toxicity. TIE approaches may also be applied to sediments and soils and to solid waste materials (see, e.g., Ankley et al. 2006).

When the chemical identity of the complex mixture is known, it becomes possible to determine whether the toxicity of the mixture follows the concepts of CA or IA. Grote et al. (2005) proposed a framework called effect-directed analysis (EDA) to analyze the toxicity of environmental samples containing complex mixtures. In this approach, applied to sediments, following extraction a fractionation was performed by chromatography, resulting in the identification of different chemical fractions. Subsequently, the total extracts, fractions, and individual chemicals identified in the extracts were tested for toxicity; Grote et al. (2005) used green algae (*Scenedesmus vacuolatus*) for this purpose. The toxicity of the total extracts was compared to that of the individual compounds, applying either the CA or the IA model. In addition, artificial mixtures were prepared by mixing the individual compounds at the ratio found in the extracts of the sediment samples. This approach can be seen as a fixed-ratio design, where the concentrations of the individual compounds are tested in the ratio found in the environmental sample, although testing other contaminant ratios is also possible. In case of pollutants with closely related or similar modes of action, such as PAHs, measured toxicity of the extracts was in good agreement with that predicted from the toxicities of the individual chemicals using the CA concept. But in case of pollution with chemicals having dissimilar modes of action, the concept of IA gave the best prediction of the toxicity of the complex mixture (Grote et al. 2005).

Elaborating further on this, De Zwart and Posthuma (2005) emphasized the importance of assessing the modes of action of chemicals in a complex mixture. Only with good insight into the chemical composition and knowledge of the modes of action of the composing chemicals, it may be possible to predict toxicity of a complex mixture (see also Chapter 3). De Zwart and Posthuma propose a

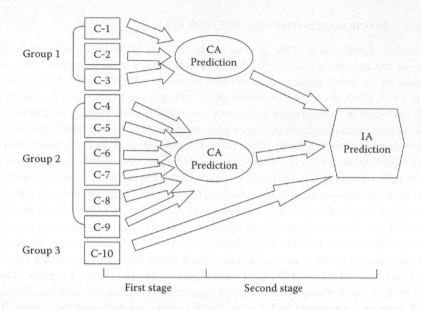

Figure 4.2 Two-step prediction model combining concentration addition (CA) and independent action (IA) models to predict the toxicity of a complex mixture of 10 chemicals (C-1 to C-10) (Redrawn from Ra et al. 2006).

combination of the CA and IA model to predict toxicity of complex mixtures. Ra et al. (2006), following the same line of reasoning, proposed a 2-step prediction (TSP) model (Figure 4.2). This TSP model uses CA to predict toxicity of (groups of) chemicals having similar modes of action and IA to predict toxicity of the complex mixture consisting of (groups of) chemicals having dissimilar modes of action.

Similar to the TIE and EDA approaches performed in ecotoxicology, fractionation of the mixture and testing the dominant or most relevant single compounds or fractions may help identify the causes of human toxicity. Figure 4.3 gives an example of an approach proposed by Groten et al. (2001). Other related approaches in human toxicology include spiking complex mixtures with single substances or lumping groups of related compounds. Examples include the fractionation of petroleum TPH (total petroleum hydrocarbon) mixtures and the assessment of toxicity of an indicator chemical for each fraction (Hutcheson et al. 1996), and the grouping of chemicals according to similar chemistry based on mode of action, for example, polychlorinated biphenyl (PCB) congeners (Andersen and Dennison 2004). Also in these cases, modes of action are taken as the starting point.

4.6.4 SIMILARITY OF MIXTURES

A characteristic aspect of human toxicology, not so frequently used in ecotoxicology, is the comparison of toxicity of a complex mixture with that of mixtures having a

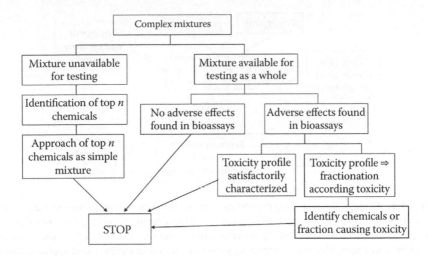

Figure 4.3 Approach to complex mixture toxicity analysis that might be used for the top-down approach. (Based on Groten et al. 2001.)

sufficiently similar composition. See Chapter 5 for a more detailed elaboration on the use of this concept in risk assessment. This concept, of course, requires a careful assessment of similarity of mixtures, taking into account similarity of mixtures' components and their relative proportions. Mixtures emitted by common sources or produced by similar processes usually may be considered similar. Nevertheless, expert judgment and statistical tools are needed to confirm that mixtures are sufficiently similar.

Bioassay-directed fractionation can facilitate the work with complex multicomponent mixtures, by treating them as simple mixtures. For example, pattern recognition and classification using multivariate statistics may be used, followed by submitting these data to a multivariate regression model. This was, for instance, done to predict mutagenicity of soot samples from chemical composition. First chemical composition of 20 soot samples was determined by gas chromatography–mass spectrometry (GC-MS), and mutagenicity determined in the Ames test. Next, PCA and partial least-squares (PLS) projections to latent structures were used for data analysis. This resulted in a PLS model containing 41 variables (chemical parameters) that with >80% accuracy predicted mutagenicity of the soot samples (Eide et al. 2002). Figure 4.4 shows the resulting strategy.

In human toxicology, more than in ecotoxicology, attention is given to the stability of the mixture. Because composition of mixtures may not be stable in time, it is important to get insight into the variability in components and their relative proportions. Refer to Chapter 1 for a discussion on mixture composition and its stability in time in relation to, for example, fate processes in the environment. Also, biotransformation may change the composition of a mixture inside the human body, and as a consequence affect its toxicity. See Chapter 2 for a discussion on this issue.

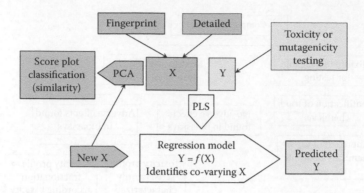

Figure 4.4 Strategy for evaluating the mutagenicity of complex mixtures applying pattern recognition. Detailed chemical analysis (fingerprinting; X) and the use of multivariate statistics, like Principal Component Analysis (PCA), provides insight into the similarity of samples containing complex mixtures. Based on mutagenicity data (Y) for these samples, and applying Partial Least-Square projections to latent structures (PLS), a regression model is developed that describes mutagenicity as a function of the chemical composition of the complex mixtures. The regression model can be applied to predict the mutagenicity of a sample from its similarity with already tested samples. Similar models may also be developed to predict toxicity of complex mixtures. (Reproduced from Eide et al. [2002], with permission from *Environmental Health Perspectives*.)

4.7 CASE STUDIES

4.7.1 CASE STUDY 1: A WHOLE MIXTURE APPROACH FROM "ECO"TOXICOLOGY

Several TRIAD-based studies are reported in the literature, for example, the study of acid mine drainage gradient analysis in southern Portugal (2000–2002) (Gerhardt et al. 2004, 2005, 2008; Janssens de Bisthoven et al. 2004, 2005, 2006). The aim was to assess and evaluate the risk of the mine effluent in a natural pH and metal gradient. A multimetric approach was chosen. The following parameters were studied: benthic macroinvertebrate community analysis was performed based on the determination of approximately 80 macroinvertebrate taxa and approximately 30 chironomid species on 3 sampling campaigns over 2 years. Different indices were calculated:

1) Pollution indices: Belgian Biotic Index (BBI), Biological Monitoring Working Party—Average Score Per Taxon (BMWP-ASPT), Ephemeroptera-Plecoptera-Trichoptera taxa of a whole macroinvertebrate sample (EPT), Saproby, and South African Scoring System—Average Score per Taxon (SASS4-ASPT).
2) Ecosystem structure: Diversity (H) and Bray-Curtis dissimilarity.
3) Indicators: Acid indicators as well as community structure of diatoms.
4) Parameters for ecosystem function: Functional feeding groups (FFGs) and index of trophic completeness.

Additionally, chemical analysis of approximately 15 substances (metals, salts) was performed as well as toxicity testing. Each assay (48 hours) was applied in the laboratory as well as directly in situ (validation), using test standard organisms (animals: *Chironomus riparius*, *Daphnia magna*; plants: *Lemna gibba*) as well as resident species (*Atyaephyra desmaresti*, *Choroterpes picteti*, *Gambusia holbrooki*), covering crustaceans, insects, and fish (whole food chain). The tests continuously recorded behavior and survival, allowing for "time to" as well as fixed-endpoint data analysis (ECx, LOEC). The results of this multifaceted study from the point of view of risk evaluation and test design 1) showed that the toxicity tests on the natural gradient (undefined) mixture described the increasing risk with decreasing pH for all species, however with different sensitivities, and 2) backed up the results from the ecological metrics in comparing risk at the different field sites. No mixture toxicity concepts have been applied on this complex natural gradient, as 1) the different components of the mixture changed independently from other components in concentration along the natural gradient, and 2) the mixture contained more than 10 compounds, hence being regarded as complex and not relevant for CA and IA testing.

4.7.2 CASE STUDY 2: A COMPONENT-BASED APPROACH FROM "HUMAN" TOXICOLOGY

An example of a well-designed component-based study is entitled "A Multiple-Purpose Design Approach to the Evaluation of Risks from Mixtures of Disinfection By-Products [DBPs]," by Teuschler et al. (2000). The researchers specifically defined a set of goals before starting their experimental work. First, they defined the risk assessment goal for the study, which is to provide data and methods for 1) estimation of human health risk from low-level multichemical DBP exposures, 2) assessment of various additivity assumptions as useful defaults for risk characterization, and 3) calculation of health risk estimates for different drinking water treatment options. For the experiments they further specified the goals: 1) to develop an efficient experimental design for the collection of data on mixtures, 2) to provide data for the development of the threshold additivity model, 3) to produce data useful in testing the proportional-response addition and interaction-based hazard index (HI) risk assessment methods, and 4) to develop an understanding of the toxicity (potency and nature of interaction) of the 4 DBPs tested. The statistical approach chosen enabled selection of single concentrations based on the model requirements, and selection of mixture ratios based on environmental relevance. The preliminary results presented in the article suggest that concentration additivity is a reasonable risk assessment assumption for the DBPs tested.

In the study of Teuschler et al. (2000), the models for analyzing the data were selected beforehand, and it was also decided to only focus on environmentally relevant mixtures. The authors indicated that these 2 factors were decisive for choosing the concentration levels to test. The concentration levels were not selected in relation to a specific endpoint, using the toxic unit approach. This may have been avoided because several different hepatotoxic endpoints have been measured simultaneously. The concentrations tested enabled the use of 3 types of models: a multiple regression CA model, the interaction-based HI, and the proportional-response addition method. A major problem with mixture toxicity research in general is the scale of the experiments, because single concentrations and mixtures preferably have to be analyzed simultaneously. The experiments in the paper of Teuschler et al. (2000) have been set up such that they enable the application of possible shortcuts in the future. Several binary mixtures were tested to investigate the possibility of predicting the effect of the mixture containing 4 chemicals from the interactions found in binary mixture experiments. In addition to this, the experimental animal was selected to explore the possibility of using single chemical data from published literature to construct the expected response of a chemical mixture rather than repeatedly generating the single chemical curves for every mixture experiment. This paper does not cover all the results, but discusses the design of the experiments. The well-defined risk assessment goals, the link between experimental design and the data analysis method, and the investigation of future experimental designs make this study a good example of a mixture concentration–response study. A criticism may be that the application of the IA model was not investigated. Also, concentration-ratio- and concentration-level-dependent deviations from CA were not studied, because of the preference of environmentally relevant concentrations. Yet, these aspects were also not mentioned in the goals. A major difficulty with measuring multiple endpoints is that the concentration ranges tested may not be suitable for all endpoints. Relevant interactions may therefore be missed. The authors do not discuss this aspect.

4.7.3 Case Study 3: A Component-Based Approach from "Eco"toxicology

If the chemical composition of the samples is known or at least partly known (in a stepwise TIE approach) or existing data allow for QSAR calculation, the samples can be ranked by TUs. Arts et al. (2006) studied, in 12 outdoor ditch mesocosms, the effects of sequential contamination with 5 pesticides in a regression design. They applied dosages equivalent with 0.2%, 1%, and 5% of the predicted environmental concentration (PEC) subsequently over 17 weeks. Endpoints recorded over 30 weeks included community composition of macroinvertebrates, plankton, and macrophytes, and leaf litter decomposition as functional ecosystem parameters. TUs were calculated in relation to acute toxicity data for the most sensitive standard species *Daphnia magna* and *Lemna minor*. Principal response curves (PRCs), a special form of constrained PCA, and Williams test (NOEC, class 2 LOEC) were used to identify the most sensitive taxa. Next to direct effects on certain species, also indirect effects, for example, how the change in abundance of a sensitive species affects the abundance of another, more tolerant species, can be detected only in mesocosm or in situ experiments. All observed effects were summarized in effect classes in a descriptive manner.

4.8   SUMMARY AND CONCLUSIONS

Mixture toxicity testing may have several aims, ranging from unraveling the mechanisms by which chemicals interact to the assessment of the risk of complex mixtures. Basically, 2 approaches can be identified: 1) a whole mixture approach in which the toxicity of (environmental samples containing) complex mixtures is assessed with a subsequent study in order to analyze which individual compounds drive the observed total toxicity of the sample, and 2) a component-based approach that is based on predicting and assessing the toxicity of mixtures of known chemical composition on the basis of knowledge on the toxicity of the single compounds. This approach is also often used to unravel the mechanisms of mixture interactions. Test design highly depends on practical and technical considerations, including the biology of the test organism, number of mixture components, and the aims of the study.

Fundamental for both the whole mixture and component-based approaches are 2 concepts of mixture toxicity, the concepts of CA and IA or response addition (RA). CA assumes similar action of the chemicals in the mixture, while IA takes dissimilar action as the starting point. In practice, this means that CA is used as the reference when testing chemicals with the same or similar modes of action, while IA is the preferred reference in case of chemicals with different modes of action.

The component-based approach usually starts from existing knowledge on the toxicity of the chemicals in the mixture, either from the literature or from a range-finding test. Several test designs may be chosen to unravel the mechanisms of interaction in the mixture or to determine the toxicity of the mixture, the CA and IA concepts serving as the reference. In addition to just testing for synergistic or antagonistic deviations from the reference concepts, focus may also be on detecting concentration-ratio- or concentration-level-dependent deviations. Experiments may be designed to determine the full concentration–response surface, often taking a full factorial design or ray design. When resources are limited or the question to be answered is more specific, the test design may be restricted to determining isoboles. Another alternative is a fixed-ratio design or a fractionated factorial design. Also, designs limited to chemical A in the presence of chemical B, or point designs, are in some cases appropriate, although less preferred, when unraveling the mechanisms of interactions in a mixture. In all cases, it is desirable to combine tests on the mixtures with tests on the single chemicals, just to ensure that even small changes in the sensitivity of the test organisms may not affect the conclusions of the mixture toxicity experiment.

The whole mixture approach generally consists of testing the complex mixture in bioassays (both in the laboratory and in situ), usually applying the same principles as used in the single chemical toxicity tests. By performing tests on gradients of pollution or on concentrates or dilutions of (extracts of) the polluted sample, concentration–response relationships may be created. However, these tests do not provide any information on the nature of the components in the mixture responsible for its toxicity. By using TIE approaches, including chemical fractionation of the sample, it may be possible to get further insight into the (groups or fractions of) chemicals responsible for toxicity of the mixture. Also, comparison with similar mixtures may assist in determining toxicity of a complex mixture. Such a comparison may be based on the chemical characterization of the mixture in combination with multivariate statistical methods. Effect-directed analysis (EDA) and the 2-step prediction (TSP) model may be used to predict toxicity when full chemical characterization of the complex mixture is possible and toxicity data are available for all chemicals in the mixture. Such a prediction can, however, only be reliable when sufficient knowledge of the modes of action of the different chemicals in the complex mixture is available. In other cases, bioassays remain the only way of obtaining reliable estimates of the toxicity and potential risk of complex mixtures.

4.9 RECOMMENDATIONS

In this paragraph, we discuss aspects of mixture experiments that need attention while analyzing and assessing the data. These aspects may be endpoint, test organism, or chemical specific:

1) Chemical measurements can change the test design. In many experiments the exposure concentrations are measured after spiking the test medium, which can be food, water, air, or soil. The measured concentrations may be different from the initial (nominal) ones. In soil and food, adsorption may occur, some chemicals in the mixture may show mutual interaction, or chemicals may be degraded or become less available (see Chapter 1 for more details). As a consequence, the exposure concentrations may be different from the starting point, and in fact, the experimental design has changed. The fixed-ratio design or isobole design may then be disrupted, which needs to be acknowledged while analyzing the data. The response surface approaches are relatively robust to shifts in concentration levels and concentration ratios. Yet, the researcher needs to investigate whether the concentration layout still supports the model parameters sufficiently.

2) Modeling hormesis. Hormesis is the finding of a stimulated rather than an inhibited response at low concentrations of a toxicant (see, e.g., Calabrese 2005). Hormesis of a single toxicant can be modeled satisfactorily by including an additional parameter in the concentration–response model (Van Ewijk and Hoekstra 1993). Technically it is possible to include this modified single concentration–response model in the CA model (Equation 4.1). However, it raises all kinds of conceptual and technical issues. For instance, it has to be decided whether 1 toxicant is expected to induce hormesis, or all toxicants in the mixture. If only 1 mixture component is inducing hormesis, what could then be expected from the mixture? Combining hormesis and concentration addition bears an odd conceptual dilemma. Concentration addition makes use of toxic units, derived by dividing the concentration of a mixture component by its own effect concentration (Equation 4.1). With hormesis the effect concentration is no longer uniquely defined, making it unclear which concentration to take. Inclusion of hormesis in a mixture concentration–response model can also lead to difficulties in model parameter estimations.

3) Responses to individual mixture components can have different end levels at high concentrations. Monotonically declining concentration–response curves may not decrease to 0, but to a minimum level. This can, for instance, happen if body size is the measured endpoint. The effect on body

size typically levels off at higher concentrations, such that a minimum body size can be identified, yet different mixture components may result in different "minimum body sizes." A similar effect can occur when increasing responses are measured, where the concentration–response maxima may be compound specific. With nonlinear response surface models it is possible to formulate the model such that an end level of the measured response at high concentrations can be estimated (Greco et al. 1990; Jonker et al. 2004). It should be realized that in this case an average end level of the response is estimated. How to model very divergent end levels for various mixture components is still an unresolved question. It is also questionable whether and how the IA concept can be applied under such circumstances, as the concept—due to its probabilistic foundation—assumes that the concentration–response curves of all compounds cover the range from 0 to 100% effect.

4) Differences in outcome between simultaneously measured endpoints. Mixture toxicity experiments are typically large, and in order to increase efficiency as well as obtain a better estimate of ecological relevance, multiple endpoints are frequently measured. The relative toxicity of the tested chemicals is usually endpoint specific (see, e.g., Cedergreen and Streibig 2005). For instance, for testing effects on reproduction one would usually use lower concentrations than those for testing effects on survival. This means that for the data analysis the test design is actually endpoint specific. It has been recognized that different endpoints show different interactions. For instance, a mixture may show synergism when its effect on reproduction is analyzed, but CA for its effect of survival. Such a difference in interaction may hold mechanistic clues.

5) Time dependence. It is widely known that effect concentrations are exposure time dependent (see, e.g., Reynders et al. 2006). This means that the experimental design for a mixture study is time dependent. It has also been reported that interactions are time dependent. For instance, it has been observed that the Cd–Cu effect on the reproduction of *Caenorhabditis elegans* changed during the course of exposure from a synergistic to a concentration-ratio-dependent deviation from CA (Jonker et al. 2004). Also, the cytotoxic effect of 4-hydroperoxycyclophosphamide (4-HC) and VP-16-213 (VP-16) on HL-60 cells changed from synergism to ratio dependent to additive (Jonker 2003). These observations indicate that time should be included in the mixture data analysis in order to make general statements about interaction. This is still an unresolved issue in mixture toxicity research, but development may benefit from a more detailed understanding of toxicokinetics and toxicodynamics (see Chapter 2).

ACKNOWLEDGMENT

Thanks are due to Geoff Hodges and Martin Scholze for their valuable contribution to the discussions at the International SETAC/NoMiracle Workshop on Mixture Toxicity in Krakow that led to this chapter.

5 Human and Ecological Risk Assessment of Chemical Mixtures

Ad M. J. Ragas, Linda K. Teuschler,
Leo Posthuma, and Christina E. Cowan

CONTENTS

5.1   INTRODUCTION

Risk assessment of chemicals is an organized process that aims to describe and estimate the likelihood of adverse outcomes from environmental exposure to chemicals (US PCCRARM 1997), and is applied for both humans and ecosystems. Risk assessment traditionally focuses on single chemicals. However, awareness is growing that exposure to single substances is the exception rather than the rule. In practice, humans and ecological receptors are often exposed to multiple chemicals that may or may not interact, that is, influence each other by physical, chemical, or biological means before or after reaching the molecular site of toxic action.

Risk assessment of chemical mixtures differs in several aspects from risk assessment of single substances. It therefore requires the development and implementation of other or additional risk assessment concepts and techniques specific for chemical mixtures, among other approaches to tackle the above-mentioned interactions. This is an evolving process that takes place at the interface between scientific research and risk assessment practice.

Environmental risks are often assessed separately for humans and ecosystems. The main reason for this is that the disciplines of human and ecological risk assessment have largely developed independently. Human risk assessment originates from the medical sciences and is strongly linked to disciplines like toxicology and epidemiology. Ecological risk assessment originates from biology and is strongly linked to ecotoxicology. However, human and ecological risks are often caused by the same sources. Management of these sources requires information about the human as well as the ecological risks. Assessing both types of risks independently may complicate decision-making processes and can result in a waste of resources. To improve efficacy, it was proposed to develop integrated risk assessment (IRA) approaches (WHO 2001; Munns et al. 2003a, 2003b).

The general goal of this chapter is to outline a consistent approach for conducting mixture risk assessments that is in line with the IRA approach and builds upon the best of the science and current practices in the respective fields of human and ecological risk assessment. It starts by comparing risk assessment of single substances with mixtures in order to identify the key features of mixture assessments. Subsequently, the current scientific state of the art in human mixture assessment is described, followed by the scientific state of the art in ecological mixture assessment. Both sections include an overview of the regulatory application of the science, especially in the United States and the European Union (EU). The state of the art contains both similarities and dissimilarities, and an analysis of these forms the basis of an integrated conceptual framework for human and ecological risk assessment of chemical mixtures outlined in the next section. The chapter concludes with a discussion of issues specific for risk assessment of chemical mixtures (e.g., the identification of sufficiently similar mixtures), and a series of conclusions and recommendations. The specific aims of this chapter are to accomplish the following:

1) Explain how mixture assessment differs from the assessment of single chemicals, and highlight the implications for risk assessment procedures (Section 5.2).
2) Summarize the scientific and practical state of the art and give an analysis of similarities and dissimilarities for human and ecological risk assessments concerning mixtures and current approaches (Section 5.3).
3) Use the analysis of the scientific and practical state of the art to derive an applicable and comparable conceptual framework for human and ecological risk assessment of mixtures, including a practical decision tree and guidance (Section 5.4).
4) Review and discuss current issues in risk assessment of mixtures (Section 5.5).
5) Identify, based on the previous analyses, gaps and key areas of research to improve both human and ecological risk assessment of mixtures, which will result in improved approaches (Section 5.6).

The approaches described in this chapter are of interest to several audiences in both human health and ecosystem risk assessments, particularly 1) research scientists who are actively involved in developing methods and expanding the knowledge base on how to conduct risk assessments of mixtures, 2) risk assessors who are charged with conducting risk assessments of mixtures, and 3) regulators who are involved in the development of risk assessment procedures and standards for mixtures.

5.2 TYPICAL FEATURES OF MIXTURE ASSESSMENT

Problems that have been solved in the risk assessment of single substances have not been solved equally well in mixture assessments. Even the most generic question in prospective risk analyses—"What is a safe level?"—poses problems. Often the mixture composition is unknown, and the mixture problem is then that the safe level would only be applicable to that particular mixture. Even if the mixture composition is well characterized, the safe exposure or concentration level would apply only to mixtures with the same or similar concentration ratios between the mixture compounds, as in cigarette smoke, diesel exhaust, or some polychlorinated biphenyl (PCB) mixtures. One option in such cases is to set a safe level for the mixture by using one of the mixture components as an indicator compound for the whole mixture. If the concentration ratios between the mixture compounds vary, there is no unique safe "mixture concentration," but an infinite number of possible safe concentration combinations.

The elementary differences between risk assessment of single substances and mixtures can best be illustrated with a schematic presentation of the risk assessment process. For single substances, this process consists of a series of consecutive steps, called the "risk assessment paradigm" (NRC 1983). It starts with the identification of a hazardous situation, then there is an exposure assessment, then an effect– or dose–response assessment, and finally a risk characterization. This process is embedded in a broader framework involving a regulatory problem formulation, risk communication, and risk management or action. Although the exact denomination and order of these steps may vary between countries and regulatory agencies (see, e.g., Van Leeuwen and Hermens 1995; Fairman et al. 1998; USEPA 1998, 2002b; WHO 2001), the essence remains more or less the same and is illustrated in Figure 5.1.

The risk assessment process for mixtures shows much similarity with that of single substances, but there are also differences. Figure 5.2 illustrates some of the issues that must be considered in addition to those for single chemicals. The problem now involves a mixture, for example, the prospective assessment of the net risks of a mixture emission, the retrospective assessment for a site contaminated with a mixture, or the regulatory wish to set a safe exposure level for a mixture. The hazard identification should account for potential toxicological interactions (e.g., initiation and promotion of tumors by different chemicals) and joint effects from exposure to combinations of chemicals causing similar or dissimilar toxic effects. Like for single substances, the exposure assessment results in an actual or predicted exposure level, but now for a mixture. Although determination of the exposure level may be more complicated for mixtures, for example, due to potential chemical interactions between the mixture components that change the mixture composition, the essence is comparable to that of single substances. However, the phase of effect assessment

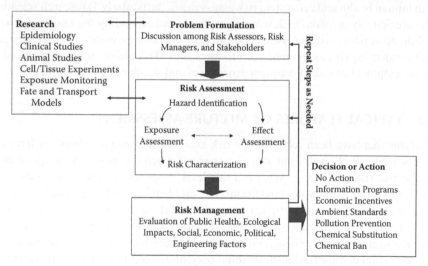

Figure 5.1 Risk assessment is traditionally organized in a series of consecutive steps—1) hazard identification, 2) exposure assessment, 3) effect assessment, and 4) risk characterization—and generally embedded in a wider framework involving research, problem formulation, risk management, and action.

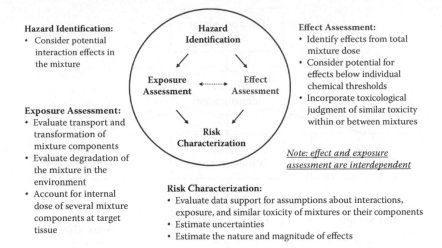

Hazard Identification:
• Consider potential interaction effects in the mixture

Exposure Assessment:
• Evaluate transport and transformation of mixture components
• Evaluate degradation of the mixture in the environment
• Account for internal dose of several mixture components at target tissue

Effect Assessment:
• Identify effects from total mixture dose
• Consider potential for effects below individual chemical thresholds
• Incorporate toxicological judgment of similar toxicity within or between mixtures

Note: effect and exposure assessment are interdependent

Risk Characterization:
• Evaluate data support for assumptions about interactions, exposure, and similar toxicity of mixtures or their components
• Estimate uncertainties
• Estimate the nature and magnitude of effects

Figure 5.2   Overview of the issues that must be considered in mixture assessment in addition to those for single chemicals.

may differ considerably between single substances and mixtures. For example, effects may occur even though the single components are below their individual threshold effect levels. Basically, 3 approaches can be followed to assess mixture effects (Figure 5.3):

1) Whole mixture approach for common mixtures. This is an option if dealing with a common, and often complex, mixture with more or less constant concentration ratios between the mixture components, for example, coke oven emissions. A reference value (e.g., PNEC) or dose–response relationship can be established for the mixture as if it were 1 (complex) compound, and a safe level can be determined like for single compounds based on toxicity data on the mixture itself or a sufficiently similar mixture. The effect data can subsequently be used in future assessments of mixtures that are identical (e.g., originating from the same source) or sufficiently similar.

2) Whole mixture approach for unique mixtures. This is an option if dealing with a mixture of completely unknown or unique origin and composition. In this case, results of previous effect studies cannot be used to assess the effects of the mixture of concern. Determination of a safe concentration level or a dose–response relationship for these mixtures is inefficient, as the effect data cannot be reused to assess the risks of other mixtures. The mixture of concern has to be tested directly in the field or the laboratory, like in the whole effluent toxicity (WET) test, resulting in a direct indication of the potential effects.

3) Component-based approach. This is an option if the mixture composition can be determined, for example, by means of chemical analysis, and if a mixture model is available that can predict the mixture effects. The mixture model can either be simple, for example, summation of PEC/PNEC ratios over all compounds into a hazard index (HI); moderately complex, for

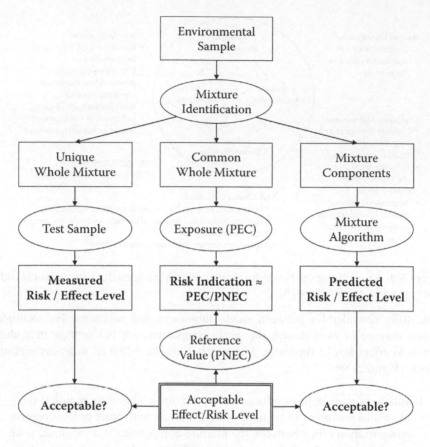

Figure 5.3 Three alternative options to assess the risk of mixtures: 1) mixtures can be tested in the field or the laboratory, particularly completely unknown mixtures; 2) if toxicity data on (sufficient) similar mixtures are available, the mixture can be evaluated using a reference value, for example, in a PEC/PNEC ratio; and 3) mixtures of which the components are known can be evaluated using component-based approaches (mixture algorithms). PEC = Predicted Environmental Concentration, PNEC = Predicted No Effect Concentration.

> example, applying the models of concentration addition (CA) or response addition (RA; or independent action) with or without taking the shape of the dose-effect curve into account; or complex, for example, like a special toxicokinetic modeling approach to predict organ-specific concentrations within organisms (see Chapter 2).

As discussed in the following sections, these 3 alternative options are widely used in human and ecological risk assessment of mixtures.

Finally, the last phase of mixture assessment, that is, risk characterization (Figure 5.2), must consider issues not addressed in single chemical risk assessments; for example, the assumptions made in order to use a particular risk assessment method (e.g., similarity in chemical composition between whole mixtures, or a shared toxic mode of action for chemical components within a mixture) must be articulated and supported with data.

5.3 STATE OF THE ART IN MIXTURE ASSESSMENT

5.3.1 HUMAN RISK ASSESSMENT OF CHEMICAL MIXTURES

Humans are exposed to multiple chemicals at any given time. Examples include air and soil pollution from municipal incinerators, leakage from hazardous waste facilities and uncontrolled waste sites, pesticides in vegetables and other consumer products, and contaminants in drinking water. Human activity patterns determine contact time with environmental pollutants; for example, the length of time spent showering or jogging outside impacts the amount of exposure to chemicals in drinking water or air, respectively. Exposures can be quite complex, involving multiple exposure routes or pathways, concurrent or sequential contact, and durations ranging from short term to episodic to lifetime. Chemical mixture composition may vary in space and time. Highly complex mixtures consist of hundreds of compounds and may be generated simultaneously from a single source or process (e.g., coke oven emissions, diesel exhaust) or produced as commercial products (e.g., gasoline, pesticide formulations). Other mixtures are composed of single components that are unrelated chemically or commercially, but are disposed of or stored together (e.g., lead, benzene, trichloroethylene, Arachlor 1254—all found at the same hazardous waste site) with the potential for combined exposure to humans (see Tables 1.1 and 1.4 in Chapter 1).

5.3.1.1 Scientific State of the Art

This section outlines the current scientific state of the art in the assessment of human health risks for chemical mixtures. It focuses on the gathering, assessment, and evaluation of effect data. The reader is referred to Chapter 1 for detailed information on exposure assessment of chemical mixtures. The section starts with an overview of methods commonly used to obtain effect data on chemical mixtures. This is followed by an overview of the current mixture approaches in human health assessments, that is, the whole mixture approach for common mixtures and the component-based approach. The section concludes with a paragraph on uncertainties in human health assessments of chemical mixtures.

5.3.1.1.1 Effect Data on Chemical Mixtures

Risk assessment of potential adverse human health effects from chemical mixture exposures may be conducted using health effects information from 1) toxicological bioassays, 2) epidemiological studies, and 3) computer models of toxicokinetic and toxicodynamic processes (i.e., in silico toxicology).

The vast majority of mixture data are obtained from toxicological bioassays on binary combinations of chemicals, with fewer studies on 3 or more chemicals or on whole mixtures (i.e., mixtures of many compounds considered together, some of which may be unidentified). Data from bioassays are useful to identify potential human health hazards, quantify dose–response relationships, and provide the basis for risk characterization, including estimates of safe levels or risk estimates for use in evaluating environmental mixture exposures. Most bioassays are done with laboratory animals like rodents and fish. Because the chemical exposures and toxicological outcomes are well controlled and documented, the information allows for extrapolation from animals to humans, introducing extrapolation uncertainty into the assessment.

Epidemiological studies on environmental mixtures are important contributors to hazard identification and, if the right data are collected, may be used in dose–response assessment or risk characterization, including estimates of safe levels and population-based risk estimates. Epidemiological data are collected on the right species, but exposures and effects are difficult to control and quantify; thus, results are highly relevant to humans, but may be difficult to interpret because of potential confounding factors in the study population. Many confounding variables are accounted for using statistical models (e.g., logistic regression models with smoking, age, or sex as a covariate), but uncertainty in the results often remains.

A final source of toxicological information has been developed through the use of computer models or in silico toxicological analyses, including biologically based models (e.g., physiologically based pharmacokinetic (PBPK) models), quantitative structure–activity relationship (QSAR) models, and models based on the simulation of biological systems (e.g., biochemical reaction network (BRN) models). In contrast to information from toxicological or epidemiological studies, information from these modeling approaches is not routinely used for human health risk assessment, because they are resource and data intensive. However, when a risk assessment issue warrants the expenditure of resources such models advance our knowledge of chemical mixtures risk assessment through a careful analysis of internal exposures, chemical reactions, or toxicity within the body. As discussed in Chapter 2, PBPK models provide internal dose estimates for multiple chemicals at the target tissues of concern, providing the risk assessor with an improved understanding of potential joint toxic action among chemicals and the nature of that action (e.g., evidence supporting the null models of dose additivity and no interaction, or of possible competitive inhibition). PBPK models have been used in human health risk assessments, for example, to derive estimates of within human variability in setting a reference value for boron (USEPA 2008) and for an analysis of the potential for metabolic interactions using internal doses of trihalomethanes in drinking water (Kedderis et al. 2006). QSAR models have mainly been used to prioritize chemicals based on their potential for toxicity when their health effects data are sparse (e.g., to prioritize drinking water disinfection by-products; see Woo et al. 2002). BRN approaches are fairly new computer algorithms that model chemical interactions using a systems biology approach (Mayeno and Yang 2005). These computer models predict the formation of metabolites from mixture exposures, including chemical and metabolic interactions, and interconnect metabolic pathways by common metabolites. BRN models are under development and have not been used in risk assessment to date. These models show promise of simulating multiple chemical interactions within the body for a larger number of chemicals than can be handled by most biologically based models.

5.3.1.1.2 Mixture Approaches

Of the 3 mixture approaches outlined in Section 5.2 (i.e., the common whole mixture, unique whole mixture, and component-based approaches), the first and the last are widely used in the assessment of human health effects. Common whole mixture approaches are generally referred to as "whole mixture approaches" in human health assessments, emphasizing the fact that the toxicity data relate to the mixture as a whole, instead of its components (USEPA 1986, 2000b; ATSDR 2004a). However, because the term "whole mixture" equally applies to the assessment of unique mixtures—which are rarely addressed in human health assessment procedures—the term "common whole mixture approach" is used here.

The state of science varies dramatically for the common whole mixture and component-based approaches. Common whole mixture approaches are most advanced for assessing carcinogenic risk, mainly because of the long use of in vitro mutagenicity tests to indicate carcinogenic potency. Numerous in vitro test procedures for noncancer endpoints continue to be developed (e.g., for developmental and reproductive endpoints, cytotoxicity) using animal and human mammalian cell lines. In contrast, the component-based procedures, particularly those that incorporate information on toxicological interactions, are most advanced for noncarcinogenic toxicity. The component approaches based on additivity are widely applied in human health risk assessment. For example, the USEPA's evaluations of contaminated sites under the Superfund program (USEPA 1989b) and of residual risk of air contaminants (USEPA 1999) routinely use the hazard index (HI) to evaluate noncancer effects; Superfund also applies RA for carcinogens to estimate the potential for human health risks. The combined use of CA and RA models has been newly introduced in human health assessments and is becoming accepted (USEPA 2003a, 2003b; Teuschler et al. 2004). Relative potency factors are used by the USEPA for the evaluation of pesticide mixtures that have a common toxic mode of action (e.g., USEPA 2002b, 2003b) and for the dioxins by the World Health Organization and other institutions in the form of toxicity equivalence factors (see Van den Berg et al. 2006).

Figures 5.4 and 5.5 illustrate the complexity of available data, assumptions, and methods currently used in assessments based on common whole mixture data (Figure 5.4) and component data (Figure 5.5). Some of the methods shown in Figures 5.4 and 5.5 are well established and frequently applied (e.g., HI, RA, epidemiological evaluations) and are endorsed by regulatory organizations (e.g., USEPA 1986, 2000b; ATSDR 2004a). Other methods are resource intensive or are not fully proven for use in the field (e.g., biological reaction network models, interaction-based HI, integrated additivity methods), but are the continued subject of risk assessment research and practical application (e.g., USEPA 2000b, 2003a, 2003b; Mayeno and Yang 2005). Details regarding the development and use of the methods shown in Figures 5.4 and 5.5 are available in many publications and are not repeated here (see USEPA 2000b; ATSDR 2004a; Teuschler 2007). Choosing among methods depends on the availability of appropriate exposure and toxicity data and on judgments concerning toxicological activity and chemical composition. The common mixture and component-based approaches in human health assessment are discussed in more detail below.

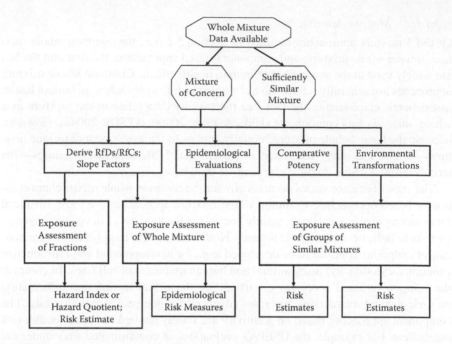

Figure 5.4 Flowchart showing various human risk assessment options for a chemical mixture based on whole mixture data. RfD = reference dose; RfC = reference concentration.

5.3.1.1.2.1 Common Whole Mixtures   The procedures based on common whole mixture data (Figure 5.4) assume that the concentration ratios between the mixture components are more or less constant or that environmental transformations in the mixture can be accounted for in the exposure assessment, for example, used to evaluate PCBs (USEPA 1996; Cogliano 1998). Based on the origin of the effect data used, distinction is often made between 1) assessments using data directly on the mixture of concern and 2) assessments using data on a "sufficiently similar" mixture as a surrogate for the mixture of concern. Both types of assessments are discussed in more detail below.

Mixture of concern. Mixture assessments based on data on the mixture of concern include direct testing of the environmental mixture itself or its concentrate, or evaluating fractions or single components of the whole mixture. When occupational, epidemiological, or bioassay data are available on the mixture of concern, a toxicological reference value, such as a reference dose (RfD),[1] reference concentration (RfC),[2] or cancer slope factor,[3] can be determined for the whole mixture using similar

---

[1] Reference dose: An estimate (with uncertainty spanning perhaps an order of magnitude) of a daily oral exposure to the human population (including sensitive subgroups) that is likely to be without an appreciable risk of deleterious effects during a lifetime (USEPA 2008).

[2] Reference concentration: An estimate (with uncertainty spanning perhaps an order of magnitude) of a continuous inhalation exposure to the human population (including sensitive subgroups) that is likely to be without an appreciable risk of deleterious effects during a lifetime (USEPA 2008).

[3] Slope factor: An upper bound, approximating a 95% confidence limit, on the increased cancer risk from a lifetime exposure to an agent (USEPA 2008).

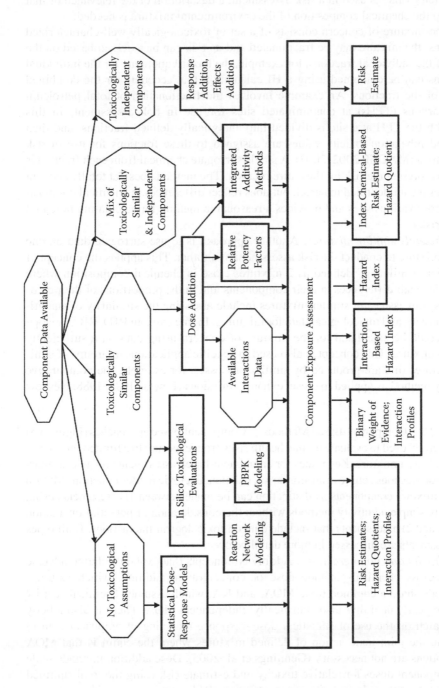

Figure 5.5 Flowchart showing various human risk assessment options for chemical mixtures based on component data. PBPK model = Physiologically Based Pharmacokinetic model.

procedures as those applied to single chemicals (see Section 5.4.3.3 for examples). If the toxicity value is used in a risk assessment, a description of the relevance of that value to the chemical composition of the environmental mixture is needed.

If the mixture of concern consists of a set of toxicologically well-characterized fractions, the mixture may be fractionated and its risk can be assessed based on the risks of the individual fractions; for example, the hazard quotients for the individual fractions can be combined using a HI calculation that accounts for the combined action of the fractions. An example involves the fractionation of total petroleum hydrocarbons (TPHs) at contaminated sites for use in risk assessment. In this approach the TPH at a site is divided into analytically defined fractions, and then oral and inhalation toxicity values are assigned to these fractions for use in risk assessment (MADEP 2002, 2003). A single surrogate chemical from each fraction is used to represent the risk for the entire fraction. The method does not totally account for all of the unidentified material, but does reflect differences in chemical composition across various sites and provides a reasonable method for calculating potential health risks.

Sufficiently similar mixtures. Another approach is to use surrogate data on one whole mixture to conduct the risk assessment of another. This applies the concept of "sufficient similarity," defined as 2 mixtures close in chemical composition where there are small differences in their components and in the proportions of their components. Key issues for similar mixtures include assessing the similarity of analytical chemistry and toxicological data for mixtures. In this case, an RfD, RfC, or slope factor could be calculated for the mixture of concern using data on a sufficiently similar mixture. This concept is also used in specific applications to groups of similar mixtures that are produced by similar processes, for example, the comparative potency method as applied to diesel exhaust emissions (Lewtas 1985, 1988; Nesnow 1990).

5.3.1.1.2.2   Component-Based Methods.   Component-based approaches (Figure 5.5) are generally used to evaluate human health risks from exposure to a limited number of chemicals as a mixture. Key issues for component-based assessments include similarity in dose–response curves and similar vs. independent toxic modes of action (MOAs) among mixture components. A distinction can be made between 1) assessments using relatively simple additivity methods without the consideration of potential interaction effects, and 2) assessments that include data on toxicological interactions. Both types of assessments are discussed in more detail below.

Additivity and no interactions. Additivity concepts that explain a shared adverse effect across chemicals include dose or concentration addition, which assumes chemicals share a common toxic MOA, and RA, which assumes chemicals act by toxicologically (and thus also statistically) independent MOA. There is also a body of research on the use of statistical dose–response modeling of empirical data to examine the joint toxic action of defined mixtures where the claim is that MOA assumptions are not necessary (Gennings et al. 2005). Dose addition methods scale the component doses for relative toxicity and estimate risk using the total summed dose, for example, using relative potency factors (RPFs), toxicity equivalency factors (TEFs), or a hazard index (HI). In contrast, RA (also named "independent action") is done by estimating the probabilistic risk of an adverse effect for each mixture component and then summing these risks, given that the risks are small. RA is derived using the statistical law of independent events, where, shown here for a 2-chemical mixture but generalizable to n chemicals, the mixture risk (R_m) is equal to 1 minus the probability of not responding to either chemical 1 (r_1) or chemical 2 (r_2):

$$R_m = 1 - (1 - r_1) \cdot (1 - r_2) \tag{5.1}$$

The algebraic simplification of this equation shows that R_m is the sum of the risks for chemical 1 (r_1) and chemical 2 (r_2) minus the probability that the toxic event from exposure to chemical 1 would overlap with the toxic event from exposure to chemical 2, as expressed in the following equation:

$$R_m = r_1 + r_2 - (r_1 \cdot r_2) \tag{5.2}$$

When risks are very low, the subtracted term is so small that its impact on R_m is negligible (e.g., for $r_1 = 0.01$ and $r_2 = 0.02$, $R_m = 0.01 + 0.02 - 0.0002 = 0.0298$, or ~0.03); thus, low risks such as those typically found in environmental risk assessments can be approximated by simple summation. Effects addition, a concept rarely applied, is a special case of RA where the biological measurements are summed across the mixture components and then a judgment is made regarding potential adverse effects based on the total measurement. Finally, when a mixture contains components with more than 1 MOA that cause the same health outcome, CA and RA methods can be integrated to assess risk (USEPA 2003b). See also Chapter 4.

Interactions. The potential for toxicological interactions to occur from coexposures is an important concern in component-based assessments. For use in risk assessment, the USEPA (2000b) defines toxicological interactions as responses that deviate from those expected under some definition of additivity. The most commonly used terms for interaction are synergism (i.e., effects are greater than additive) and antagonism (i.e., effects are less than additive). The analysis of toxicological interactions is complicated by the dose dependence of interactions, the vast amount of chemical and dose level combinations, and the lack of toxicity data for higher-order combinations beyond binary mixtures. In recent years, approaches for evaluating interactions have been proposed that use available binary toxicity data to adjust the value of the additive HI:

1) ATSDR has developed a qualitative binary weight of evidence (BINWOE) for each chemical pair in a mixture, considering mechanistic evidence, strength of interactions data, influence of exposure duration, and route and sequence of exposure (Pohl et al. 2003; ATSDR 2004a); and
2) USEPA has developed a similar but quantitative method for the evaluation of interaction data, the interaction-based HI, to numerically modify the HI using binary interactions data (USEPA 2000b; Hertzberg and Teuschler 2002).

Newer methods for evaluating toxicological interactions for component-based assessments include the use of PBPK models and BRN modeling. For example, Krishnan et al. (2002) used PBPK modeling to extrapolate the occurrence and magnitude of interactions from binary to more complex chemical mixtures. The goal of this research is to predict the kinetics of chemicals in complex mixtures by accounting for binary interactions alone within a PBPK model, thus providing a method for developing interaction-based risk assessment for chemical mixtures. BRN approaches predict the formation of metabolites from mixture exposures, including chemical and metabolic interactions, and interconnect metabolic pathways by common metabolites. As such models continue to be developed and validated, they support risk assessors in predicting metabolism and toxicity and in understanding MOA. The advantage of this approach is to simulate interactions for mixtures composed of a large number of components and to link computer simulation techniques with PBPK models.

Uncertainties and probabilistic assessments. Mixture risk assessments usually involve substantial uncertainties. If the mixture is treated as a single complex substance, these uncertainties range from inexact descriptions of environmental exposure, including its variability, to inadequate toxicological characterization. When viewed as a simple collection of a few component chemicals, the uncertainties include a generally poor understanding of the magnitude and nature of the toxicological interactions, especially those interactions involving 3 or more chemicals. Because of these uncertainties, the assessment of health risk from chemical mixtures should include a thorough discussion of all assumptions and the identification, when possible, of the major sources of uncertainty.

Whenever appropriate, and in line with the probabilistic concept of risk, probability distributions are used in human risk assessments for mixtures. Exposure modeling efforts typically result in the production of exposure distributions for various segments of the population. Chemical mixture exposure distributions have been produced for the dioxins (Lorber et al. 1994), drinking water disinfection by-products (USEPA 2006a), and several classes of pesticides, including the organophosphates (USEPA 2002c). Exposure distributions are produced fairly often, particularly in the evaluation of contaminated sites, but when a probabilistic exposure distribution is produced, it is usually compared with a single toxicity reference value to determine if and by how much a safe level could be exceeded.

5.3.1.2 Regulatory State of the Art in Different Nations

This section gives a nonexhaustive overview of the regulatory state of the art on human risk assessment of chemical mixtures in the United States, the European Union (EU), and other nations and (inter)national agencies. The focus is on regulations for environmental pollution, but when available and relevant, mixture regulations from other policy areas have also been included in the overview, for example, for food quality and the workplace. The reader is referred to McCarty and Borgert (2006) and Monosson (2005) for a more extensive overview of mixture toxicity regulations related to human health.

5.3.1.2.1 United States of America

The United States has several laws that authorize regulatory agencies to address risks of chemical mixtures, for example, the Comprehensive Environmental Response, Compensation, and Liability Act (CERCLA); the Food Quality Protection Act (FQPA); the Safe Drinking Water Act Amendments; and the Occupational Safety and Health Act (OSHA). Guidance is provided by different agencies and for different risk assessment areas, for example, for occupational exposure (NIOSH 1976; ACGIH 1984, 2000; OSHA 1993, 2001) and drinking water (NRC 1989). Most of these agencies recommend an HI approach in which the ratios between exposure and threshold levels are summed for similar endpoints. If this sum exceeds 1, the exposure limit for the mixture is exceeded. The most extensive guidance for human health assessment of mixtures has been developed by the Agency for Toxic Substances and Disease Registry (ATSDR) and the US Environmental Protection Agency (USEPA). This guidance is described in more detail below.

5.3.1.2.1.1 ATSDR.

CERCLA, of the United States, determines that, where feasible, the ATSDR shall develop methods to determine the human health effects of substances in combination with other substances with which they are commonly found. ATSDR developed a guidance manual that outlines the latest methods used in risk assessment of chemical mixtures (ATSDR 2004a). Based on this guidance, a series of documents, called Interaction Profiles, have been compiled for priority mixtures, for example, 1) persistent chemicals in breast milk (ATSDR 2004b); 2) arsenic, cadmium, chromium, and lead (ATSDR 2004c); and 3) chlorpyrifos, lead, mercury; and methyl mercury (ATSDR 2006). The purpose of the Interaction Profiles is to evaluate data on the toxicology of the mixture and on the joint toxic action of the chemicals in the mixture in order to assess the potential hazard to public health. A tiered approach is followed to prioritize the available data. Data on the mixture of concern or a similar mixture are the preferred basis for an assessment. If these are lacking, a component-based approach is undertaken. This approach focuses on the mixture components that are present at toxicologically significant exposure levels; that is, at least 2 components must exceed a hazard quotient of 0.1 for noncancer effects or a 10^{-6} lifetime risk level for cancer effects. PBPK or other biologically based models or joint toxic action studies are the preferred information sources to evaluate the potential health hazard of the components of concern. If these are unavailable, a HI method is used to screen for potential additivity of the components causing noncancer health hazards. Potential additivity of carcinogenic components is screened by summing the cancer risks of the individual components. A weight-of-evidence method based on data from binary mixtures is proposed to evaluate the potential impact of interactions between chemicals on noncancer as well as cancer health effects. A qualitative binary weight of evidence (BINWOE) is determined for each chemical pair, considering mechanistic evidence, strength of interactions data, influence of exposure duration, and route and sequence of exposure. For HI > 0.1, ATSDR recommends a qualitative weight of evidence to be used to assess potential consequences of toxic joint action (Pohl et al. 2003).

5.3.1.2.1.2   USEPA.   The first USEPA guidance document on risk assessment of chemical mixtures was published in 1986 (USEPA 1986). It follows a tiered approach. Priority is given to 1) data on the mixture of concern, 2) similar mixtures, and 3) components-based approaches, respectively. The component-based approach consists of 1) the calculation of an HI for systemic toxicants that influence the same endpoint, 2) summation of the risks for carcinogens, and 3) an assessment of the effects of potential interactions. In 1989, the *Risk Assessment Guidance for Superfund* (RAGS) contaminated sites was published, which expanded on the use of the HI and RA approaches in contaminated site assessments. A supplementary guidance was published in 2000 (USEPA 2000b). This guidance introduces a new mixture assessment category, that is, assessment based on a group of similar mixtures. This category refers to chemically related classes of mixtures that act by a similar MOA, have closely related chemical structures, and occur together routinely in environmental samples, usually because they are generated by the same commercial process. Methods proposed to assess the risk of a chemical mixture based on a group of similar mixtures include the comparative potency approach and environmental transformation methods. The comparative potency approach uses the relative potency of a group of mixtures tested in different assays to assess the human health risk of the mixture of concern. The supplementary guidance of the USEPA (2000b) also elaborates on components-based approaches in greater detail than the 1986 guidelines. It outlines an interaction-based HI based on the evaluation of the weight of evidence for interaction effects, which is ultimately expressed in numerical scores. The supplementary guidance also outlines the use of RPFs to assess the risk of mixtures of related chemicals that are assumed to be toxicologically similar, for example, dioxins and polycyclic aromatic hydrocarbons (PAHs). In 2002, the USEPA's Office of Pesticide Programs published additional guidance on the evaluation of pesticide mixtures that have a common mechanism of toxicity using common mechanism groups and an RPF approach to evaluating potential health risks (USEPA 2002b).

5.3.1.2.2   European Union

Like the United States, the EU has legislation that enables regulatory agencies to address mixture risks. Examples are the directives on risk assessment of existing substances (1488/94/EC), the placing of biocidal products on the market (98/8/EC), safety and health of workers at work (89/391/EEC) and contaminants in food (315/93/EEC). However, unlike the United States, explicit guidance on the assessment and regulation of chemical mixtures is often lacking. An exception is an annex of the EU Technical Guidance Document on Risk Assessment (EC 2003) that provides explicit guidance on human risk assessment of petroleum mixtures.

The limited EU guidance on risk assessment of chemical mixtures does not mean that the issue is not addressed by the EU member states. However, the level of detail and the status of the guidance vary considerably between the member states and the different policy areas. The UK Food Standards Agency established a special working group to formulate advice on risk assessment of multiple residues of pesticides and veterinary medicines in food, and of multiple sources of exposure to these substances (COT 2002). They concluded that there is limited exposure of humans to multiple residues, but that the nature and extent of combined exposure and the likelihood of the resulting adverse effects should be evaluated when carrying out a risk assessment. An extensive Danish study on the combined actions and interactions of chemicals in mixtures concluded that it is not advisable to recommend rigid use of any single approach for the risk assessment of all chemical mixtures (Binderup et al. 2003). The HI approach of Reffstrup (2002) is proposed for the assessment of the relatively simple mixtures, and those of Groten et al. (2001) and ATSDR (2004a) for the assessment of more complex mixtures.

5.3.2 ECOLOGICAL RISK ASSESSMENT OF CHEMICAL MIXTURES

Like humans, ecosystems are exposed to multiple chemicals at any given time. This implies exposure of individuals, species, and ecosystems. The exposure conditions vary considerably, and are determined by the characteristics of the exposure medium as well as by physiological characteristics and behavior of the exposed species and individuals. The object of concern can be separate species, like protected birds, mammals, or butterflies, or it can be assemblages or communities of different organisms or ecosystems. The concern can pertain to the structure or the function of the system, or both. Stressors like mixture exposures may trigger a multitude of responses in ecosystems, including indirect effects due to changes in competition and predator-prey relationships.

Ecological risk assessment of chemical mixtures thus has to deal with a variety of field phenomena, a possible range of assessment endpoints, and a variety of assessment approaches. Moreover, there exists a huge variety in the regulatory questions and problem formulations addressed in ecological risk assessment of chemical mixtures. Examples include the protection of specific species against well-defined mixtures (like PCBs and PAHs), the protection of an undefined concept like "the ecosystem," and retrospective assessments for highly or diffusely contaminated systems.

5.3.2.1 Scientific State of the Art

This section outlines the current scientific state of the art in ecological risk assessment of chemical mixtures. It focuses on the gathering, assessment, and evaluation of ecological effect data. The reader is referred to Chapter 1 for detailed information on exposure assessment. The section starts with an overview of methods commonly used to obtain ecological effect data on chemical mixtures. This is followed by an overview of the current mixture approaches in ecological risk assessments, that is, the common whole mixture, unique whole mixture, and component-based approaches. The section concludes with a paragraph on uncertainties in ecological assessments of chemical mixtures.

5.3.2.1.1 Effect Data on Chemical Mixtures

Ecological risk assessment of chemical mixtures may be conducted using the same types of data sources and approaches as in human risk assessment of mixtures. Available data and approaches are, however, different in kind and numbers. The vast majority of data are from laboratory toxicity tests with mostly binary mixtures and sometimes higher-order mixtures. The tests are performed most often with cultured species that have also frequently been tested with single compounds. Further data originate from tests with field-collected substrates or effluents (bioassays). A relatively new trend is the development of epidemiological approaches, that is, eco-epidemiological analyses of empirical data from field inventories, with diagnosis of the role of mixtures, and disentangling mixture effects from the effects of other (confounding) factors. The least data are available on approaches that apply complex mathematical modeling of interactions (like the PBPK and BRN models being developed for human risk assessment).

5.3.2.1.2   Mixture Approaches

All 3 mixture approaches identified in Section 5.2 (i.e., the common whole mixture, unique whole mixture, and component-based approaches) are widely used in the ecological risk assessment of mixtures. Unlike human health assessment of mixtures, ecological mixture assessment does not distinguish between assessments of common whole mixtures based on effect data from the mixture of concern and those based on effect data from similar mixtures. Furthermore, data on mixture interactions are rarely used in ecological assessments. There are tendencies to develop adapted models (with extra shape parameters) to be fitted to experimental data sets if those data contain specific deviations of, for example, a sigmoid concentration–effect relationship (e.g., Jonker 2003). However, bioassays or whole mixture studies do not normally focus on "mechanistic understanding" of the observed effects in ecotoxicology. When the concept of mode or mechanism of action is better understood, there might be latitude for use of more sophisticated models. Finally, contrary to human health assessments, there is a wide range of techniques available to assess the ecological risks on unique whole mixtures. The 3 main mixture approaches in ecological risk assessment are discussed in more detail below.

5.3.2.1.2.1   Common Whole Mixtures   There are few systematic studies of mixtures that are strictly based on the approach of the "mixture of concern" or "similar mixtures" as defined under human risk assessment of mixtures. Most ecological effect studies have more characteristics in common with a component-based or unique whole mixture approach than with the common mixture approach. A rare example of the common whole mixture approach in ecological risk assessment is the hydrocarbon block method. In this case, mixture effects are predicted on the basis of partial characterization of hydrocarbon mixtures. The hydrocarbon block method is used to determine the risks of a total hydrocarbon mixture on the basis of discriminating different chain length fractions of hydrocarbons, for each of which toxicities are known (King et al. 1996).

5.3.2.1.2.2   Unique Whole Mixtures   Many techniques are available to directly quantify total toxic impacts of mixtures. The risk assessment question in this case is related to the regulatory evaluation of emissions, like in the case of effluents, or it is related to the concept of good ecological status (GES), introduced in the EU Water Framework Directive (EU Directive 2000/60/EC).

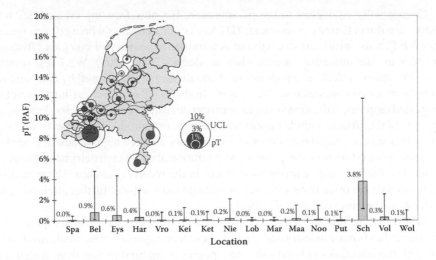

Figure 5.6 An example of a spatially explicit monitoring of mixture risks. Dutch surface waters at various sampling stations (X-axis names) were taken, concentrated, and analyzed as to determining the local toxic potency (pT-value) of the waters. For further explanation: see text and De Zwart and Sterkenburg (2002). (Reproduced from De Zwart and Sterkenburg [2002], with permission.)

The techniques that have been applied to quantify overall mixture impacts in ecosystems vary considerably, in relation to a diversity of problem formulations. An example of a spatially explicit measurement technique that is applied to mixture impact monitoring is the so-called "toxic potency" (pT) approach (Slooff and De Zwart 1991; De Zwart and Sterkenburg 2002; Struijs 2003). In this case, no attempts are made to quantify the concentrations of all compounds that could possibly contribute to the toxic impact. In the pT approach attention is focused on whole water samples, and the approach aims at quantifying the toxic pressure of unknown mixtures, that is, predicting the fraction of species that may occur at a site and that would be affected when exposed to the mixture (varies between 0% and 100%). Surface water samples are taken at various spots, and all samples are concentrated, for example, 1000-fold, with appropriate techniques. Thereafter, these concentrated samples are diluted in a dilution series. In various short-term microtests (like the *Daphnia*-IQ test), it is investigated at which level of concentration or dilution some specified level of effect is observed. Higher pT values imply higher fractions of species being likely affected by a local mixture. The analyses of the data finally result in a spatially explicit quantification of (relative) mixture risk (Figure 5.6). The pT value is conceptually closely related to the multisubstance probably affected fraction (msPAF) approach, in which also multisubstance (mixture) probably affected fractions of species are predicted, but in that case based on known mixture composition (De Zwart and Posthuma 2005). For a set of artificial mixtures, the measured pT values correlated well with modeling results obtained on the basis of species sensitivity distribution (SSD) modeling (Struijs 2003).

Conceptually equivalent to the pT approach, but now focusing on aquatic "hot spots," are direct toxicity assessments (DTAs), better known as whole effluent toxicity (WET) tests, which are executed to determine the toxicity of complex effluents in order to take immediate action when needed (SETAC 2004). WET test results describe adverse effects to a population of aquatic organisms caused by exposure to an effluent (or contaminated surface water). In the WET approach, effluent samples are tested (applying a dilution series as appropriate) using sentinel species, looking at impacts of the effluent sample on survival, growth, or reproductive capacity, and the data obtained are analyzed by either of 2 regulatory adopted approaches. The analyses provide an estimate of the effluent concentration above which detrimental impact from the effluent would be predicted to occur in the receiving stream. The results of such methods are often used to reduce or manage risks without further attempting to understand the observed mixture effects.

In other whole mixture approaches, overall mixture impact levels are quantified and thereafter further investigated. In fact, these techniques address whole mixtures, but with the intention to identify the most potent compound (going from undefined total mixture of concern to identified mixtures). The unraveling is usually not done to investigate the accuracy of mixture predictions, but attempts are made to identify the most potent compound groups within the mixture, by either chemical methods (toxicity identification evaluation (TIE), (e.g., Mount and Anderson-Carnahan 1988, 1989; Coombe et al. 1999); or biological response methods (BDF), (e.g., Brack et al. 1999). See Chapter 4 for more detailed descriptions of these methods. In these methods, physical and chemical approaches are used to collect mixture subsamples, and these are tested on one or more species. By looking at correlations between toxic effects and a specific subsample, or at compound-specific symptoms or species sensitivities, or finally by spiking water with a hypothesized key compound, the most potent fraction or compounds can be identified.

Retrospective whole mixture assessments may also be founded on a weight-of-evidence (WOE) approach. Such methods are partly oriented on whole mixtures and partly on risk modeling of identified mixtures. The combination of approaches is based on the principle that the studied phenomena need not be addressed from a mechanistic perspective due to a lack of comprehensive scientific models that address all relevant phenomena, but can also be addressed pragmatically. The pragmatic approach is taken, that various lines of evidence (LOEs) can provide better insight for decision making than a single LOE, without necessarily providing a full scientific explanation of observed phenomena. A specific example of that method is the Dutch soil TRIAD approach as developed by De Zwart et al. (1998, Figure 5.7), and applied and further developed by Mesman et al. (2003) and Jensen and Mesman (2006). In this approach, 3 independent, retrospective mixture impact quantification attempts (LOEs 1 to 3) are undertaken: 1) the toxic pressure approach (mixture risk modeling resulting in an estimate of the msPAF), 2) bioassay testing with sentinel organisms exposed in field-collected substrate, and 3) field observations on local organism groups. Note that various LOEs pertain to different levels of biological organization (community, test species, and assemblage, respectively). Irrespective of that, all measures of impact are scaled between the values of 0 and 1, to allow compilation of the 3 LOEs, and comparisons in the WOE. The WOE approach in this case

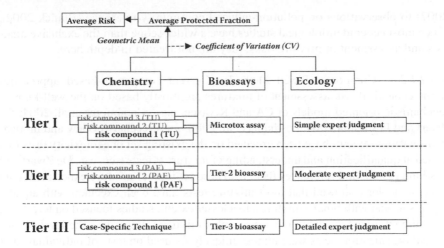

Figure 5.7 Weight-of-evidence (WOE) approach in assessing mixture effects. Three Lines of Evidence (LOE) are considered, in a tiered fashion (simple and quick methods in the tier-1, more complex methods in higher tiers). For each LOE, the response is scaled between 0 and 1. The LOEs can indicate a similar type of response, which results in a low Coefficient of Variation (CV), or differences (high CV). When the CV is lower than a predefined value, the assessment is stopped, and a decision (e.g., on remediation) is taken. TU = Toxic Unit, PAF = Probably Affected Fraction of species.

operates through setting a trigger value for the coefficient of variation (CV) between these 3 scaled quantities. Below the trigger level, all 3 signals are pointing in the same direction, being either no, slight, or large impacts, and the process is stopped. When a high CV is found at a lower tier, a next tier (of more complex analyses) is pursued. The formalized approach has been proposed and used for both aquatic or sediment systems (Chapman et al. 1992) and terrestrial systems (Jensen and Mesman 2006), in which contexts they serve for risk management.

Ecoepidemiological analyses focus on the diagnosis of impacts, and diffuse mixture pollution may be one of the stress factors. An example is the Causal Analysis/Diagnosis Decision Information System (CADDIS) developed by the USEPA in order to identify the main stressors in biologically impaired water bodies (see www.epa.gov/caddis/). This kind of system can help to determine whether mixtures have a detrimental impact on a water body. However, apart from establishing the presence of toxicant-induced changes (mostly by mixtures), ecoepidemiological studies do not usually contain specific analyses on the role and origin of mixtures and its components. Recognizing the impacts of individual chemicals or mixtures by field experts (ecotoxicologists or ecologists) is extremely complex at realistic ambient concentrations, especially when there is no clear clue to the types of compounds that might play a role in causing the observed impacts. A vast amount of literature is available to suggest that unknown mixtures do cause adverse effects in field ecosystems. Differences between exposed and nonexposed conditions were found at all levels of biological organization, ranging from differential gene responses to metal exposure in springtail populations (Roelofs et al.

2007) to observations on pollution-induced community tolerance (Blanck 2002). Since most ecoepidemiological studies have a wider scope than the exclusive analysis and assessment of mixture risks, they are not treated in depth here.

5.3.2.1.2.3 Component-Based Methods   Current component-based approaches in the ecological risk assessment of mixtures are mostly based on the well-known, mechanistic-oriented models of CA and RA (see, e.g., Altenburger et al. 2004; De Zwart and Posthuma 2005; Posthuma et al. 2008, for a recent overview and motivation). Various reviews have been executed on the existing set of ecotoxicity tests that aimed at quantification and understanding of mixture effects (see, e.g., De Zwart and Posthuma 2005, and references therein). These reviews of mixture experiment data in ecotoxicology showed that most mixture tests have been executed with aquatic organisms and with binary mixtures; to a lesser extent, studies focused on terrestrial or sediment organisms and on more complex mixtures. Regarding the level of biological organization, tests were almost uniquely focused on tests of individuals, not on communities. The latter poses problems when risk assessment questions focus on the level of communities and ecosystems. The pattern of results obtained in the reviews suggests the following:

1) Concentration additivity (CA) for similarly acting chemicals as a mathematical null model for testing observed responses (associated with the pharmacological concept of simple similar action) often fits well or quite well to the data, but misfits do occur, and when they occur, they are often in the tails of the response curves.
2) Alternatively, response additivity (RA) for independently acting chemicals as a mathematical null model for testing observed responses (associated with the pharmacological concept of independent joint action) and with an assumed correlation of sensitivities of 0 also often fits the data well. Again, misfits occur (e.g., when the test mixture consists of compounds with the same MOA at concentrations below the individual compound's no-effect concentrations), and when they occur, they are often in the tails of the response curves.
3) Well-designed tests, with powerful test designs and clear working hypotheses on the (differences in) MOA of the compounds in the mixture, allow for differentiation between the different null models: assumed similar MOA mixtures fit best to CA predictions, and assumed dissimilar MOA mixtures fit best to RA predictions.
4) No pattern can be derived from a scientific review of the (few) available data for the community level, but there (again) the pharmacological null models may be of help in the analysis and characterization of the observed responses in relation to the expectations from the classical toxicological null models (e.g., Backhaus et al. 2004).

De Zwart and Posthuma (2005) proposed a stepwise approach to quantify the probable impact of mixtures. First, interactions in the exposure medium are taken into account. This is an issue usually addressed as exposure analysis, and which often results in differences in availability between different substrates. Except for very high concentrations, or in the case of direct chemical interactions like precipitation, the most relevant interactions are dependent on the interaction of each compound separately with the local substrate. Second, toxicokinetic interactions can be taken into account. Third, toxicodynamic interactions need to be taken into account (for these issues see Chapter 2). This procedure results in a parameter that is known as the toxic pressure. It quantifies the fraction of species that is probably directly affected by a compound or the mixture at a given ambient concentration; that is, when the calculated toxic pressure equals 15%, then it is to be expected that 15% of the test species would directly suffer from the mixture. When the test species are representative for the local species assemblage, the toxic pressure also estimates the fraction of species that is probably affected in the local ecosystem.

The toxic pressure of each of the compounds in a mixture is calculated using the species sensitivity distribution (SSD) concept. In this concept, laboratory toxicity data for various species are collected from a database, for example, the USEPA's Ecotox database (USEPA 2005) or the RIVM e-toxBase (Wintersen et al. 2004), and compiled for each compound. A statistical distribution of these data, called the SSD, is derived. Each SSD describes the relationship between exposure concentration (X) and toxic pressure (Y), whereby the latter is expressed as the probably affected fraction (PAF, %) per compound (Posthuma et al. 2002). Depending on the test endpoint chosen for deriving SSDs, there is the option to derive chronic and acute toxic pressures, based on SSD_{NOEC}s and SSD_{EC50}s, respectively.

Various approaches can be used to aggregate the per-compound risk estimates to a single mixture risk estimate, that is, the multisubstance PAF (msPAF). This can be done using the models of CA or RA (with $r = 1$), but also with a mixed-model approach. In a mixed-model approach, which is similar to the 2-step prediction (TSP) model described in Figure 4.2, mixtures are treated as follows: 1) CA is used for quantifying the net effects within a subgroup of compounds in the mixture for which similar MOAs are assumed (e.g., all narcotic acting compounds or all organophosphorus insecticides), and 2) RA is used to aggregate to the net effect of the whole mixture. In the latter action, the toxic pressures posed by the subgroups of compounds with similar MOAs are aggregated over the different MOAs. This yields the msPAF. Posthuma and De Zwart (2006) have shown that the predicted loss of species attributable to mixture exposure (msPAF) was linearly associated with observed species in the field, whereby this loss was attributable to mixture exposure (Figure 5.8). In other words, for that monitoring data set, msPAF was linearly related to species loss.

The msPAF has been proposed as a method for assemblage-level mixture risk assessment in ecotoxicology, and has been used for various purposes (see, e.g., Mulder et al. 2005; Mulder and Breure 2006; De Zwart et al. 2006; Harbers et al. 2006). One example concerns a recent analysis of (bio)monitoring data that combined the study of site-specific impacts on species assemblages with toxic mixture modeling, the latter according to the approach described by De Zwart and Posthuma (2005). The aim was to detect and quantify the net impact of the local mixtures relative to the probable local impacts of other stressors, like changes in classical water chemistry parameters (pH, dissolved oxygen) and physical changes in habitats (like

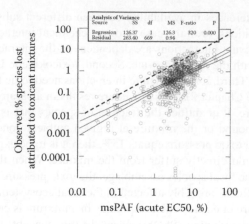

Figure 5.8 Interpretation of the concept of (acute) toxic pressure (multisubstance probably affected fraction of species (msPAF), based on species sensitivity distributions (SSDs) made from EC50s) using fish species census data from a large monitoring data set. The predicted msPAF (X-axis) is linearly associated to the observed impact of local mixtures on fish assemblages (species loss) in Ohio surface waters (approx. 700 sampling sites). (Study data from Posthuma and De Zwart [2006])

canalization). The study showed a significant role of toxic mixtures in shaping local species assemblages of fish, with large differences in the relative contribution of mixture stress to the observed impacts between sites (De Zwart et al. 2006). Another example of this kind, based on almost similar data and the same mixture assessment approach, but different statistical and geospatial analysis techniques, is provided by Kapo and Burton (2006).

For predictions of toxic pressures of mixtures on the level of assemblages, the concept of MOA again plays a role. An insecticide is designed for killing insects, and its primary mode (or even mechanism) of action may be well understood. However, the compound can also affect vital functions of nontarget organs, and thus the vitality of other organisms, due to secondary or tertiary modes of action. One compound can just interact with 1 major, but also many other receptor sites in organisms. This phenomenon is the subject of an investigative study into options to improve the concept of MOA in ecotoxicology (Jager et al. 2007; Figure 5.9). The notion of multiple MOAs for 1 substance implies that mixture effects depend on the endpoint considered; that is, 2 substances may show simple similar action for one endpoint and independent action for another. A statement about the joint action of 2 substances (e.g., simple similar action) should therefore always include a specification of the endpoints considered in the assessment.

An aspect not covered by the models discussed so far is the extra level of integration for species assemblage-level problems, that is, ecological interactions. When quantitative risks can be established for separate subgroups, it might become possible to address this issue further. When a mixture of photosynthesis inhibitors would affect 100% of the plant species, simple reasoning suggests that other species will go extinct

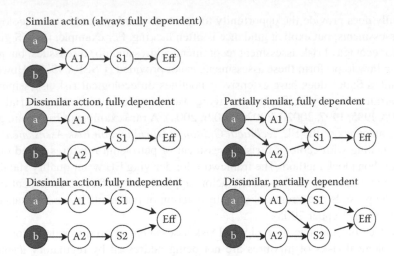

Figure 5.9 A limited array of possible (dis)similar actions of compounds at the target site of intoxication, including the relationship between the toxicological interaction and the final effect that is observed. In this figure, a and b are compounds, A1 and A2 are sites of action, S1 and S2 are affected subsystems, and Eff is an effect that can be observed (e.g., reproduction).

too as a consequence of absence of primary production, even when the other species are completely insensitive for the direct toxicological effects of the compounds.

5.3.2.1.3 Uncertainties and Probabilistic Assessments

Like human health assessments of chemical mixtures, ecological assessments usually involve substantial uncertainties. Examples include the assumptions of simple similar action or independent joint action, the extrapolation from laboratory data to field circumstances, and the inclusion of ecological interactions in the assessment. Because of these uncertainties, the assessment should always include a thorough discussion of all assumptions and the identification, when possible, of the major sources of uncertainty.

Whenever appropriate, and in line with the probabilistic concept of risk, probability distributions are used in ecological risk assessment of mixtures. This applies to the assessment of exposure (e.g., the probabilistic application of multimedia fate models; see Hertwich et al. 1999; Ragas et al. 1999; MacLeod et al. 2002), as well as to the assessment of effects, especially the SSD approach. Recent developments (both conceptually and practically) suggest that joint probability assessments (looking at exposure and effects distributions simultaneously) are applied more frequently. This relates to the refined questions being posed, but also to theory development (e.g., Aldenberg et al. 2002) and technical facilitation by software (e.g., Van Vlaardingen et al. 2004).

5.3.2.2 Regulatory State of the Art

In contrast to the situation for human risk assessment, regulatory rules to deal with mixture risks for ecological endpoints are limited in number. US and EU legislation generally does provide the opportunity to account for mixture effects in ecological risk assessments, but explicit guidance is often lacking. For example, the US guidelines for ecological risk assessment recommend addressing mixture issues, but guidance on how to perform these assessments is not provided (USEPA 1998). However, the United States does have extensive guidelines on ecological risk assessment of the overall toxicity of effluents, receiving waters, sediments, and terrestrial sites (USEPA 1995, 1997, 2002d, 2002e, 2002f, 2004). A rare example of European guidance is an annex of the EU *Technical Guidance Document on Risk Assessment* (EC 2003) dealing with ecological risk assessment of petroleum mixtures based on the hydrocarbon block method. The framework for deriving EU water quality standards states that it is assumed that safety factors applied in the effects assessment phase cover the possible occurrence of combined action of pollutants in most instances to a great extent (Lepper 2005).

The lack of guidance on ecological risk assessment of mixtures does not imply that ecological risks of mixtures are not being addressed by regulatory agencies. The TEF-TEQ approach has been applied by various agencies to assess the mixture risks for specific species, such as birds and mammals (Van den Berg et al. 1998). Sometimes a safety factor is used to account for the possible occurrence of mixtures, for example, in the derivation of Dutch environmental quality standards (VROM 1989). Mixture theories have also been used to derive environmental quality criteria for compound groups for which mixtures usually are present with more or less constant concentration ratios such as PAHs. The msPAF approach has recently been adopted in the Netherlands for retrospective assessments of the possible impact of mixtures in sediments.

In summary, various regulations address the issue of mixture toxicity in one way or another, but there are many specific solutions, and few generally adopted ones. Among the generally adopted ones, the TEF/TEQ approach for species-level assessments is well established, like the HI-like approaches for species assemblages—in this case often limited for use within groups of compounds with the same MOA. Note that, for practical assessments of contaminated sites, all approaches can be mechanistically wrong or not justified to a certain extent, but that neglecting mixture effects would be even worse (Posthuma et al. 2008).

5.3.3 SIMILARITIES AND DISSIMILARITIES BETWEEN HUMAN AND ECOLOGICAL MIXTURE ASSESSMENT

The state of the art presented in the 2 previous sections illustrates that human and ecological mixture assessments are similar in many ways, but there are also dissimilarities. This section summarizes the most important similarities and dissimilarities.

5.3.3.1 Similarities

An important similarity between human and ecological assessment of mixtures is the structure of the assessment procedure. Both procedures are often organized along a series of consecutive steps, that is, problem definition, hazard identification, exposure assessment, effect– or dose–response assessment, and risk characterization

(Figure 5.1). A second major similarity is that mixtures do pose realistic management problems in both fields; that is, in both fields an array of problem definitions has been formulated. Examples are the evaluation of mixture emissions, mixture evaluations at contaminated sites, mixture assessments in food, and the establishment of safe exposure or concentration levels for common mixtures such as PAHs. A third major similarity is that both fields deal with the toxic action of mixtures in living organisms. The processes involved in this action are similar for humans and other biological receptors because they share the same basic building blocks (e.g., DNA, proteins, membranes, cells) and many physiological processes (respiration, transport, signaling). This notion is reflected in the common terminology, theoretical concepts, toxicity measures, and disciplinary origins. The availability of data on mixtures is limited, for human as well as ecological test systems. Component-based models such as CA, RA, or mixed models are being used in both fields, although the names and details of the methods may vary. Examples are the use of RPFs, TEFs, an HI, or a cumulative risk index. Finally, both fields have to deal with uncertainties, especially when it comes to mixtures. Application of mixture models and assumptions is necessary because our knowledge about the processes and dynamics involved is limited, resulting in risk predictions that can have considerable uncertainty.

5.3.3.2 Dissimilarities

There is much dissimilarity between the fields of human and ecological risk assessment, but many of the differences are not typical for mixtures. Examples are differences in assessment endpoints (individuals vs. species or communities), in exposure routes and media (oral, inhalatory, and dermal for humans vs. aquatic or terrestrial for ecosystems), and in the level of mechanistic understanding (generally larger in human than in ecological studies).

When it comes to mixtures, an important development is the use of the internal dose as a dose metric, particularly in human assessments. The internal dose is either measured directly or modeled using PBPK models, for example, as a blood or a target tissue concentration. Application of an internal dose metric makes it possible to account for 1) interindividual variability in toxicokinetics, 2) temporal variations in exposure patterns, and 3) interactions between substances during absorption, metabolism, and transport. In ecological risk assessment, internal doses are sometimes measured but rarely modeled with PBPK models. The awareness is growing that the internal dose is a useful metric but the use in formal risk assessment procedures is still limited, for separate compounds as well as for mixtures.

Many similar mixture techniques and approaches are being used in human as well as ecological risk assessment, but whole mixture approaches for unique mixtures are generally limited to the field of ecological risk assessment, for example, WET tests and the pT approach. This is probably due to the fact that ecological test species can often directly be tested in the laboratory, whereas human assessments have to rely on tests with stand-in species such as rats and mice, and extrapolate the results to humans. Assessment of mixture risks based on testing a sufficiently similar mixture is typical for human risk assessment. The human field is clearly leading in the development and application of interaction-based methods for mixture assessment such as BRN models and BINWOE methods. The human field is also leading

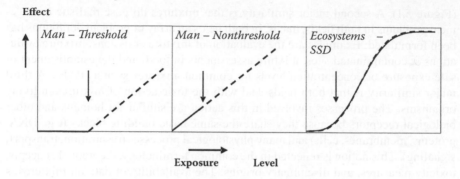

Effect

Exposure   Level

Figure 5.10 Difference in the dose-effect models for humans and species assemblages (species sensitivity distribution [SSD], right). Threshold-type curves are used for many compounds; it is assumed that below a certain daily intake there will be no effects. Nonthreshold chemicals (i.e. certain types of carcinogens) lead to increased probability of cancer, and for this a linear model is assumed in the relevant concentration range. Species sensitivities are assumed to follow a non-linear curve (the SSD), relating the exposure to the fraction of species affected, with a maximum of 100% of the species affected.

in the development of regulatory guidance for mixture assessments. Finally, typical for ecological mixture assessment are the application of the msPAF concept, validation of mixture predictions with field studies, and combination of different lines of evidence to assess mixture effects, like in a formalized TRIAD approach. These typical features can be explained by the fact that ecological assessments include an extra level of biological organization, but have better opportunities to study the effects of realistic exposure concentrations under field conditions. A unique aspect of the msPAF approach is that the whole, nonlinear, concentration–effect curve is considered. In human risk assessment, the models are assumed linear in the relevant concentration range when a linear MOA is assumed (i.e., not a threshold effect). This difference implies that CA and RA are different models in ecotoxicology, whereas both models can be referred to as "additive" in human risk assessment, since they are equal in the concentration range causing responses (see Figure 5.10).

5.4   CONCEPTUAL FRAMEWORK FOR HUMAN AND ECOLOGICAL RISK ASSESSMENT OF MIXTURES

The state of the art presented in the previous section illustrates that the methodologies for human and ecological assessment of chemical mixtures have similar conceptual foundations in the disciplines of toxicology, ecotoxicology, pharmacology, and (to a lesser extent) epidemiology, but they developed more or less along independent lines. However, there is a growing awareness that integration of human and ecological effect assessment may improve assessment quality and efficiency (WHO 2001; Suter et al. 2003). A first step toward such integration is the development of a common conceptual framework for mixture assessment in which the different human and ecological assessment methods that were reviewed in the previous section are placed together. Such a framework reveals the relationships between the different methods and highlights (dis)similarities. It is a useful tool to support the selection of suitable procedures, methods, and techniques for effect assessment of chemical mixtures depending on the problem at hand, data availability, and desired level of assessment detail. The aim of this section is to outline such a common conceptual framework.

The first step of the framework is a clear description of the mixture problem at hand, including the assessment goals and strategy (Section 5.4.1). The next step is the choice of one or more suitable methods for assessing mixture effects. This choice depends on the mixture problem at hand, for example, whether the mixture composition is known, its frequency of occurrence, the variation in concentration ratios, and the availability of toxicity data (Section 5.4.2). A distinction is made between assessment methods that estimate the toxicity of the mixture as a whole and component-based methods. Different methods for whole mixture assessment are discussed, varying from inaccurate to accurate and from poorly characterized to well characterized (Section 5.4.3). The component-based methods are discussed within the framework of a tiered approach, varying from rough methods that likely produce a conservative estimate of mixture risk to sophisticated methods that likely produce more accurate estimates (Section 5.4.4).

5.4.1 MIXTURE PROBLEM FORMULATION

Which effect assessment method should be applied in a particular situation depends on the nature of the mixture problem at hand. Because the diversity in assessment methods is large, it is important to clearly describe the problem. For example, derivation of a safe level for a proposed industrial mixture emission requires a different approach than the prioritization of a number of sites contaminated with mixtures. The former problem requires the assessment of realistic risks, for example, by the application of a suite of fate, exposure, and effect models, whereas the application of a simple consistent method suffices to address the latter problem, for example, a toxic unit approach. A successful and efficient assessment procedure thus starts with an unambiguous definition of the mixture problem at hand. The problem definition consists of the assessment motive, the regulatory context, the aim of the assessment, and a structured or stepwise approach to realize the aim. Elaboration of the problem definition is an iterative process (Figure 5.1) that strongly depends on factors such as resources, methods, data availability, desired level of accuracy, and results of previous studies.

The motive and regulatory context are 2 important factors of the problem definition. A possible motive could be that the use of a particular mixture by industry is rapidly increasing, and there is a need for safe exposure levels at the workplace and safe concentration levels in the environment. In this case, the regulatory context would be legislation or policy guidelines in the areas of working conditions and environmental protection. Another reason to perform a mixture assessment could be the discovery of a mixture contamination at a new building site. In this case, the aim would be to assess the risks for humans and nature, and the policy context (to interpret mixture risk estimates for the site) would be provided by environmental legislation such as a soil protection act.

It is noteworthy that the regulatory context of most mixture problems consists of legislation and regulations that focus on either (individual) substances or compartments. For example, widely used common mixtures such as PAHs and PCBs are often evaluated within the context of substance-specific regulations such as the Toxic Substances Control Act in the United States and the new REACH program in the EU. Mixtures encountered at polluted sites are evaluated within the context of specific regulations such as CERCLA in the United States, dealing with improperly deposited wastes, and the Water and Soil Framework Directives in the EU. All these regulations allow for assessment of mixture risks, but the problem definition is generally limited to the context of these specific regulations, for example, exposure to contaminated soil or to a specific mixture. However, the nature of a mixture problem can be much broader, for example, a farmer who works with pesticides and lives on a metal-polluted soil in a house with high radon exposure levels. Identification and definition of this type of mixture problems requires an approach that differs from the identification of traditional substance- or compartment-oriented problems. Receptor-oriented approaches (and thus often system-oriented approaches) are required, for example, based on analyzing consumption and living patterns or based on combining exposure data from different sources (Loos et al. 2010). We conclude that the identification and regulation of mixture risks requires a broader regulatory context than covered by most current regulations. The focus should not be on a specific substance, mixture, or compartment, but on the integrated protection of human or ecological receptors from exposure through different routes and at different moments in time. This implies a broader assessment perspective than currently in use.

The aim of the mixture assessment should include a specification of the relevant effect endpoints and exposure durations. This is particularly important because empirical studies have shown that the results of mixture tests are dependent on the endpoints considered (Chapter 4) and the duration of a mixture test (Chapter 2). It should be carefully evaluated whether the tested endpoints and exposure durations are relevant for the assessment at hand.

Furthermore, it is important to consider the accuracy and available resources when defining a mixture problem. It can be difficult to determine the accuracy and costs beforehand, because accuracy often trades off with costs, and it is influenced by data availability and the actual risk level. Efficiency considerations imply that an assessment should be as cheap as possible, but sufficiently detailed to make a motivated decision. For site-specific assessments, a detailed and costly assessment is necessary only when the actual risks are close to the level that is considered unacceptable. For other outcomes, for example, when the actual risk is much lower or higher than the unacceptable level, a more crude and cheap assessment may suffice. Similar reasoning applies when deriving safe exposure or concentration levels. If limited data are available, a conservative value may be derived. If implementation of this conservative value turns out to be too costly, it is worthwhile to invest resources in gathering additional data and subsequently deriving a more accurate value. This is reflected in the loops of the risk assessment approach (Figure 5.1), which has resulted in many tiered approaches, such as those discussed in Section 5.4.4.

5.4.2 Exploration of the Mixture Assessment Options

A risk assessor has different options to evaluate the potential effects of a mixture. This raises the question of which option should be preferred in a particular situation. The answer depends not only on the level of determinacy of the mixture, but also on other factors, for example, the capacity to further chemically characterize the mixture, whether it is a common or rare mixture, whether the concentration ratios between the components are more or less stable, and whether data on a sufficiently similar mixture are available. For example, it does not seem sensible to put much effort into full characterization of the mixture components if dealing with a rare mixture of unique composition. In this situation, it is more efficient to test a mixture sample directly in the laboratory or the field.

Based on this type of insight, Table 5.1 provides elementary guidance on how to identify a suitable assessment option for a mixture assessment in a particular situation. The guidance is based on answering 4 basic questions:

- Is the composition of the mixture of concern known, partially known, or unknown?
- Is the frequency of occurrence of the mixture common, rare, or unknown?
- Are the concentration ratios between the mixture components fixed, unique, or unknown?
- Are toxicity data available about a (sufficiently) similar mixture?

Table 5.1 provides the following assessment options:

1) Test the toxicity of the mixture of concern directly in the laboratory or the field and use this information in determining the risk.
2) Conduct the assessment based on data from (sufficiently) similar mixtures.
3) Conduct a component-based evaluation.
4) Determine a dose–response relationship or safe mixture level for the mixture of concern (or for its fractions) for future evaluation of sufficiently similar mixtures, for example, based on laboratory or field testing in combination with the use of one or more indicator substances.

Option 1 corresponds to the unique whole mixture approach introduced in Section 5.2. Options 2 and 4 can be considered communicating vessels; that is, option 4 generates the toxicity data necessary to realize option 2. Both options are examples of a common whole mixture approach. Options 1, 2, and 4 are elaborated in more detail in Section 5.4.3. Option 3 refers to the component-based methods, which are further discussed in Section 5.4.4.

If more than 1 assessment option is possible in a particular situation, the preferential option for the majority of such cases is listed first in Table 5.1, and alternative options are listed in parentheses. For example, the effects of a mixture of known composition are most likely to be assessed using component-based methods, but depending on the assessment context it is also possible to test the mixture of concern in the laboratory or the field.

Table 5.1   Overview of possible mixture problems and associated approaches for practical risk assessments

Composition?	Occurrence?	Concentration ratios?	Similar mixture Data available?	Assessment option[a]	Data gathering options
Unknown or partially known	Unknown	Not applicable	Not applicable	1. Test mixture	A. Composition? B. Occurrence?
	Rare	Unknown	Not applicable	1. Test mixture	A. Composition? C. Ratios?
		Unique	Not applicable	1. Test mixture	A. Composition?
		Fixed	No	1. Test mixture (4)	A. Composition?
			Yes	2. Similar mixture (1)	A. Composition?
	Common	Unknown	Not applicable	1. Test mixture	A. Composition? C. Ratios?
		Unique	No	1. Test mixture	A. Composition?
		Fixed	No	4. Safe level (1)	A. Composition?
			Yes	2. Similar mixture (1)	A. Composition?
Known	Unknown	Not applicable	Not applicable	3. Component based (1)	B. Occurrence?
	Rare	Unknown	Not applicable	3. Component based (1)	C. Ratios?
		Unique	Not applicable	3. Component based (1)	—
		Fixed	No	3. Component based (1, 4)	—
			Yes	2. Similar mixture (3, 1)	—
	Common	Unknown	Not applicable	3. Component based (1)	C. Ratios?
		Unique	Not applicable	3. Component based (1)	—
		Fixed	No	4. Safe level (3, 1)	C. Ratios?
			Yes	2. Similar mixture (3, 1)	—

Note: The numbers (1, 2, 3, 4) and letters (A, B, C) in the last two columns correspond to approaches described in the text.

[a] If more than one assessment option is possible, the preferential option is listed first and alternative options are listed in parentheses.

Table 5.1 also identifies options to gather additional data, which enable the application of alternative assessment options. For example, a risk assessor may opt to test the toxicity of an unknown mixture in the field or the laboratory, but alternatively the assessor can decide to determine the mixture composition and subsequently perform a component-based assessment. The following actions for gathering additional data are listed:

1) Determine mixture composition and reassess (composition?).
2) Determine whether the mixture is common or rare (e.g., based on the origin of the mixture) and reassess (occurrence?).
3) Determine whether the concentration ratios of the mixture are fixed or unique (e.g., based on the origin of the mixture) and reassess (ratios?).

The set of questions leads to a logical approach for all potential mixture assessment situations. For example, mixtures of which the composition is unknown or partially known and which have an unknown frequency of occurrence and unknown or unique concentration ratios should be assessed on a case-by-case basis using toxicity tests with whole mixture samples in the field or the laboratory. If this is not feasible, the risk assessor can opt to gather additional data about the mixture composition or its occurrence and subsequently reassess the mixture. It also leads to a logical approach for mixtures of which the composition is known, namely, the application of component-based methods or the use of appropriate data from similar cases.

5.4.3 Whole Mixture Approaches

Whole mixture approaches cover all methods that aim to assess the overall risk of the mixture based on testing the mixture as a whole or partially (e.g., a mixture fraction or a selection of its components). A whole mixture may be thought of as a mixture that contains a large number of chemical components, some of which may be unidentified. The relationships between the components may be unknown, or the mixture may contain unidentified substances considered important because of known or suspected toxic properties. Examples of whole mixtures include diesel exhaust, PAHs, drinking water disinfection by-products (DBPs), PCBs, TPHs, and coke oven emissions. The complexity of the evaluation depends on the goal of the assessment and the sophistication of the chemical characterization of the mixture. Some risk assessment approaches are conducted as a screen for potential risks (conservative estimates that are protective), while others are conducted to more accurately assess risk (central tendency estimates that are predictive). Full chemical characterization and toxicological testing of a mixture may be prohibitive for several reasons. Such analyses may be analytically difficult, expensive to conduct, or highly variable across exposure scenarios or testing protocols. Thus, factors to consider in choosing how to assess risk include the goal of the assessment, the tolerable amount of uncertainty, and the limits on available resources.

Many different methods for whole mixture assessment are available, and they vary substantially in level of accuracy and uncertainty. Figure 5.11 provides an

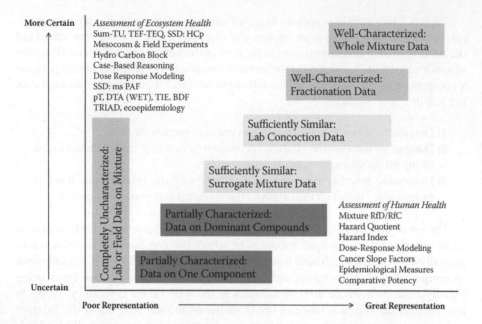

Figure 5.11   An overview of the approaches and methods used in the assessment of human (lower right corner) and ecological (upper left corner) effects of whole mixtures. Methods for assessing quantitative risks are Sum-TU through Case-Based Reasoning in Ecosystem Health, and Mixture RfD/RfC through Hazard Index for Human Health; the remainder are methods for determining safe levels. Moving from left to right in the diagram (horizontal axis), information availability about the mixture increases; a mixture on the right-hand side will be better characterized than a mixture on the left-hand side. Running from the lower to the higher part of the diagram (vertical axis), the uncertainty in the risk assessment decreases. TU = Toxic Unit, TEF = Toxicity Equivalency Factor, TEQ = Toxicity Equivalent, HCp = Hazardous Concentration for p% (5%) of the species, SSD = Species-Sensitivity Distribution, msPAF = multi-substance Probably Affected Fraction of species, pT = Toxic Potency, DTA = Direct Toxicity Assessment, WET = Whole Effluent Toxicity, TIE = Toxicity Identification and Evaluation, BDF = Bioassay-Directed Fractionation, TRIAD = Integrated approach, applying a combination of chemical analysis, bioassays and ecological field observations, RfD = Reference Dose, RfC = Reference Concentration.

overview. Moving from left to right in the diagram (horizontal axis), information availability about the mixture increases; a mixture on the right-hand side is better characterized than a mixture on the left-hand side. In general, better characterization implies that more information is available about the mixture's chemical composition. However, this does not mean that the exact composition of the mixture is known. For example, a mixture can also be well characterized because it is a common mixture with stable concentration ratios that originates from a well-known source, such as coke oven emissions. Running from the lower to the higher part of the diagram (vertical axis), and assuming equally good toxicity data for all levels of the figure, the uncertainty in the risk assessment decreases.

Figure 5.11 distinguishes 4 assessment situations:

1) Completely Uncharacterized Mixtures: Mixtures that are completely unknown in terms of composition, toxicity, and origin (completely uncharacterized mixtures).
2) Partially Characterized Mixtures: Mixtures of which the toxicity can be predicted based on one or more of their individual components (partially characterized mixtures).
3) Sufficiently Similar Mixtures: Mixtures of which the toxicity can be predicted based on their similarity with other, known mixtures (e.g., because they have a similar origin).
4) Well-Characterized Mixtures: Mixtures that are well characterized in terms of chemical composition and origin, and for which extensive dose–response data are available, for either the mixture itself or its fractions.

Figure 5.11 implies that the certainty of a whole mixture risk assessment generally increases as the characterization improves from a complete lack of characterization to partial characterization to a similar mixture to a well-characterized mixture (i.e., mixture of concern). In other words, based on available mixture insights, it is considered likely that refinement of information on the mixture can be reached by characterization of the mixture.

Different techniques are currently available in human and ecological risk assessment to help in assessing the risks of the 4 mixture types. These techniques are indicated in Figure 5.11. Ecological techniques are placed in the upper left part of the diagram, and human techniques in the lower right part. Distinction is made between techniques that are used to derive safe levels and techniques that are used for quantitative assessment of risks. The techniques are discussed in more detail in the following sections.

5.4.3.1 Completely Uncharacterized Mixtures

When a mixture is completely uncharacterized and it is infeasible or inefficient to determine its composition, the potential effects can be determined by testing a field sample. This is highly representative for the problem, and the outcomes of the toxicity test itself are relatively certain, but the reproducibility depends strongly on the similarity of collecting the same mixture next time. There is no "understanding" and no "generalization" to other samples or mixtures. Many uncertainties still exist; for example, it is unknown if the mixture will change rapidly as it sits in the field, refined and long-term effects such as bioaccumulation and carcinogenicity may not be detected in the field test, and in some cases, there may be insufficient samples across the geographic area to ensure the toxicity test represents the variance of the field sample test. So, the uncertainties are still large, though not necessarily larger than for assessments based on partially characterized or similar mixtures, which explains the position of the completely uncharacterized mixtures in Figure 5.11. Accuracy could be improved if characterization were to be done, in which case it would move into the upper-right-hand corner.

Whole mixture bioassays are rare in human health assessments. One exception is the testing of DBP mixture concentrates for developmental and reproductive effects in animal bioassays (Simmons et al. 2002). In ecological risk assessments, direct

toxicity studies with completely uncharacterized mixtures are conducted on a regu-
lar basis. This is likely because, unlike human health evaluations, ecological toxicity
assessments can be conducted on relevant test species or sentinel species using whole
mixture techniques, often conducting the toxicity test directly on the mixture of
concern. These approaches in ecological risk assessment are possible, practical, and
often more accurate than the application of modeling on partial mixture techniques
that require many assumptions. This is probably why many types of direct testing of
whole mixtures have been developed in ecotoxicology, like WET testing, TIE procedures, BDF techniques, the TRIAD approach, and the pT approach (see Section
5.3.2.1 and Chapter 4).

5.4.3.2 Partially Characterized Mixtures

In Figure 5.11, methods are shown that use only a partial chemical characterization
of the whole mixture. These mixtures are characterized by only one or a few of its
components, whereby it is assumed that the toxicity of those compounds is sufficient
to act as proxy for the toxicity of the whole mixture. This may range from very simply
using a single chemical to represent the whole mixture (e.g., assuming all of the TPH
is really benzene and evaluating it using the cancer slope factor for benzene when
assessing human risks of TPH) to using a simple, defined mixture (e.g., assessing a
complex mixture composed of hundreds of DBPs by using only 4 trihalomethanes
and 5 haloacetic acids). The former is relatively easy to do, but may be highly inaccurate; the latter provides a better exposure characterization and evaluation of toxicity, but still fails to take into account the risk from exposure to the chemicals not
included in the assessment. When 2 or more components are involved, toxicity is
often assessed based on component-based additivity models (Section 5.4.4).

In human health risk assessment, the partial characterization of mixtures (e.g., using
a single chemical or a simple, defined mixture to represent a whole mixture) may
be evaluated for safety using a hazard quotient (i.e., exposure level divided by a
safe level) or HI (sum of hazard quotients), respectively. Alternatively, risk estimates for these same entities could be made using RA based on summing of risks
(e.g., using cancer slope factors) or using CA (e.g., by applying RPFs).

In ecological risk assessment, partially characterized mixtures are usually not
evaluated using partial chemical information. Instead, such mixtures are often treated
as completely uncharacterized mixtures and tested directly (see Section 5.4.3.1).

5.4.3.3 Sufficiently Similar Mixtures

As shown in Figure 5.11, a sufficiently similar mixture may be used as a surrogate
to assess the risks of the mixture of concern. In this case, the evaluation may use
surrogate exposure and toxicity information on a mixture of the same chemical class
(e.g., PAHs) or on a group of similar mixtures produced by the same process (e.g.,
from a diesel engine in the laboratory). The risk of the mixture of concern can be
estimated based on the degree of similarity between mixtures for which the dose–
response relationships are well characterized for at least one of the mixtures. In this
case, the mixture composition can be relatively clear, represented by all compounds,
by a large number of major components, or by chemical measures that represent
the entire mixture. An example of the latter is the percentage of brominated vs.

chlorinated compounds found in the total organic halide material that comprises a mixture of drinking water DBPs. The mixture's toxicity is often not measured itself (on an environmental sample), but extrapolated using data on the known mixture as surrogate data. This introduces reproducible model outcomes (same mixture + same model delivers reproducible outcomes), but that outcome is uncertain and may be imprecise, depending on the accuracy of the models used and the certainty that the tested mixture represents the true mixture composition.

In human toxicology, toxicity values in the form of RfDs, RfCs, or cancer slope factors may be derived from these surrogate data. For example, USEPA (2000b) discusses the "comparative potency approach" that has been applied to diesel engine emissions. This approach builds a regression model using in vivo and in vitro data on several different combustion mixtures to estimate human cancer slope factors for diesel emissions (Lewtas 1985, 1988; Nesnow 1990). Transformation of a mixture in the environment (e.g., chemical degradation due to weathering) may also be addressed using a similar mixtures approach. For example, Cogliano (1998) suggested that cancer slope factors derived from commercial bioassay data on PCBs can be adjusted for changes in potency based on the chlorine content of the PCBs found in the environment.

In ecological risk assessment, similar mixture methods have not been applied widely, but there are options that could be elaborated further. It would imply mixture-to-mixture extrapolation (see Solomon et al. 2008). It can be imagined that the similar mixture principle could be worked out according to the principle of case-based reasoning (CBR). So far, this method only exists for predicting complex community responses to single pesticides (Van den Brink et al. 2002). In a formalized software program, pesticide effects data are stored, and by CBR one can predict the risks of an untested pesticide, based on similarity of conditions (e.g., pesticide group, concentration, water type). Posthuma et al. (2006) proposed this approach for a decision support system for handling contaminated sediments: by gross characterization of sediments on the basis of few proxy chemicals, the similarity to well-known and well-analyzed mixtures would be established, and the decision on sediment risk management would be based on limited mixture characterization and a mixture risk database. Furthermore, it is possible to execute mesocosm and sometimes field experiments with partially similar mixtures.

5.4.3.4 Well-Characterized Mixtures

The 2 boxes in the upper right corner of Figure 5.11 reflect a situation in which the mixture of concern is well characterized, for example, because a lot of information is available about its composition, its origin, and its dose–response relationship. A well-characterized mixture can be thought of as a commonly occurring mixture with a stable chemical composition, which is more or less known, for example, coke oven emissions. It is often infeasible to determine the exact chemical composition of the mixture at hand because the mixture contains hundreds or thousands of different components. This is also unnecessary because dose–response data on the mixture of concern are available from previous studies, for example, epidemiological data on coke oven emissions. A mixture is also considered well characterized if it can be divided into fractions of which the dose–response relationship is known, whereby the fractions together represent the total mixture. The line between a sufficiently similar and well-characterized mixture is somewhat arbitrary; the main difference is that available toxicity data are considered directly applicable to a well-characterized mixture, whereas some adjustment may be necessary in the case of a sufficiently similar mixture. Well-characterized mixtures should not be confused with mixtures of which the toxicity has been tested directly in the laboratory or the field (Section 5.4.3.1). Although the latter mixtures may be relatively well characterized in terms of toxicity, uncertainties remain and their composition and origin often remain unknown.

In human risk assessment, an inhalation cancer slope factor for coke oven emissions can be found on the USEPA's Integrated Risk Information System (IRIS), which was derived from an occupational study (USEPA 2008). Similarly, risk estimates (e.g., odds ratios, relative risks) using human epidemiological data represent a direct evaluation of the mixture of concern. The approach of mixture fractionation has been applied in human risk assessment by the Massachusetts Department of Environmental Protection (MADEP 2002) to conduct site-specific assessments of TPH mixtures, whose chemical composition is highly variable across locations. This approach divides the PH at a site into analytically defined PH fractions and then assigns oral and inhalation toxicity values to these fractions for use in risk assessment (MADEP 2003), using a single surrogate chemical from each fraction to represent the risk for the entire fraction. For example, this information could be used to calculate hazard indices for the TPH. Fractionating the PH mixture still requires the risk assessor to determine the similarity of the surrogate chemical to the rest of its PH fraction, and it does not totally account for all of the unidentified material. However, this approach illustrates a flexible method for characterizing PH exposures that reflects differences in chemical composition across various sites and provides a reasonable method for calculating potential health risks.

In ecological risk assessment, methods for well-characterized mixtures are rarely applied in practice, mainly because it is often more practical to test the mixture of concern directly in the laboratory or the field (i.e., corresponding to a unique mixture approach or the assessment of a completely uncharacterized mixture; see Section 5.4.3.1). Fractionation of whole mixtures is sometimes done, for example, the fractionation of TPH mixtures (King et al. 1996).

5.4.4 COMPONENT-BASED APPROACHES

Component-based approaches are based on the assumption that the toxicity of a mixture can be assessed based on knowledge of the individual components in combination with the application of a mixture model. A component-based approach is generally applied to mixtures with relatively few chemical components that have all or partially been identified. If only partial information is available, the approach corresponds to the partially characterized boxes in Figure 5.11.

The use of component-based techniques is widespread in human and ecological risk assessment. What is usually done is based on the fact that many binary or more complex mixture tests have been done, and compiled in databases. Various authors have reflected on those data, and have been drawing generalized conclusions on the mixture models that explain the data best. Many authors have concluded that the CA model (related to simple similar action as a mechanism) is a good model, because it reasonably fits many data sets. Alternative and more sophisticated models, such as RA, mixed models, and PBPK models, have also been applied to the same data to make accuracy comparisons (see sections on scientific state of art). From those enterprises, (eco)toxicologists usually address mixture problems for known mixtures by combining measured or expected environmental concentrations and the preferred model to predict or quantify the mixture effect.

Many different component-based techniques have been developed, varying from very simple and rough to highly sophisticated and accurate. A tiered approach can be applied when various scientific techniques can be used to address 1 problem definition in risk assessment, like the HI and the msPAF index. Tiered approaches are widely used in risk assessment of single substances (Cowan et al. 1995; Van Leeuwen and Hermens 1995; USEPA 1998; Solomon et al. 2008). As the assessment moves from lower to higher tiers it becomes more refined and requires more data and resources. The tiered approach provides a systematic way of determining what level of investigation is appropriate for the situation at hand, minimizing unnecessary investigations and allowing an efficient use of resources. It requires an explicit description of the management questions involved and decision criteria to proceed from one tier to the next.

Posthuma et al. (2008) proposed a tiered system for component-based methods in ecotoxicological risk assessment, which has been elaborated further here and extended to include human techniques. It is presented schematically in Figure 5.12. Ecological techniques are placed in the upper left part of the diagram, and human techniques in the lower right part. The different techniques have been discussed in part in Section 5.3.2 and are discussed in more detail below. Table 5.2 summarizes the underlying assumptions and data needs of the different tiers. An important issue in using tiered approaches is that there is a need for a defined trigger to go to the next tier. In the context of this chapter it is only possible to generally mention this subject, since local regulations have defined or may define such triggers specifically, in the context of specific problem definitions.

5.4.4.1 Tier 0: Safety Factors

In tier 0, mixture effects are considered potentially relevant for the assessment, but detailed data on mixture effects are lacking. In such cases, a nonmixture data- or theory-driven safety factor is used, whereby the mere presence of this factor in the assessment reflects uncertainty on various issues, including mixture impacts. Three different situations in which safety factors may be applied can be distinguished:

  1) Mixture assessment is not possible given the assessment problem. In such cases, safety factors can be applied, but these are not derived from mixture assessment data. An example is the application of an extra safety factor when setting standards for single compounds in order to account for

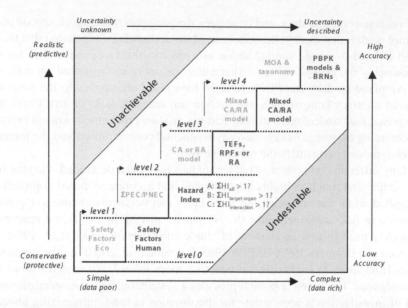

Figure 5.12 The principle of tiering in risk assessment: simple questions can be answered by simple methods that yield conservative answers, and more complex questions require more sophisticated methods, more data, and more accurate risk predictions. PEC = Predicted Environmental Concentration, PNEC = Predicted No Effect Concentration, HI = Hazard Index, CA = Concentration Addition, RA = Response Addition, TEF = Toxicity Equivalency Factor, RPF = Relative Potency Factor, MOA = Mode of Action, PBPK = Physiologically Based Pharmacokinetic, BRN = Biochemical Reaction Network.

potential mixture effects. This is common practice in the Netherlands when setting environmental quality standards to protect human health and eco-systems (see Section 5.3.3.2).

2) Component-based models are applied to a mixture for which only partial information about its composition is available. In this case, the safety factor is supposed to cover for the components for which information is lacking. The value of the safety factor should reflect the extent of the missing information.

3) There are indications that interaction effects may occur, but quantitative data are lacking. In this case, the height of the safety factor depends on the nature of the suspected effects (e.g., synergistic or antagonistic) and the quality of the available information.

Please note that the safety factors that cover for situations B and C can be used in combination with the methods of tiers 1, 2, and 3.

5.4.4.2 Tier 1: Hazard Indices

Tier 1 involves the application of a simplified form of CA. It boils down to calculat-ing the ratio between the exposure concentration of each component and a point estimate from its concentration–effect curve (e.g., the EC10, EC50, or NOEC), and the summation of these ratios. The human variant of this approach is the HI, and

Table 5.2   Major tiers that can be distinguished in combined effect extrapolation

Tier	Model	Assumptions/model	Information required
0	Safety factors	A. Single substance standard does not cover mixture effects	None
		B. Partial information	Assess information extent
		C. Interaction effects	Likelihood of interactions
1	Hazard indexes (optional: modify HI based on binary interactions data)	Point estimates on concentration–effect curves (optional: assume binary interaction data represent higher-order interactions)	Toxicological reference values for the mixture components, for example, EC50, NOEC (optional: binary interaction data)
2	Concentration addition or response addition	Full-curve-based approaches, modes of action assumed fully similar or fully dissimilar	Concentration-response relationships for the components
3	Mixed models	Similar and dissimilar action allowed, full-curve-based approach	Concentration-response relationships for the components, mode of action information
4	Sophisticated mechanistic models, including interaction data and data on different characteristics in the set of receptor species (assemblage-level assessments only)	Similar and dissimilar action allowed, full-curve-based approach, kinetics and dynamics of mixture components and toxicological interactions are included	Concentration-response relationships for the components, mode of action information, kinetics and dynamics data, receptor species information

the ecological equivalent is the summation of PEC/PNEC or PEC/ECx ratios. These methods are often applied in cases where the assessment problem is either simple or vaguely formulated (no clear, identified endpoint). Ideally, a separate index should be calculated for each specific effect caused by the mixture. However, these preconditions for application are often relaxed because CA has been shown to result in relatively conservative risk estimates (see Drescher and Bödeker 1995). So, the approach is often applied as a conservative tool to address all mixture effects. In human risk assessment, a tiered HI approach is sometimes followed. In the first tier, the hazard ratios of all substances in the mixture are summed, irrespective of the effect they cause. If this sum exceeds 1.0, target-organ-specific HIs are calculated. One step further, separate HIs for specific molecular receptors can be calculated if detailed information about these receptors is available. Finally, qualitative information about potential interaction effects can be included, resulting in an interaction-based HI. In ecotoxicology, clear examples are the use of the TEF-TEQ approach to assess the risks of PCBs for birds and mammals, or HI-type approaches to quantify the magnitude of exceedance of regulatory quality standards.

5.4.4.3   Tier 2: Concentration or Response Addition

Tier 2 assumes either a uniform MOA for all compounds (i.e., concentration addition) or a complete nonuniform set of modes of action (i.e., response addition). Limited information on the MOA is typically available to use in the assessment, and the techniques are relatively simple. CA in tier 2 differs from that in tier 1 by using the full-dose–response curve. First, the concentration of the components is expressed in comparable units. Subsequently, these units are summed and a dose–response model is applied to predict the response. Examples include the application of RPFs, TEFs, and toxic units. These techniques are commonly used in human as well as in ecological risk assessment of mixtures, though the use of whole curve estimates is by far less common than the use of point estimates (tier 1).

5.4.4.4   Tier 3: Mixed Models

Tier 3 involves the use of both CA and RA models together (mixed-model approaches). This approach differs from the previous tiers by using detailed information on the modes of action for the different mixture components as well as full-curve-based modeling approaches. Mixed models are used in human as well as ecological assessment. An example of mixed-model approaches in ecological risk assessments is the approach proposed for assemblages (De Zwart and Posthuma 2005); a similar approach has been proposed by Ra et al. (2006); see Chapter 4 and Figure 4.2.

5.4.4.5   Tier 4: Sophisticated Mechanistic Models

Tier 4 includes all methods that go beyond CA or RA and attempt to provide some kind of mechanistic explanation for the mixture effects, including potential inter-actions between the mixture components. It requires detailed information on the toxicokinetic and toxicodynamic processes involved. The diversity of models that belong to this category is huge. Examples from human mixture assessment include the application of PBPK and BRN models. In ecological risk assessment, it may involve the consideration of multiple modes of action per mixture component as well as the assumed characteristics of sets of receptor species. Therefore, tier 4 methods only apply to problems that are defined in a very specific way (regarding site, species, compounds), and where an accurate result is preferred over a conservative one.

In human risk assessment, tier 4 approaches would require PBPK modeling, BRN models, or other highly specific approaches. In ecological risk assessment, emphasis can be put on modes of action in relation to subgroups of exposed species (e.g., insecticidal action and focus on impacts on insects). This type of approach was proposed some years ago (Posthuma et al. 2002), and has been applied in the evalu-ation of pesticide risks (De Zwart 2005; Henning-de Jong et al. 2008).

It is evident that some techniques do not have conceptually similar equivalents with various levels of complexity. Hence, tiering is not (yet) possible for all problem definitions. Moreover, it is clear that human risk assessment can sometimes operate on a higher conceptual tier than ecological risk assessment, for example, when BRN modeling and PBPK models are used. On the other hand, ecological risk assessment approaches may be sometimes more diverse, and can be better tailored to a risk assessment problem and its context.

5.5 ISSUES IN HUMAN AND ECOLOGICAL RISK ASSESSMENT OF MIXTURES

In the previous sections, different methods for human and ecological risk assessment of mixtures have been reviewed and discussed. Some typical mixture issues were briefly mentioned within the context of this methodological review, but they have not been extensively discussed. Examples of such issues are 1) exposure assessment, 2) sufficient similarity, 3) interactions vs. additivity, 4) QSARs, 5) uncertainties, and 6) risk perception of mixtures. These topics are discussed in more detail in the following sections.

5.5.1 EXPOSURE ISSUES

Exposure is a critical part of both human and ecological risk assessment for mixtures. Exposure assessment for mixtures is discussed in detail in Chapter 1. As with single chemicals, ideally the magnitude and duration of exposure for the mixture should be determined. In addition, a mixture assessment must consider the interactions of the individual components with the environment and with each other that impact the composition of the mixtures to which the human or ecosystem is exposed. Thus, the chemical characteristics, the release pattern, and the environmental characteristics must all be considered jointly to determine the final exposure to the mixture. Chemical characteristics that are important include volatilization, sorption, solubility, and especially degradation within environmental compartments. Relevant release pattern characteristics include the environmental compartments into which the mixture is released and the presence of other releases in the vicinity. Environmental compartment characteristics that can impact the exposure include pH, percentage organic matter, and dilution. The interactions of all these factors as well as interactions among the mixture components affect the composition of the mixture in space and time. One approach to predict the composition of the mixture in space and time, and thus the exposure, is to use environmental fate models, but often these are limited in their capability to look at all these parameters simultaneously as well as the interactions among the mixture components and degradation products of the mixture components.

In addition to these factors that impact the external concentration of the components in the mixture, the bioavailability of the mixture components must be considered to predict the transfer of the mixture components into the organisms and within food chains. Bioavailability represents the interaction of the environment and the organism that impacts the degree and rate at which the chemical is absorbed into a living system. Bioavailability is particularly difficult to predict for components of mixtures because the presence of other chemicals within the mixture may affect the bioavailability in ways that are not easily predicted based on the individual chemical characteristics. The various factors that can impact bioavailability are discussed in detail in Chapter 1.

The final factor in estimating the exposure to the mixture is the understanding of the habits and practices of humans and the distribution of organisms in the environment, which can increase or decrease the potential contact with the mixture. For example, consumption of various foods, time spent outdoors vs. indoors, occupation, and life stage can impact the potential exposure to a mixture.

Given all these interactions and potential confounding factors, there has been an increased interest in using biomonitoring data and internal dose measurements to assess exposure. However, usually these data cannot be linked to exposure history, resulting in a limited understanding of the sources contributing to the exposure. The usefulness of biomonitoring data is hampered by the limited understanding of the internal fate of the mixture components and analytical techniques to detect these components in biological matrices.

5.5.2 (SUFFICIENT) SIMILARITY OF MIXTURES

The risk assessment of whole mixtures by modeling and based on fully appropriate data is difficult because the required mixture toxicity data are never available for all combinations of chemicals and dose levels within such a mixture. Direct toxicity testing of the environmental mixture of concern provides the most relevant data. However, even when environmental samples are tested, variability exists from one sample to another for different sampling locations, weather conditions, or sampling dates. Thus, sufficient similarity is a key concept for evaluation of a whole mixture. It is applied when toxicity data are inadequate to directly evaluate a mixture of concern, but toxicity data are available on another mixture composed of similar chemical components in similar proportions. If the 2 mixtures are judged to be sufficiently similar, then the toxicity data for the latter can be used as surrogate data in conducting a quantitative risk assessment for the former. Although this general concept for evaluating complex mixtures has been generally accepted as a risk assessment method (USEPA 1986), specific guidance is lacking regarding statistical and toxicological criteria for determining with confidence whether a pair of chemical mixtures is actually sufficiently similar for use in risk assessment.

Currently, there are no generally applied methods available for determining sufficient similarity, and only a few published studies on the subject (e.g., Eide et al. 2002; Stork et al. 2008). Eide et al. (2002) developed a method to predict the mutagenicity of a sufficiently similar mixture. They used multivariate regression modeling of variables identified using pattern recognition techniques to characterize the chemistry and mutagenicity of organic extracts of exhaust particles. Variables were identified from gas chromatography–mass spectrometry (GC-MS) chromatograms of the complex mixture, and then multivariate analysis was used to build a model that identified the peaks that covaried with mutagenicity. Thus, the mutagenicity of a new mixture of exhaust particles can be predicted from its GC-MS chromatogram as input to the statistical model. See also Chapter 4. From such an analysis, the risk assessor is provided with information on both chemical composition and toxicity for use in determining sufficient similarity of 2 or more extracts of soot particles. Stork et al. (2008) approaches sufficient similarity using a statistical analysis of the variance in dose–response curves based on changes in chemical mixture composition. In this research, statistical equivalence testing, logic, and mixed-model theory are used to develop a methodology to define sufficient similarity in dose–response for mixtures of many chemicals containing the same components with different ratios.

Several methods are also being developed that key off decisions made regarding chemical and toxicological measures of DBP mixtures (Rice et al. 2009). When comparing 2 drinking water DBP mixtures, for example, an important factor to consider is the relative amount of chlorinated or brominated total organic halide material, because brominated compounds are generally considered to be more toxic. It is also known that various disinfectants (e.g., ammonia, chlorine, ozone) and source water combinations (e.g., high vs. low bromide waters) form quite different DBP mixtures. For example, different carcinogens may be formed; nitrosamines are formed with use of ammonia, bromate results from ozone application, and neither is generally found from chlorination. Other important chemical composition factors that affect the toxicology include similar concentrations and proportions of individual DBPs with known toxicity data on the same endpoint, a similar unidentified fraction of total organic halide material, and similar toxicity outcomes from whole mixture testing, such as mutagenicity measures. Statistical techniques such as linear regression models and pattern recognition methods can be applied to determine whether mixtures are similar based on such measures. Important DBP mixture summary measures for sufficient similarity that can be used in statistical modeling include mutagenicity data, total organic carbon, total organic halides, total trihalomethanes, total haloacetic acids, total haloacetonitriles, and the percent of such chemical measures that are made up of brominated compounds.

Sufficient similarity assumes that the toxicological consequences of exposure to 2 mixtures (i.e., a mixture of concern and a mixture for which toxicological data are available) are identical or indistinguishable from one another. In practice, ad hoc procedures require some degree of chemical similarity, or at least an understanding of how chemical differences between the mixtures affect their toxicology. Professional judgments of sufficient chemical similarity are based on an evaluation of the composition of 2 mixtures by noting if significant and systematic differences exist in the quantities of components and their proportions in the mixture. In addition, if information exists on differences in environmental fate, uptake and pharmacokinetics, bioavailability, or toxicological effects for either of these mixtures or their components, it should be considered in the determination of sufficient similarity.

5.5.3 INTERACTIONS VERSUS ADDITIVITY

The potential for toxicological interactions to occur from coexposures to multiple chemicals is a common concern for evaluating chemical mixtures, particularly when environmental dose levels are elevated above recommended toxicity reference values for the single components. Toxicological interactions are generally considered to occur as chemical-chemical reactions, and as toxicokinetic and toxicodynamic interactions. Toxicological interactions are defined by federal agencies in the United States as responses that deviate from those expected under a specified definition of additivity (USEPA 2000b; ATSDR 2004a). Thus, interactions are generally referenced as effects that are greater than additive, or synergistic (e.g., increased carcinogenicity for coexposures to asbestos and tobacco smoke), or less than additive, or antagonistic (e.g., decreased cadmium toxicity through coexposure to dietary zinc that reduces cadmium absorption). More specific terms are in use, including "inhibition," "potentiation," and "masking." Inhibition occurs when a component that does not have a toxic effect on a certain organ system decreases the apparent effect of a second chemical on that organ system; potentiation is the opposite, in that it increases the apparent effect of the second chemical (ATSDR 2004a). The term "masking" is used when the components produce opposite or functionally competing effects on the same organ system, and diminish the effects of each other, or one overrides the effect of the other (ATSDR 2004a).

The character of a toxicological interaction depends on the doses and mixing ratios of the chemicals involved. A specific example of this is provided by Sharma-Shanti et al. (1999), who showed concentration-dependent antagonistic, nonadditive, and synergistic effects of metal mixture exposures to root growth in the plant *Silene vulgaris*. Interactions may include changes in the type of adverse effect that is manifested as doses increase or as ratios change, as well as changes in the type of joint toxic action (e.g., synergism vs. antagonism). Interactions typically range from additivity at low doses to synergism in the midrange of the dose–response curve to antagonism at high doses close to a maximum biological response. For use in risk assessment, it is recommended that experimental doses in animal bioassays mimic environmentally relevant dose levels and ratios as far as is feasible (Teuschler et al. 2002). Interactions data can be used in risk assessment based on the strength of evidence for these requirements: 1) adequate toxicity data are available on dose response and MOA, 2) data on the same route of exposure should be used across components or similar mixtures, 3) data on components should be from comparable studies (e.g., same species, endpoint, study duration), and 4) observed interaction effects should be toxicologically significant (USEPA 2000b).

5.5.4   QSARs for Mixtures

When no data on the physical, chemical, transformation, and toxicological properties of a chemical are found, QSARs or SARs are commonly used to fill in the missing properties. When any of these types of data on a similar mixture are available, they are often used to represent the mixture of interest.

Altenburger et al. (2003) reviews and discusses current QSAR methodology for analysis of aquatic mixture toxicity and some physical and chemical properties. Much of the following discussion is based on that document. It is recognized that components in complex mixtures can interact and cause substantial changes in the apparent properties of the components from the situations where they exist individually. These changes can relate to physical–chemical properties as well as to the toxicological properties. When the precise composition of the mixture (i.e., components and relative abundance) is known, then a pragmatic approach for estimating a property of the mixture is to add up the properties of individual components weighted by their molar fraction in the mixture. This approach is limited to ideal mixtures and perhaps to only a few physical–chemical properties (e.g., K_{ow} and aqueous solubility), but can be applied to nonideal mixtures when activity coefficients can be estimated and are used as correction factors. However, this precise composition of the mixture is the exception, not the norm, for most mixtures, and thus this approach has limited application. For a mixture of chemicals with identical modes of interaction, additive joint action estimated through CA may be appropriate. For example, a QSAR-type approach can be applied to mixtures based on the argument that all the components of the mixture (known or unknown) act as baseline narcotics even if at high concentrations and at specific sites in the organism they may exhibit specific effects. This gives an impression of the minimum toxicity since the narcotic MOA is the least toxic MOA for organic chemicals. Thus, if the octanol–water partition coefficient, K_{ow}, can be estimated for the mixture, then application of a baseline toxicity QSAR using this K_{ow} yields an impression of the minimum toxicity of the mixture.

Recently, Riviere and Brooks (2007) published a method to improve the prediction of dermal absorption of compounds dosed in complex chemical mixtures. The method predicts dermal absorption or penetration of topically applied compounds by developing quantitative structure-property relationship (QSPR) models based on linear free energy relations (LFERs). The QSPR equations are used to describe individual compound penetration based on the molecular descriptors for the compound, and these are modified by a mixture factor (MF), which accounts for the physical–chemical properties of the vehicle and mixture components. Principal components analysis is used to calculate the MF based on percentage composition of the vehicle and mixture components and physical–chemical properties.

However, it must be recognized that due to the complex interactions of components in the mixtures, quantitative applications of QSARs to mixtures in environmental risk assessment are limited. For exploratory analysis and interpretation of environmental observations and test results, QSARs can be useful for developing a basic understanding of the mixture and its deviation from the theory for ideal mixtures.

5.5.5 UNCERTAINTIES

Our limited knowledge about chemical mixtures, their fate, and their interactions with biological systems means that estimates of mixture risks are uncertain. This uncertainty and uncertainties on other aspects of the assessment are important ingredients in the risk management process. For example, highly uncertain risk estimates may result in other management decisions and interventions than relatively certain risk predictions. This notion is reflected in the precautionary principle (PP), which states that strategies and policies must be based on the best available scientific information while erring on the side of caution.

Many sources of uncertainty in mixture assessment are comparable to those in the assessment of single substances. Examples are uncertainties about emission loads, the fate of the substances in the environment, exposure scenarios, variability in individual characteristics, extrapolation of toxicity data between species, and the shape of the dose–response curve. But there are also uncertainties that are typical for mixture assessment. An example is the composition of the mixture, which is often unknown or only partly known. This is irrelevant if the mixture of concern can be tested directly in the laboratory or the field, but it results in uncertainty if the assessment is based on sufficient similarity or on the known individual components. It is therefore important to explicitly state the arguments and assumptions in the evaluation process of sufficient similarity or in the application of component-based methods, including an indication of their likelihood. Application of an extra safety factor may be considered if the unknown or partly known composition is expected to result in high uncertainty (see Section 5.4.4.1).

Another typical source of uncertainty in mixture assessment is the potential interaction between substances. Interactions may occur in the environment (e.g., precipitation after emission in water), during absorption, transportation, and transformation in the organism, or at the site of toxic action. Interactions can be either direct, for example, a chemical reaction between 2 or more mixture components, or indirect, for example, if 1 mixture component blocks an enzyme that metabolizes another mixture component (see Chapters 1 and 2). Direct interactions between mixture components are relatively easy to predict based on physical–chemical data, but prediction of indirect interactions is much more difficult because it requires detailed information about the processes involved in the toxic mechanisms of action. One of the main challenges in mixture risk assessment is the development of a method to predict mixture interactions. A first step toward such a method could be the setup of a database, which contains the results of mixture toxicity tests. Provided such a database would contain sufficient data, it could be used to predict the likelihood and magnitude of potential interaction effects, that is, deviations for CA and RA. This information could subsequently be used to decide whether application of an extra safety factor for potential interaction effects is warranted, and to determine the size of such a factor. The mixture toxicity database could also support the search for predictive parameters of interaction effects, for example, determine which modes of action are involved in typical interactions.

Application of the concepts of CA and RA also involves uncertainty. Both concepts are based on assumptions of which the validity should be explored for the mixture of concern and the endpoints of concern. CA assumes that substances have a similar toxic MOA, that is, act on the same receptors in the target organs. In practice, this assumption is often relaxed to acting on the same target tissue or organ, which introduces uncertainty in the assessment. RA assumes that the toxic MOA of the substances involved is independent. The arguments used to choose a particular model, or combination of models, should always be explicitly stated to allow an assessment of their validity. However, in practice, both models often produce similar results when the slopes of the response curves are not very steep or shallow, implying that the choice between both models is frequently and quantitatively not a very important source of uncertainty. On the other hand, discussions on this issue continue, since it is frequently argued that a conceptually wrong model should not be applied despite being numerically rather accurate.

Finally, CA and RA both apply algorithms that combine the results of single substance evaluations to produce an estimate of the mixture risk. The uncertainty in the estimate is a combination of the uncertainties in the individual components. Calculation rules for evaluating the overall uncertainty based on the uncertainties in the individual components are provided in Appendix 1.

5.5.6   PERCEPTION OF MIXTURE RISKS

Risk perception is an important ingredient of risk management. Even if actual risks are relatively low, people may perceive something as a serious threat and risk managers may decide to take action in order to reassure people. Risk perception thus is a social and political reality, which must be accounted for in the management process. It is therefore important to know how people perceive mixture risks and which factors influence mixture risk perception.

It is remarkable how little research has been done on the perception of mixture risks. This may be explained by the fact that in real life people are generally exposed to mixtures and not to single substances. Perception studies that deal with environmental pollution implicitly include perception of mixture risks. Where scientists tend to consider mixtures as an extra complicating factor in the risk assessment process, laypersons consider mixtures a fact of life. They find it difficult to understand why scientists study effects of single substances, while in real life they are exposed to mixtures; they fail to understand the complexity of research on chemical mixtures.

One of the few studies that dealt with the perception of mixture risks was a survey of 1500 adults in Baden-Württemberg, Germany (Zwick and Renn 1998; Renn and Benighaus 2006). The participants were confronted with the following statement on mixtures:

> Just as combined consumption of tablets and alcohol can cause serious health problems, relatively harmless substances in the environment can also cause serious damage to one's health when they interact.

Two-thirds of the respondents agreed with this statement, 23.8% were indifferent, and only 9.6% disagreed. Although these results may not be fully representative because the question is framed in a tendentious way, they suggest that people are concerned about potential super-additive effects of substances. This is remarkable since the probability of super-additive effects is generally considered low by experts (COT 2002; Hertzberg and MacDonell 2002). The elevated concern under laypersons may be explained by the gap that exists between the reality of mixture exposure and being provided with incomplete risk information, often based on single substances. This gap indicates uncertainty, and it is well known that uncertainty generally results in a more negative assessment of risks (Siegrist and Cvetkovich 2001). It may also explain why people tend to show a preference for conservative management options if environmental pollution is involved (Renn and Benighaus 2006).

One of the main challenges for risk assessors and managers is to make people aware that our knowledge on chemicals is incomplete and uncertainty is inherent to the risk assessment process, especially if mixtures are involved. This is a difficult task since the notion of uncertainty is in conflict with the traditional image of the natural sciences as a provider of unambiguous answers. Not only the general public suffers from this view, but also administrators and policy makers, who often consider uncertainty a nuisance in their search for clear answers and effective policy actions. Here lies an important task for risk assessors and communicators, that is, the development of adequate tools to identify, describe, quantify, and communicate uncertainties in such a way that they can be understood and handled by policy makers and the general public. A promising way to realize this is by involving stakeholders in the risk assessment and management process (US PCCRARM 1997). An important advantage of stakeholder involvement is that potential sources of conflict can be identified in an early stage, leaving sufficient room to anticipate and readjust the assessment process. Stakeholders can contribute not only to the identification of suitable risk reduction measures, but also to the formulation of an adequate problem definition, the selection of suitable research approaches, and the identification of important sources of uncertainty. Especially in location-specific assessments, they can bring to the table a wealth of practical knowledge and expertise. Through their involvement the stakeholders also gain a better understanding and appreciation of the complexity of mixture issues and the limitations of scientific research.

5.6 DISCUSSION, CONCLUSIONS, AND RECOMMENDATIONS

5.6.1 Discussion

The review presented in the previous sections shows an enormous diversity in risk assessment methods and procedures for chemical mixtures. This diversity is characteristic for the current state of the art. The awareness that mixtures may cause risks that are not fully covered by single compound evaluations is growing, but the knowledge required to accurately assess the risks of chemical mixtures is still limited. The scientific community is attempting to unravel the mechanisms involved in mixture exposure and toxicity, and over recent decades, a multitude of new techniques to assess mixture risks have been developed. However, a comprehensive and solid conceptual framework to evaluate the risks of chemical mixtures is still lacking. The framework outlined in Section 5.4 can be considered a first step toward such a conceptual framework. The framework recognizes that the problem definitions vary greatly (between protective and retrospective assessments, for humans and ecosystems), and that each question has resulted in a different type of approach.

Within a regulatory context, mixture risks can only be adequately addressed if there are laws and guidelines that account for mixtures. The review of mixture regulations (Sections 5.3.1.2 and 5.3.2.2) shows that most laws on chemical pollution do account for mixtures, but explicit regulatory guidelines to address mixture risks are scarce. Moreover, the available laws do not cover all potential exposure situations in which mixtures may cause problems. Particularly, sequential exposures to different compounds and exposures through multiple pathways are generally poorly covered, for example, simultaneous or sequential exposure through food, drinking water, indoor air, and consumer products. An important reason for this lack of regulation is the enormous diversity of potential multiple and sequential exposure situations in real life. It is a challenge to identify those real-life exposure situations that require management priority because of potential high risk. There is a clear need for instruments that can support the identification of such high-risk exposure situations, for example, based on an analysis of food consumption and behavioral patterns, and the occurrence of common mixture combinations.

Acknowledgment that mixtures may cause risks that are not fully covered by single compound evaluations does not automatically imply that mixture assessments should be performed for all potential mixture exposures. This depends on the problem at hand, the specific exposure situation, and the available information. If a regulator must make a decision about the remediation of 1 contaminated soil plot, and it is already known that one of the mixture components exceeds the remediation threshold, and this fact results in a need for remedial action itself, then a mixture assessment is redundant. Mixture assessment for contaminated soils (and other compartments and exposure routes) is useful if the known individual components do not exceed their respective thresholds, but if it is suspected that the overall mixture may still cause unacceptable adverse effects, and in case risk managers have a limited budget for a large number of contaminated sites, that is, when they have to prioritize the most hazardous sites to be remediated first, while other sites should possibly be subject to simple risk reduction measures. The situation is different for mixture emissions. Here, potential mixture effects should always be addressed because the aim was and is to establish an emission level that is lower than a certain (acceptable) effect level.

Compared to per-compound risk assessment, mixture risk assessment is based on relatively little data, relatively weaker scientific underpinning, and relatively few and simple null models. In contrast to these "simpler as compared to" characteristics is the presence of all kinds of mixtures in the field, with exposure levels ranging from very low to very high, and pertaining to an innumerable set of possible mixture compositions. Therefore, it is hard to provide proof that mixture risk assessments are scientifically fully justifiable for a case, and it is hard to obtain the relevant data for the most justifiable approach. Despite this, mixture assessments are frequently requested, and—when it is adopted as principle that risk assessors should provide the best possible answer to a problem—the question is whether the approaches that are proposed are "better" than neglecting mixtures, and handling each situation compound by compound. Our firm belief, founded in both review of the field and practical experience, is that executing mixture risk assessments is feasible, at least partially, and often indeed provides better answers than those generated by applying a per-compound approach. This particularly holds for relatively well-characterized common mixtures such as PCBs, PAHs, and coke oven emissions. It also holds for the default models of the component-based approaches, that is, CA and RA, particularly if the MOA of the chemicals is known. But even if the MOA is unknown, there are mathematical resemblances between predictions generated by both mixture models, as long as the slopes of the separate concentration–effect curves are more or less similar (Drescher and Bödeker 1995). For problems where a good assessment is crucial, it is often even feasible to apply multiple approaches, and follow a multicriteria-like analysis, like in TRIAD (see Section 5.3.2.1). When absolute predictions of mixture risks might be out of reach (since those require proof that is not there), "relative answers" may be of high value to the risk manager. In many cases the relative answer is already sufficient for improving the risk management, for example, to determine the remediation priority of a set of contaminated soils.

Notwithstanding the useful application of current mixture assessment methods, it must be acknowledged that current methods are not always accurate and estimates of mixture risks can be highly uncertain. One area where current methods can be improved is the application of the concept of sufficient similarity. This concept is applied to assess whole mixture risks based on toxicity data from other mixtures, which are considered "sufficiently" similar. Basically, each assessment that involves the application of toxicity data from mixtures, which may differ from the mixture of concern, either in terms of compounds or in terms of concentration ratios, should be evaluated for sufficient similarity. However, explicit guidelines for sufficient similarity are currently lacking. Such guidelines should be based on a qualitative or quantitative evaluation of the mixture composition or the processes from which the mixture originates. Analytical techniques to characterize complex mixtures could support the development of such guidelines, for example, GC-MS and high-performance liquid chromatography (HPLC) coupled to mass spectrometry (HPLC-MS) or high-field nuclear magnetic resonance (HPLC-NMR; Levsen et al. 2003).

As noted above, the default models of CA and RA generally perform well as a tool for risk assessment. Based on a review of available literature data, Warne (2003) concludes that approximately 10% to 15%, 70% to 80%, and 10% to 15% of mixtures show antagonistic, additive, and synergistic toxicity, respectively. Analyses by Ross (1996) and Ross and Warne (1997) indicated that 5% of the mixtures had toxicity values that differed more than a factor of 2.5 from CA, and 1% of the mixtures had toxicity values that were different by more than a factor of 5. These empirical figures indicate that application of the default models of CA and RA generally results in risk estimates that are sufficiently accurate for most regulatory purposes, even if the underlying mechanisms might not be adequately represented by both models.

In view of the above-mentioned findings, the main challenge for mixture assessment is not so much to debate the average cases (for which the solution appears empirically right), but to identify and assess those mixtures that significantly deviate from the default model (see Chapter 3). In order to realize this, we need to better understand the mechanisms involved in mixture toxicity, particularly those mechanisms that may result in high risk. McCarty and Borgert (2006) advocate the development of a comprehensive, generally accepted mechanism or mode of toxic action classification system to improve mixture assessment. Although such a system could be highly beneficial to distinguish between the default models of CA and RA, it remains to be seen whether such a system would significantly improve mixture assessment. Most current classification schemes (Amdur et al. 1991; Verhaar et al. 1992; McCarty and Mackay 1993; Klaassen 1996) are based on receptor-toxicant interactions or target organ. However, there are many more processes involved in mixture toxicity, which these simple classification schemes fail to capture, for example, absorption, transport, metabolism, and excretion. A more promising approach is the development of process-based models, which aim to describe the chain of processes leading to toxic effects, for example, PBPK and BRN models. However, the results of these models are highly specific for particular mixtures, and extrapolation to other mixtures is not always possible. What is needed is a system that is generally applicable and sufficiently detailed to account for mixture interactions at different process levels, such as absorption, metabolism, target site, physiological processes, and effects (Jager et al. 2006, 2007). Classification of mixtures based on their potential interaction with these basic biological processes could significantly reduce the huge number of different chemicals and species to a limited set of potential interactions. This requires a thorough understanding of the physiological processes in living organisms. As Yang et al. (2004, p 65) stated: "Once we know the normal biological processes, all external stressors are merely perturbations of these processes."

When looking at test data reviews, it appears that most tests have been performed with few frequently studied compounds or compound groups, and that extrapolation to other groups of compounds cannot be avoided. Also, when looking at exposure levels, it is the acute or chronic exposure levels that induce visible effects that are represented most—and not the concentration levels that may more frequently occur in the environment (that is, ambient levels just above some quality standard). Finally, when looking at the data, it appears that binary mixtures are overrepresented and other mixtures underrepresented. The situation could, in the longer term, be improved by testing various crucial mixture working hypotheses that are defined beforehand, and that (as a set) would result in a set of experiments that critically look into the role of the different interaction levels (ambient, uptake, intoxication) and the role of the different pharmacological models, and in improved "rules of thumb" for practical assessments. Toxicity testing of mixtures should move beyond the standard tests for deviations from the default models of CA and RA, toward a more mechanistic understanding of the process involved in mixture toxicity. Within this context, toxicogenomic studies with mixtures could contribute significantly to the identification of mechanisms of toxicity and potential interaction effects (Sterner et al. 2005). Mechanistic and toxicogenomic studies should focus not only on the processes and effects involved in concurrent exposure to multiple substances (i.e., cocktails), but also on those involved in sequential exposure to multiple substances.

Tiered risk assessment approaches have been designed to be scientifically "wrong"—because they yield conservative results in the lower tiers—but are practically a cost-effective "right" approach. This implies that validation of mixture risk assessment methods should show conservatism as to real field effects when considering lower-tier approaches. Regarding human risk assessment, the validation of methods by comparing predicted risks to observed impacts is problematic, because this would require detailed and successful epidemiological analyses. Validation of exposures is possible, but often not different from the validation of per-compound exposure models. Regarding ecological risk assessment, validation of predicted (relative) impacts to observed impacts is technically feasible, and within reach. By mimicking ecosystem structure, mixture tests can be executed, as, for example, in microcosms or mesocosms (Cuppen et al. 2002). In these cases, the experimental design and statistical analyses can be similar, as in single species mixture tests. In the field, mixture issues can be studied profitably on pollution gradients around point sources. Further mixture issues can be studied by investigating large monitoring databases, like done by De Zwart et al. (2006) and Kapo and Burton (2006). In these studies, the predicted mixture toxic pressure (msPAF) for a large set of monitoring sites correlated significantly with the local impacts (species loss) on fish communities resulting from mixture exposure levels.

Comparisons between human and ecological risk assessments have elucidated many similarities and dissimilarities. Conceptually similar approaches are present for both subdisciplines, but some approaches are more common in one subdiscipline than the other. For example, toxicity tests for uncharacterized whole mixtures, such as WET tests, are more common in ecological risk assessment. In contrast, refined mechanistic approaches for mixture assessment are further developed in the context of human risk assessment, like BRN and PBPK models, in silico approaches, in vitro techniques, and the inclusion of binary interaction data in mixture assessments. But there is also criticism on the application of these techniques. For example, McCarty and Borgert (2006) argue that not all interaction data are suitable for mixture evaluation, for example, because they have been gathered at dose or concentration levels or for species that have little relevance for human and/or environmental risks. Although this also applies to many single substance data, it must be acknowledged that the use of this type of uncertain data can result in over- or underestimation of mixture risk, which could lead to risk management decisions that do not protect public health and/or the environment, or to remedial action where the risk in fact is more limited than calculated. It is clear that further refinement of mixture assessment approaches is a scientific target in both human and ecological mixture toxicology. But it is also clear that practical risk assessments should deliver useful answers, before those systems have been developed further. The challenge for human and ecological risk assessors is to develop those techniques that are scientifically sound and useful in practice.

5.6.2 CONCLUSIONS

Based on the review and discussion of human and ecological assessment techniques for chemical mixtures, several conclusions can be drawn. These conclusions are listed below and related to the aims of this chapter, outlined in Section 5.1. The recommendations to improve human and ecological risk assessment of mixtures (Aim 5) are provided in the next section.

Aim 1: Typical features of mixture assessment
- There is a clear need for mixture risk assessment, since most environments are characterized by mixture exposure situations. This means that risk assessments should pay specific attention to all aspects of mixture exposures and effects in order to make accurate risk predictions.
- The establishment of a safe dose or concentration level for mixtures is useful only for common mixtures with more or less constant concentration ratios between the mixture components and for mixtures of which the effect is strongly associated with one of the components. For mixtures of unknown or unique composition, determination of a safe concentration level (or a dose–response relationship) is inefficient, as the effect data cannot be reused to assess the risks of other mixtures. One alternative is to test the toxicity of the mixture of concern in the laboratory or the field to determine the adverse effects and subsequently determine the acceptability of these effects. Another option is to analyze the mixture composition and apply an algorithm that relates the concentrations of the individual mixture components to a mixture

risk or effect level, which can subsequently be evaluated in terms of acceptability.

Aim 2: State of art and comparison of human and ecological approaches

* There are many concepts in use for the assessment of risks or impacts of chemical mixtures, for both human and ecological risk assessment. Many of these concepts are identical or similar, for example, whole mixture tests, (partial) mixture characterization, mixture fractionation, and the concepts of CA and RA.
* Regulatory application and implementation of toxicity tests for uncharacterized whole mixtures is typical for the field of ecological risk assessment.
* The human field is leading in the development and application of process-based mixture models, such as PBPK and BRN models, and BINWOE methods.
* Most national laws on chemical pollution do account for mixture effects, but explicit regulatory guidelines to address mixtures are scarce. Only the United States has fairly detailed guidelines for assessing mixture risks for humans.
* The multitude of different mixture assessment techniques is typical for the current state of the art in mixture assessment. There is a clear need for a comprehensive and solid conceptual framework to evaluate the risks of chemical mixtures.

Aim 3: Toward a conceptual framework for mixture risk assessment

* The framework outlined in Section 5.4 can be considered a first step toward a conceptual framework for integrated assessment of mixture risks. The framework is proposed as a possible line of thinking, not as a final solution.
* Distinction is made between approaches for assessment of whole mixtures and component-based approaches. The most accurate assessment results are obtained by using toxicity data on the mixture of concern. If these are not available, alternatives can be used, such as the concept of sufficient similarity, (partial) characterization of mixtures, and component-based methods. Which method is most suitable depends on the situation at hand. A single mixture assessment method that always provides accurate risk estimates is not available.
* Tiering is proposed as an instrument to balance the accuracy of mixture assessments with the costs. When lower tiers do not provide sufficiently accurate answers for the problem at hand, the option exists to go to a higher tier, for example, by more detailed characterization of the mixture or application of more sophisticated mixture models.

Aim 4: Current issues in mixture risk assessment

* In both human and ecological risk assessment, there is considerable scientific latitude to develop novel methods (e.g., those that exist in only one of the subdisciplines could be useful in the other one) and to refine approaches (e.g., by considering complex reaction networks and more specific attention for modes of action). The refinements are needed to

improve the scientific evidence that is available for underpinning risk assessments.

- Several key issues in risk assessment of chemical mixtures were identified, that is, exposure assessment of mixtures (e.g., mixture fate and sequential exposure), the concept of sufficient similarity, mixture interactions, QSARs, uncertainty assessment, and the perception of mixture risks. Resolving these key issues will significantly improve risk assessment of chemical mixtures (see next section).

5.6.3 RECOMMENDATIONS

To further improve human and ecological risk assessment of mixtures, the following recommendations are suggested:

- Toxicity testing of mixtures should move beyond the standard tests for deviations from the default models of CA and RA, toward a more mechanistic understanding of the process involved in mixture toxicity. These studies should focus not only on the processes and effects involved in concurrent exposure to multiple substances (i.e., cocktails), but also on those involved in sequential exposure to multiple substances.
- Tools should be developed to support the identification of mixture exposure situations that may cause unexpectedly high risks compared to the standard null models of CA and RA, for example, based on an analysis of food consumption and behavioral patterns, and the occurrence of common mixture combinations that cause synergistic effects.
- The general framework proposed for organizing problem definitions and associated mixture assessment approaches should be critically tested and improved.
- Mixture assessment methods from human and ecological problem definition contexts should be further compared for improving the methods.
- Criteria should be developed further to support appropriate application of the concept of "sufficient similarity" as formalized in practical mixture assessments.
- Criteria should be developed for the inclusion of interaction data in mixture assessments.
- National authorities should develop legislation that enables the assessment and management of potential high-risk situations caused by sequential exposures to different compounds and exposures through multiple pathways, with specific emphasis on a systems approach rather than on approaches focusing solely on chemicals, or on water or soil as compartments.

Appendix 1: Uncertainty of Concentration and Response Addition

Concentration (or dose) addition (CA) and response addition (RA) both apply algorithms that combine the results of single substance evaluations to produce an estimate of the mixture risk. The uncertainty in the estimate is a combination of the uncertainties in the individual components. The hazard index (HI), which is a specific case of CA, adds the ratios between the exposure and reference values of the individual substances:

$$HI = \sum_{i=1}^{n} \frac{E_i}{RfD_i}$$

Different safety factors may have been used in the derivation of the reference values of the individual substances (RfD_i). A deterministic HI thus sums risk ratios that may reflect different percentile values of a risk probability distribution. Assessment and interpretation of the uncertainty in the HI may be severely hampered by this summation of dissimilar distribution parameters. In a probabilistic risk assessment, the uncertainty in the exposure and reference values is often characterized by lognormal distributions. The ratio of 2 lognormal distributions also is a lognormal distribution. The variance in a quotient of 2 random variables can be approximated as follows (Mood et al. 1974, p 181):

$$\text{var}\left[\frac{E}{RfD}\right] \approx \left(\frac{\mu_E}{\mu_{RfD}}\right)^2 \left(\frac{\text{var}[E]}{\mu_E^2} + \frac{\text{var}[RfD]}{\mu_{RfD}^2} - \frac{2\,\text{cov}[E,RfD]}{\mu_E \mu_{RfD}}\right) \quad \text{(A1.1)}$$

which, under the assumption of no significant covariation between exposure and reference values, can be simplified to

$$\text{var}\left[\frac{E}{RfD}\right] \approx \left(\frac{\mu_E}{\mu_{RfD}}\right)^2 \cdot \left(\frac{\text{var}[E]}{\mu_E^2} + \frac{\text{var}[RfD]}{\mu_{RfD}^2}\right) = \left(\frac{\mu_E}{\mu_{RfD}}\right)^2 \cdot \left(CV_E^2 + CV_{Rfd}^2\right) \quad \text{(A1.2)}$$

where CV stands for the coefficient of variation. The variance in the HI, the summation of the individual risk ratios, can subsequently be calculated by (Mood et al. 1974, p 178):

$$\text{var}\left(\sum_i^n \left[\frac{E_i}{RfD_i} \right] \right) = \sum_i^n \text{var}\left[\frac{E_i}{RfD_i} \right] + 2\sum_i \sum_j \text{cov}\left[\frac{E_i}{RfD_i}, \frac{E_j}{RfD_j} \right]$$

(A1.3)

which can also be written as (substitution of Equation A1.2 in Equation A1.3):

$$\text{var}\left(\sum_i^n \left[\frac{E_i}{RfD_i} \right] \right) = \sum_i^n \left[\left(\frac{\mu_E}{\mu_{RfD}} \right)^2 \left(CV_E^2 + CV_{Rfd}^2 \right) \right] + 2\sum_i \sum_j \text{cov}\left[\frac{E_i}{RfD_i}, \frac{E_j}{RfD_j} \right]$$

(A1.4)

Closer inspection of Equation A1.4 shows that substances with a high expected risk ratio (μ_E/μ_{RfD}) contribute most to the uncertainty (or variance) in the HI. If 1 or 2 components dominate the mixture, it seems sufficient to base the uncertainty assessment on these dominant components. However, mixtures are often dominated by more than 2 components. Furthermore, the covariance between the individual risk ratios should not be ignored, since exposure estimates (E_i) of individual mixture components can be (positively) correlated, as well as their reference values (RfD_i). The uncertainty in the HI may be severely underestimated if these correlations are not accounted for, which is evident from the last part of Equation A1.4. The central limit theorem states that the final HI will approach a normal distribution when the number of substances in the mixture becomes large or if no single risk ratio dominates the sum (De Groot 1986).

RA makes use of the following equation:

$$R_{\text{mix}} = 1 - \prod_i^n (1 - R_i)$$

(A1.5)

where R_i represents the probability of a response of an individual, the percent responding in a population or the percentage of affected species in an ecosystem, as a consequence of exposure to substance i. The response values represent probabilities or fractions and can take values between 0 and 1. Rewriting Equation A1.5 for a mixture of 3 components results in

$$R_{\text{mix}} = R_1 + R_2 + R_3 - R_1 R_2 - R_1 R_3 - R_2 R_3 + R_1 R_2 R_3$$

(A1.6)

which shows that the overall risk consists of the summation of the individual responses minus a correction factor ($R_1 R_2 + R_1 R_3 + R_2 R_3 - R_1 R_2 R_3$) that accounts for the fact that individuals or species that respond to more than 1 substance should not be counted double or triple in the response calculations. If response values are low, for example, for cancer risk estimates, the response products in Equation A1.6 can be ignored:

$$R_{mix} \approx R_1 + R_2 + R_3 \tag{A1.7}$$

In ecological mixture assessments, the response values are often above 0.1, and in this situation the response products have a significant impact on the overall risk.

If RA is applied in a deterministic assessment, the uncertainty of the end result is determined by the likelihood of the individual response values. Summation of conservative values, such as the upper bounds of cancer estimates, results in even more conservative estimates of the mixture risk. However, this source of uncertainty is generally considered small compared to other sources of uncertainty in carcinogenic risk assessment (Kodell and Chen 1994; Cogliano 1997). Several procedures have been proposed to reduce the conservativeness in deterministic risk estimates of carcinogenic mixtures, for example, dividing the sum by a factor of 2 (Cogliano 1997) or using the upper bound of the most uncertain mixture component (Putzrath 2000). A more appropriate way to prevent overly conservative risk estimates is to perform a probabilistic assessment. At low response levels, the uncertainty in the mixture risk is given by a simplified version of Equation A1.3:

$$\mathrm{var}\left(\sum_i^n R_i \right) = \sum_i^n \mathrm{var}\left[R_i \right] + 2 \sum_i \sum_j \mathrm{cov}\left[R_i, R_j \right] \tag{A1.8}$$

Because it is unlikely that the risk estimates of substances with an independent mode of action are correlated, the covariance can be ignored and the overall variance is simply calculated by adding the variances of the individual components. For a many-compound mixture, the central limit theorem states that the uncertainty of the mixture risk is approximately normally distributed. At high response levels, the uncertainty in the mixture risk depends not only on the variance in the individual responses, but also on the variances of the response products. Calculation procedures are available but are rather complex (Mood et al. 1974).

Glossary

AChE: Acetylcholinesterase.

Acute: Occurring within a short period in relation to the life span of the organism (usually 4 days for fish). It can be used to define either the exposure (acute test) or the response to an exposure (acute effect).

Acute toxicity: The harmful effects of a chemical or mixture of chemicals occurring after a brief exposure to relatively high concentrations. *See also* chronic toxicity.

ADI: Acceptable daily intake. The amount of a substance that could be consumed daily for an entire life span without appreciable risk. *See also* TDI.

ADME: Accumulation, distribution, metabolism, and excretion.

AMAP: Arctic Monitoring and Assessment Program.

AMD: Acid mine drainage.

Analysis of effects: A phase in an ecological risk assessment in which the relationship between exposure to contaminants and effects on endpoint entities and properties is estimated along with associated uncertainties.

Analysis of exposure: A phase in an ecological risk assessment in which the spatial and temporal distributions of the intensity of the contact of endpoint entities with contaminants are estimated along with associated uncertainties.

ANOVA: Analysis of variance.

Antagonism: A deviation from a chosen mixture toxicity reference model (e.g., CA or IA), suggesting a mixture is less toxic than expected from the toxicities of the individual chemicals.

Assessment endpoint: An explicit expression of the environmental value to be protected. An assessment endpoint must include an entity and specific property of that entity.

ATSDR: Agency for Toxic Substances and Disease Registry (USA).

Background concentration: The concentration of a substance in environmental media that are not contaminated by the sources being assessed or any other local sources. Background concentrations are due to natural occurrence or regional contamination.

Battery toxicity testing: The parallel application of a range of different toxicity tests.

BBI: Belgian Biotic Index.

BDF: Bioassay-directed fractionation. Fractionation of mixture samples followed by toxicity testing of the fractions in a bioassay. Often performed to identify the most toxic fraction for further chemical characterization.

BINWOE: Binary weight of evidence. A procedure to include qualitative judgment of pairwise interaction data in mixture assessments.

Bioaccessibility: Indication of the fraction of the total amount of a chemical present in ingested food, water, or soil and sediment particles that at maximum can be released during digestion (adopted from Peijnenburg and Jager 2003).

Bioaccumulation: The net accumulation of a substance by an organism due to uptake from all environmental media.

Bioassay: Commonly used as synonymous with toxicity test; its use may, however, also be restricted to the application of a toxicity test to an environmental sample (diagnosis).

Bioavailability: Indication of the fraction of the total amount of a chemical present in a specific environmental compartment that, within a given time span, is or can be made available for uptake by organisms (adopted from Peijnenburg and Jager 2003). Also defined as the extent to which the form of a chemical is susceptible to being taken up by an organism. A chemical is said to be bioavailable if it is in a form that is readily taken up (e.g., dissolved) rather than a less available form (e.g., adsorbed to solids or to dissolved organic matter).

Bioconcentration: The net accumulation of a substance by an organism due to uptake from an aqueous solution.

Biosensors: Analytical devices that incorporate biological material or a material mimicking biological membranes associated with or integrated within a physical–chemical transducer or transducing system. Biosensors are based on biochemical biomarker responses at the suborganism level.

BLM: Biotic ligand model.

BMWP-ASPT: Biological monitoring working party—average score per taxon.

BRN: Biochemical reaction network. A system of biochemical reactions that interact with each other.

CA: Concentration addition. Concept used to calculate the risk of chemical mixtures based on the assumption that the mixture components act through a simple similar mode of action (SSA). *See also* DA (dose addition).

CBR: Case-based reasoning. The process of solving new problems based on the solutions of similar past problems.

CBR approach: Critical body residue approach.

CDC: Centers for Disease Control and Prevention (USA).

CERCLA: Comprehensive Environmental Response, Compensation, and Liability Act (USA), commonly known as Superfund.

Chronic: Occurring after an extended time relative to the life span of an organism (conventionally taken to include at least 1/10 of the life span). Long-term effects are related to changes in metabolism, growth, reproduction, or the ability to survive.

Chronic toxicity: The harmful effects of a chemical or mixture of chemicals occurring after an extended exposure to relatively low concentrations. *See also* acute toxicity.

CLEA: Contaminated Land Exposure Assessment model.

Community: The biotic community consists of all plants, animals, and microbes occupying the same area at the same time. However, the term is commonly used to refer to a subset of the community, such as the fish community or the benthic macroinvertebrate community.

Conceptual model: A representation of the hypothesized causal relationship between the source of contamination and the response of the endpoint entities.

Contaminant: A substance that is present in the environment due to release from an anthropogenic source and is believed to be potentially harmful.

CV: Coefficient of variation. The ratio between the standard deviation and the mean.

DA: Dose addition. Concept used to calculate the risk of chemical mixtures based on the assumption that the mixture components act through a simple similar mode of action. *See also* CA (concentration addition).

DAM: Damage assessment model.

DBP: Disinfection by-product.

DBTK: Data-based toxicokinetics.

DCE: 1,1-Dichloroethylene.

DDT: Dichlorodiphenyl trichloroethane.

DEB: Dynamic energy budget.

DEN: Diethylnitrosamine.

DGT: Diffusive gradients in thin films.

Direct effect: An effect resulting from an agent acting on the assessment endpoint or other ecological component of interest itself, not through effects on other components of the ecosystem. *See also* indirect effect.

D_{lipw}: Liposome-water distribution coefficient.

DOC: Dissolved organic carbon.

DOM: Dissolved organic matter.

DTA: Direct toxicity assessment. Biological effect test for whole mixture samples.

DTPA: Diethylene triamine penta acetate.

EC: European Commission.

ECB: European Chemicals Bureau.

EC/Dx: Concentration (dose) that affects designated effect criterion (e.g., a behavioral trait) by x% compared to the control. The EC/D50 is known as the median effective concentration (dose). The EC/D values and their 95% confidence limits are usually derived by statistical analysis of effects in several test concentrations, after a fixed period of exposure. The duration of exposure must be specified (e.g., 96-hour EC50). *See also* LC/Dx.

EC50: Median effective concentration, the concentration causing 50% reduction of a certain response parameter (e.g., reproduction, growth).

Ecosystem: A collection of populations (microorganisms, plants, and animals) that occur in the same place at the same time and that can therefore potentially interact with each other as well as their physical and chemical environment, and thus form a functional entity.

Ecosystem function: A biological, chemical, or biochemical process taking place in an ecosystem.

Ecosystem structure: The composition of the biological community in an ecosystem and the interrelationships between the individual populations of species (e.g., food web structure).

Ecotoxicity: The property of a compound to produce adverse effects in an ecosystem or one of its components.

Ecotoxicology: The study of toxic effects of chemical and physical agents in living organisms, especially on populations and communities within defined ecosystems; it includes transfer pathways of these agents and their interaction with the environment.

EDA: Effect-directed analysis.

EDTA: Ethylene diamine tetra acetic acid.

Effect criterion: The type of effect observed in a toxicity test (e.g., immobility).

EM: Extensive metabolizer.

EMEP: European Monitoring and Evaluation Program.

Endpoint entity: An organism, population, species, community, or ecosystem that has been chosen for protection. The endpoint entity is 1 component of the definition of an assessment endpoint.

EPER: European Pollutant Emission Register.

EPT: Ephemeroptera–Plecoptera–Trichoptera taxa of a whole macroinvertebrate sample.

EQC: Environmental quality criterion. The concentration of a potentially toxic substance, which can be allowed in an environmental medium over a defined period. The term is used in this book generally, for which also EQO (objective) and EQS (standard) are used in different contexts.

EQO: Environmental quality objective. *See* EQC.

EQS: Environmental quality standard. *See* EQC.

ERA: Ecological risk assessment. A process that evaluates the likelihood that adverse ecological effects may occur or are occurring as a result of exposure to one or more agents.

ESD: Emission Scenario Document.

EU: European Union.

EXAFS: X-ray absorption fine structure methodology for the analysis of metals in solid materials.

EXAMS: Exposure Analysis Modeling System.

Exposure: The contact or co-occurrence of a contaminant with a receptor organism, population, or community.

Exposure assessment: The component of an ecological risk assessment that estimates the exposure resulting from a release or occurrence in a medium of a chemical, physical, or biological agent. It includes estimation of transport, fate, and uptake.

Exposure pathway: The physical route by which a contaminant moves from a source to a biological receptor. A pathway may involve exchange among multiple media and may include transformation of the contaminant.

Exposure route: The means by which a contaminant enters an organism (e.g., inhalation, stomatal uptake, ingestion).

Extrapolation: 1) An estimation of a numerical value of an empirical (measured) function at a point outside the range of data which were used to calibrate the function, or 2) the use of data derived from observations to estimate values for unobserved entities or conditions.

FA: Factor analysis.

FFG: Functional feeding group.

FQPA: Food Quality Protection Act (USA).

Frequency distribution: The organization of data to show how often certain values or ranges of values occur.

Fugacity: The tendency of a chemical to escape a certain compartment.

GC-MS: Gas chromatography–mass spectrometry.

GES: Good ecological status. Concept used in the EU Water Framework Directive to refer to water bodies that have biological and chemical characteristics expected under sustainable conditions.

GIS: Geographic Information System. System that allows for the interrelation of quality data (as well as other information) from a diversity of sources based on multilayered geographical information processing techniques.

GSH: Glutathione.

GST: Glutathione-S-transferase.

GST-P: Glutathione-S-transferase placental form.

Half-life: Time it takes to degrade, transform, or eliminate a chemical to 50% of its initial concentration.

Hazard: The set of inherent properties of a chemical substance or mixture, which make it capable of causing adverse effects in man or the environment when a particular degree of exposure occurs. *See also* risk.

Hazard assessment: Comparison of the intrinsic ability to cause harm with expected environmental concentration. In Europe it is typically a comparison of PEC with PNEC. It is sometimes loosely referred to as risk assessment.

Hazard index: Summation of the hazard quotients for all chemicals to which an individual is exposed. A hazard index value of ≤1.0 indicates that no adverse human health effects (no cancer) are expected to occur (from http://web.ead.anl.gov/uranium/glossacro; last accessed July 2010).

Hazard quotient: Ratio of the estimated chemical intake (dose) to a reference dose level below which adverse health effects are unlikely. The value is used to evaluate the potential for noncancer health effects, such as organ damage, from chemical exposures (from http://web.ead.anl.gov/uranium/glossacro; last accessed July 2010).

HazDat: Hazardous Substance Release/Health Effects Database (ASTDR).

$HC_{Percentage}^{Endpoint}$: General notation to identify that an environmental quality criterion is a percentile derived from an SSD, explicitly stating the endpoint and the chosen percentage p as cutoff value, for example, to identify in general notation what is known as the HC5 based on NOEC toxicity data. Optionally, one could add a superscript prefix to show the number of toxicity data from which the model was derived, like for an HC_p based on 4 data points.

HC_p (HC5): Hazardous concentration for $p\%$ (5%) of the species.

HI: Hazard index. A summation of the hazard quotients for all chemicals to which an individual is exposed.

HQ: Hazard quotient. The ratio between the estimated chemical intake of a substance (dose or exposure concentration) and a reference value below which adverse health effects are unlikely.

HRA: Human risk assessment. A process that evaluates the likelihood that adverse human health effects may occur or are occurring as a result of exposure to one or more agents.

Hydrocarbon block method: Method for environmental risk assessment of petroleum substances based on calculating and combining PEC/PNEC ratios for "blocks" of constituent hydrocarbons.

IA: Independent action. Concept used to calculate the risk of chemical mixtures based on the assumption that the mixture components act through independent joint action (IJA), also named RA (response addition).

IC: Industrial category.

IJA: Independent joint action. The joint action of chemicals in case they act independently, that is, if their mode of action is dissimilar (the primary sites of action of the chemicals are not the same) and noninteractive (the presence of one substance does not influence another).

Indirect effect: An effect resulting from the action of an agent on components of the ecosystem, which in turn affect the assessment endpoint or other ecological components of interest. Indirect effects of chemical contaminants include reduced abundance due to toxic effects on food species or on plants that provide habitat structure. *See also* direct effect.

Interaction: Chemicals influence each other by physical, chemical, or biological means before or after reaching the molecular site of toxic action. Toxicological interactions are responses that deviate from those expected under some definition of additivity (e.g., following the concepts of IA or CA). The most commonly used terms for interaction are "synergism" and "antagonism."

IRA: Integrated risk assessment. A science-based approach that combines the process of risk estimation for humans, biota, and natural resources in 1 assessment.

IRIS: Integrated Risk Information System. A database of human health effects that may result from exposure to various substances found in the environment, maintained by the USEPA.

Joint action: Two or more chemicals exerting their effects simultaneously.

K_{aw}: Air–water partition coefficient.

K_d: Partition coefficient, describing the distribution of chemicals over the soil or sediment solid phase and the corresponding pore water.

K_{oa}: Octanol–air partition coefficient.

K_{ow}: Octanol–water partition coefficient.

LC/Dx: The concentration or dose of a substance in water that is estimated to be lethal to x% of the test organisms. The LC50 is known as the median lethal concentration. The LC values and their 95% confidence limits are usually derived by statistical analysis of mortalities in several test concentrations, after a fixed period of exposure. The duration of exposure must be specified (e.g., 96-hour LC50).

LC50: Lethal concentration 50%, the concentration killing 50% of the test organisms.

LOAEC/L: Lowest-observed-adverse-effect concentration/level. The lowest level of exposure to a chemical in a test that causes statistically significant differences from the controls in a measured negative response.

LOE: Line of evidence. A set of data and associated analysis that can be used, alone or in combination with other lines of evidence, to estimate risks. Each line of evidence is qualitatively different from any others used in the risk characterization. In ecotoxicological assessments, the most commonly used lines of evidence are based on 1) biological surveys, 2) toxicity tests of contaminated media, and 3) toxicity tests of individual chemicals.

LOEC/L: Lowest-observed-effect concentration/level. The lowest concentration of an agent used in a toxicity test that has a statistically significant effect on the exposed population of test organisms, compared with the controls.

LRTP: The potential for long-range transport of organic chemicals.

MADEP: Massachusetts Department of Environmental Protection (USA).

MDS: Multidimensional scaling.

Measure of effect: A measurable ecological characteristic that is related to the valued characteristic chosen as the assessment endpoint (equivalent to the earlier term "measurement endpoint").

Measure of exposure: A measurable characteristic of a contaminant or other agent that is used to quantify exposure.

Mechanism of action: The molecular sequence of events from absorption of an effective dose to production of a specific biological response.

Mechanistic model: A mathematical model that simulates the component processes of a system rather than using simple empirical relationships.

Median lethal concentration: A statistically or graphically estimated concentration that is expected to be lethal to 50% of a group of organisms under specified conditions. *See* LC50.

MOA: Mode of action. A set of physiological and behavioral signs characterizing an adverse biological response.

Model uncertainty: The component of uncertainty concerning an estimated value that is due to possible misspecification of a model used for the estimation. It may be due to the choice of the form of the model, its parameters, or its bounds.

MPC: Maximum permissible concentration. Environmental quality standard used in the Netherlands.

msPAF: Multisubstance probably affected fraction. The fraction of species that is expected to be affected by the exposure to multiple substances.

NHANES: National Health and Nutrition Examination Survey (USA).

NOAEC/L: No-observed-adverse-effect concentration/level. The highest level of exposure to a chemical in a test that does not cause statistically significant differences from the controls in any measured negative response.

NOEC/L: No-observed-effect concentration or level. The highest concentration of a test substance to which organisms are exposed that does not cause any observed and statistically significant effects on the organism, compared with the controls. For example, the NOEC might be the highest tested concentration at which an observed variable such as growth did not differ significantly from growth in the control. The NOEC customarily refers to the most sensitive effect unless otherwise specified. NEL, NOAEL, NEC, and NOEC are used as equivalent terms.

OECD: Organisation for Economic Co-operation and Development.

OP: Organophosporous pesticide.

OSH Act: Occupational Safety and Health Act (USA).

PAF^{Endpoint} : General notation to identify that the probably affected fraction (PAF) of species is based on an SSD with a specific type of endpoint, for example, PAF^{NOEC} to identify a PAF based on SSD^{NOEC}. Optional: One could add a superscript prefix to show the number of toxicity data from which the model was derived, like ^{4}PAF^{Endpoint} for a PAF based on 4 data points.

PAH: Polycyclic aromatic hydrocarbon.

PBB: Polybrominated biphenyl.

PBDE: Polybrominated diphenyl ether.

PBPD: Physiologically based pharmacodynamics.

PBPK model: Physiologically based pharmacokinetic model. Physiologically based compartmental model used to characterize pharmacokinetic behavior of a chemical. Available data on blood flow rates, and metabolic and other processes, which the chemical undergoes within each compartment are used to construct a mass-balance framework for the PBPK model.

PBTD: Physiologically based toxicodynamics; synonymous to PBPD.

PBTK: Physiologically based toxicokinetics; synonymous to PBPK.

PCA: Principal component analysis.

PCB: Polychlorinated biphenyl.

PCDD: Polychlorinated dibenzodioxin.

PCDF: Polychlorinated dibenzofuran.

PD: Pharmacodynamics.

PEC: Predicted environmental concentration. The concentration in the environment of a chemical calculated from the available information on certain of its properties, its use and discharge patterns, and the associated quantities.

PLS: Partial least-squares projections to latent structures.

PM: Poor metabolizer.

PNEC: Predicted no-effect concentration or level. The maximum level (dose or concentration) that, on the basis of current knowledge, is likely to be tolerated by an organism or ecosystem without producing any adverse effect.

POP: Persistent organic pollutant.

Population: An aggregate of interbreeding individuals of a species, occupying a specific location in space and time.

Potentiation: The situation in which a chemical becomes more toxic when applied together with another chemical that by itself is nontoxic. Potentiation is a special case of synergism; in case of synergism, both chemicals in the mixture show a dose-related toxicity when applied individually.

P_{ov}: Overall persistence.

PP: Precautionary principle. A moral and political principle that states that if an action or policy might cause severe or irreversible harm to the environment or the public, in the absence of a scientific consensus that harm would not ensue, the burden of proof falls on those who would advocate taking the action.

PRC: Principal response curve.

Problem formulation: The phase in a risk assessment in which the goals of the assessment are defined and the methods for achieving those goals are specified.

PRTR: Pollution Release and Transfer Register.

pT: Toxic potency. A measure indicating the relative toxic response of a surface water sample during a short-term microtest (e.g., the *Daphnia*-IQ test).

QSAR: Quantitative structure–activity relationship. Quantitative structure-biological activity model derived using regression analysis and containing as parameters physical–chemical constants, indicator variables, or theoretically calculated values.

QSPR: Quantitative structure-property relationship.

RA: Response addition. Concept used to calculate the risk of chemical mixtures based on the assumption that the mixture components act through independent joint action (IJA or IA).

RAGS: Risk Assessment Guidance for Superfund (USA).

RAM: Rate of metabolism.

REACH: Registration, Evaluation, Authorization and Restriction of Chemicals (EU legislation).

Receptor: In toxicology: Molecular structure in or on a cell, which specifically recognizes and binds to a compound and acts as a physiological signal transducer or mediator of an effect.

In exposure assessment: Organism, population, or community that is exposed to contaminants. Receptors may or may not be assessment endpoint entities.

RfC: Reference concentration. An estimate (with uncertainty spanning perhaps an order of magnitude) of a continuous inhalation exposure to the human population (including sensitive subgroups) that is likely to be without an appreciable risk of deleterious effects during a lifetime.

RfD: Reference dose. An estimate (with uncertainty spanning perhaps an order of magnitude) of a daily oral exposure to the human population (including sensitive subgroups) that is likely to be without an appreciable risk of deleterious effects during a lifetime.

Risk (toxic): The predicted or actual probability of occurrence of an adverse effect on humans or the environment of exposure to a chemical substance or mixture. *See also* hazard.

Risk assessment: An organized process that aims to describe and estimate the likelihood of adverse health outcomes from environmental exposure to chemicals.

Risk characterization: A phase of the risk assessment process that integrates the exposure and stressor response profiles to evaluate the likelihood of adverse effects associated with exposure to the contaminants.

Risk management: The process of deciding what regulatory or remedial actions to take, justifying the decision, and implementing the decision.

RIVPACS: River Invertebrate Prediction and Classification System.

RPF: Relative potency factor. A factor that expresses the toxic potency of a mixture component relative to an index compound. In the RPF approach, RPF values of mixture components are summed and the risk of the whole mixture is estimated using dose–response data of the index compound.

Safety factor: A factor applied to an observed or estimated toxic concentration or dose to arrive at a criterion or standard that is considered safe. The terms "safety factor" and "uncertainty" factor are often used synonymously. *See also* uncertainty factor.

SAR: Structure–activity relationship. The relationship between the chemical structure of a compound and its biological or pharmacological activity.

SASS4-ASPT: South African Scoring System—Average Score per Taxon.

SCALE: Science, Children, Raising Awareness, Legal Instruments, and Evaluation. An EU program to reduce disease burden caused by environmental factors.

Slope factor: An upper bound, approximating a 95% confidence limit, on the increased cancer risk from a lifetime exposure to an agent.

SPE: Solid-phase extraction.

SPME: Solid-phase microextraction.

SSA: Simple similar action. The joint action of chemicals in case they act in the same way, by the same mechanisms, and differ only in their potencies.

SSD: Species sensitivity distribution. A frequency distribution of the toxicity of a certain compound or mixture to a set of species that may be defined as a taxon, assemblage, or community. Empirically, the distribution is estimated from a sample of toxicity data for the specified species set.

SSD^{Endpoint}: General notation for a species sensitivity distribution based on a specific type of input data, for example, NOECs (SSD^{NOEC}), EC50s (SSD^{EC50}), or LC50s (SSD^{LC50}). A superscript prefix can be added to SSD^{Endpoint} to show the number of toxicity data from which the model was derived, like ^{4}SSD^{Endpoint} for a SSD based on 4 data points.

Susceptibility: The condition of an organism or other ecological system lacking the power to resist a particular disease, infection, or intoxication. It is inversely proportional to the magnitude of the exposure required to cause the response.

Synergism: A deviation from a chosen mixture toxicity reference model (e.g., CA or IA), suggesting a mixture is more toxic than expected from the toxicities of the individual chemicals.

TCDD: 2,3,7,8-Tetrachlorodibenzo-*p*-dioxin.

TCE: Trichloroethylene.

TD: Toxicodynamics.

TDI: Tolerable daily intake. *See also* ADI.

TEAM: Total Exposure Assessment Methodology.

TEF: Toxicity equivalency factor. Ratio of the toxicity of a chemical to that of another structurally related chemical (or index compound) chosen as a reference. TEFs are toxicity potency factors used to evaluate the toxicities of highly variable mixtures of dioxin-like compounds. The most toxic members, 2,3,7,8-TCDD and 1,2,3,7,8-pentachlorodibenzo-*p*-dioxin, are assigned a TEF of 1. The values of individual TEFs indicate how closely family member compounds resemble the toxicity of 2,3,7,8-TCDD.

TEQ: Toxicity equivalent. Contribution of a specified component (or components) to the toxicity of a mixture of related substances relative to a reference toxicant. TEQs are calculated by multiplying the actual grams weight of each dioxin and dioxin-like compound by its corresponding TEF (e.g., 10 grams × 0.1 TEF = 1 gram TEQ) and then summing the results. The result is referred to as grams TEQ.

Test endpoint: A response measure in a toxicity test, that is, the values derived from a toxicity test that characterize the results of the test (e.g., NOEC or LC50).

TIE: Toxicity identification evaluation. A process used to determine the compounds responsible for toxicity in ambient waters, effluents, and sediments.

TK: Toxicokinetics.

Toxic strength: Potential toxicity of a mixture expressed as the sum of the number of toxic units of each of the individual chemicals present in the mixture.

Toxic unit: Concentration scaled to a measure of toxicity (e.g., EC50 or LC50).

Toxicity test: The determination of the effect of a substance on a group of selected organisms under defined conditions. A toxicity test usually measures either 1) the proportions of organisms affected (quantal) or 2) the degree of effect shown (graded or quantitative), after exposure to specific levels of a chemical or mixture of chemicals.

TPH: Total petroleum hydrocarbon.

TRIAD approach: Integrated approach to determine ecological effects of pollution by a complex mixture, applying a combination of chemical analysis, bioassays, and ecological field observations.

TSCA: Toxic Substances Control Act (USA).

TSP: Two-step prediction model. Model that predicts mixture toxicity by grouping of chemicals according to similar chemistry based on mode of action, and applying CA for chemicals with similar modes of action and IA or RA for (groups of) chemicals with different modes of action.

TU: Toxic unit. The concentration of a chemical expressed as a fraction or proportion of its effective concentration (measured in the same units). It is calculated as toxic unit = actual concentration of chemical in solution/LC50. If this number is greater than 1.0, more than half of a group of aquatic organisms are killed by the chemical. If it is less than 1.0, less than half the organisms are killed.

TV: Target value. Environmental quality standard used in the Netherlands to indicate the targeted quality of air, water, soil, and sediments.

UK-PBMS: UK Predatory Bird Monitoring Scheme.

Uncertainty: Imperfect knowledge concerning the present or future state of the system under consideration; a component of risk resulting from imperfect knowledge of the degree of hazard or of its spatial and temporal pattern of expression.

Uncertainty factor: A factor applied to an exposure or effect concentration or dose to correct for identified sources of uncertainty. *See also* safety factor.

US-AMP: The Arctic Monitoring Program of the USEPA.

USDA: US Department of Agriculture.

USEPA: US Environmental Protection Agency.

VOC: Volatile organic compound.

WET: Whole effluent toxicity. The total toxic effect of an effluent measured directly with a toxicity test or bioassay.

WHO: World Health Organization.

WOE: Weight of evidence.

References

Abadin H, Hibbs B, Pohl H. 1997. Breast-feeding exposure of infants to environmental contaminants—a public health risk assessment viewpoint. II. Cadmium, lead, mercury. Toxicol Ind Health 13:495–517.

Abou-Donia MB, Makkawy HM, Campbell GM. 1985. Pattern of neurotoxicity of n-hexane, methyl-n-butyl ketone, 2,5-hexanediol, and 2,5-hexanedione alone and in combination with o-4-nitrophenyl phenylphosphonothionate in hens. J Toxicol Environ Health 16:85–100.

ACGIH. 1984. Threshold limit values—discussion and thirty-five year index with recommendations. Cincinnati (OH): American Conference of Governmental Industrial Hygienists.

ACGIH. 2000. 2000 TLVs and BEIs. Threshold limit values for chemical substances and physical agents and biological exposure indices. Cincinnati (OH): American Conference of Governmental Industrial Hygienists.

Ahlborg UG, Becking GC, Birnbaum LS, Brouwer A, Derks H, Feeley M, Golor G, Hanberg A, Larsen JC, Liem AKD, Safe SH, Schlatter C, Waern F, Younes M, Yrjanheikki E. 1994. Toxic equivalency factors for dioxin-like PCBs. Chemosphere 28:1049–1067.

Albers EP, Dixon KR. 2002. A conceptual approach to multiple-model integration in whole site risk assessment. In: Rizzoli AE, Jakeman AJ, editors, Integrated assessment and decision support. Proceedings of the First Biennial Meeting of the International Environmental Modelling and Software Society. Part 1. Manno (CH): iEMSs. p 293–298.

Alcock RE, Boumphrey R, Malcolm HM, Osborn D, Jones KC. 2002. Temporal and spatial trends of PCB congeners in UK Gannet eggs. Ambio 31:202–206.

Alda Álvarez O, Jager T, Kooijman SALM, Kammenga JE. 2005. Responses to stress of *Caenorhabditis elegans* populations with different reproductive strategies. Funct Ecol 19:656–664.

Alda Álvarez O, Jager T, Marco Redondo E, Kammenga JE. 2006b. Assessing physiological modes of action of toxic stressors with the nematode *Acrobeloides nanus*. Environ Toxicol Chem 25:3230–3237.

Alda Álvarez O, Jager T, Nuñez Coloa B, Kammenga JE. 2006a. Temporal dynamics of effect concentrations. Environ Sci Technol 40:2478–2484.

Aldenberg T, Jaworska JS, Traas TP. 2002. Normal species sensitivity distributions and probabilistic ecological risk assessment. In: Posthuma L, Suter GW, II, Traas TP, editors, Species sensitivity distributions in ecotoxicology. Boca Raton (FL): Lewis Publishers. p 49–102.

Alexander M. 1995. How toxic are toxic chemicals in soil? Environ Sci Technol 29:2713–2717.

Ali N, Tardif R. 1999. Toxicokinetic modeling of the combined exposure to toluene and n-hexane in rats and humans. J Occup Health 41:95–103.

Allen BC, Kavlock RJ, Kimmel CA, Faustman EM. 1994. Dose–response assessment for developmental toxicity II. Comparison of generic benchmark dose estimates with no observed adverse effect levels. Fund Appl Toxicol 23:487–495.

Allen HE, editor. 2002. Bioavailability of metals in terrestrial ecosystems: importance of partitioning for bioavailability to invertebrates, microbes, and plants. Metals and the Environment Series. New York: SETAC.

Altenburger R, Nendza M, Schüürmann G. 2003. Mixture toxicity and its modeling by quantitative structure–activity relationships. Environ Toxicol Chem 22:1900–1915.

Altenburger R, Schmitt H, Schüürmann G. 2005. Algal toxicity of nitrobenzenes: combined effect analysis as a pharmacological probe for similar modes of interaction. Environ Toxicol Chem 24:324–333.

229

Altenburger R, Walter H, Grote M. 2004. What contributes to the combined effect of a complex mixture? Environ Sci Technol 38:6353–6362.

Amdur MO, Doull J, Klaassen CD, editors. 1991. Casarett & Doull's toxicology: the basic science of poisons. 4th ed. New York: McGraw-Hill.

Amweg EL. 2006. Effect of piperonyl butoxide on permethrin toxicity in the amphipod *Hyalella azteca*. Environ Toxicol Chem 25:1817–1825.

Andersen ME. 1995. Development of physiologically based pharmacokinetic and physiologically based pharmacodynamic models for applications in toxicology and risk assessment. Toxicol Lett 79:35–44.

Andersen ME, Birnbaum LS, Barton HA, Eklund CR. 1997. Regional hepatic CYP1A1 and CYP1A2 induction with 2,3,7,8-tetrachlorodibenzo-p-dioxin evaluated with a multicompartment geometric model of hepatic zonation. Toxicol Appl Pharmacol 144:145–155.

Andersen ME, Clewell HJ III. 1983. Pharmacokinetic interaction of mixtures. In: Proceedings of the 14th Annual Conference on Environmental Toxicology. AFAMRL-TR-83-099. Dayton (OH): AFAMRL p 226–238.

Andersen ME, Dennison JE. 2004. Mechanistic approaches for mixture risk assessments—present capabilities with simple mixtures and future directions. Environ Toxicol Pharmacol 16:1–11.

Andersen ME, Gargas ML, Clewell HJ III, Severyn KM. 1987. Quantitative evaluation of the metabolic interactions between trichloroethylene and 1,1-dichloroethylene *in vivo* using gas uptake methods. Toxicol Appl Pharmacol 89:149–157.

Ankley GT, Dierkes JR, Jensen DA, Peterson GS. 1991. Piperonyl butoxide as a tool in aquatic toxicological research with organophosphate insecticides. Ecotoxicol Environ Safety 21:266–274.

Ankley GT, Schubauer-Berigan MK, Hoke RA. 2006. Use of toxicity identification techniques to identify dredged material disposal options: a proposed approach. Environ Manage 16:1–6.

Archer V. 1985. Enhancement of lung cancer by cigarette smoking in uranium and other miners. Carcinogenesis 8:23–37.

Arrhenius Å, Grönvall F, Scholze M, Backhaus T, Blanck H. 2004. Predictability of the mixture toxicity of 12 similarly acting congeneric inhibitors of photosystem II in marine periphyton and epipsammon communities. Aquat Toxicol 68:351–367.

Arts GHP, Buijse-Bogdan LL, Belgers JDM, Van Rhenen-Kersten CH, Van Wijngaarden RPA, Roessing I, Maund SJ, Van den Brink PJ, Brock TCM. 2006. Ecological impact in ditch mesocosms of simulated spray drift from a crop protection program for potatoes. IEAM 2:105–125.

Ashauer R, Boxall A, Brown C. 2006. Predicting effects on aquatic organisms from fluctuating or pulsed exposure to pesticides. Environ Toxicol Chem 25:1899–1912.

Ashauer R, Boxall ABA, Brown CD 2007a. Modeling combined effects of pulsed exposure to carbaryl and chlorpyrifos on *Gammarus pulex*. Environ Sci Technol 41:5535–5541.

Ashauer R, Boxall ABA, Brown CD. 2007b. New ecotoxicological model to simulate survival of aquatic invertebrates after exposure to fluctuating and sequential pulses of pesticides. Environ Sci Technol 41:1480–1486.

Ashford JR. 1981. General models for the joint action of mixtures of drugs. Biometrics 37:457–474.

ATSDR. 1997. Toxicological profile for chlorpyrifos. Agency for Toxic Substances and Disease Registry. Atlanta (GA): Department of Health and Human Services, Public Health Service. Available from: http://www.atsdr.cdc.gov/toxpro2.html

ATSDR. 1998a. Toxicological profile for chlorinated dibenzo-p-dioxins (CDDs). Agency for Toxic Substances and Disease Registry. Atlanta (GA): US Department of Health and Human Services, Public Health Service. Available from: http://www.atsdr.cdc.gov/toxpro2.html

ATSDR. 1998b. Toxicological profile for radon 1990. Agency for Toxic Substances and Disease Registry. Atlanta (GA): US Department of Health and Human Services, Public Health Service. Available from: http://www.atsdr.cdc.gov/toxpro2.html

ATSDR. 1999. Toxicological profile for lead. Agency for Toxic Substances and Disease Registry. Atlanta (GA): US Department of Health and Human Services, Public Health Service. Available from: http://www.atsdr.cdc.gov/toxpro2.html

ATSDR. 2004a. Guidance manual for the assessment of joint toxic action of chemical mixtures. Agency for Toxic Substances and Disease Registry. Atlanta (GA): US Department of Health and Human Services. Available from: http://www.atsdr.cdc.gov/interaction-profiles/ipga.html

ATSDR. 2004b. Interaction profile for persistent chemicals found in breast milk (chlorinated dibenzo-p-dioxins, hexachlorobenzene, p,p'-DDE, methylmercury and polychlorinated biphenyls). Agency for Toxic Substances and Disease Registry. Atlanta (GA): US Department of Health and Human Services, Public Health Service.

ATSDR. 2004c. Interaction profile for arsenic, cadmium, chromium and lead. Agency for Toxic Substances and Disease Registry. Atlanta (GA): US Department of Health and Human Services, Public Health Service.

ATSDR. 2006. Interaction profile for chlorpyrifos, lead, mercury, and methylmercury. Agency for Toxic Substances and Disease Registry. Atlanta (GA): US Department of Health and Human Services, Public Health Service.

Aylward LL, Hays S, Finley B. 2002. Temporal trends in intake of dioxins from foods in the U.S. and Western Europe: issues with intake estimates and parallel trends in human body burdens. 22nd Int Symp Halogenated Environ Organic Pollutants POPs 55:235–238.

Aylward LL, Hays SM. 2002. Temporal trends in human TCDD body burden: decreases over three decades and implications for exposure levels. J Expo Anal Environ Epidemiol 12:319–328.

Baas J, Jager T, Kooijman SALM. 2009. A model to analyse effects of complex mixtures on survival. Ecotoxicol Environ Safety 72:669–676.

Baas J, Van Houte BPP, Van Gestel CAM, Kooijman SALM 2007. Modelling the effects of binary mixtures on survival in time. Environ Toxicol Chem 26:1320–1327.

Bachmann KA, Ghosh R. 2001. The use of *in vitro* methods to predict *in vivo* pharmacokinetics and drug interactions. Curr Drug Metab 2:299–314.

Backhaus T, Altenburger R, Boedeker W, Faust M, Scholze M, Grimme LH. 2000a. Predictability of the toxicity of a multiple mixture of dissimilarly acting chemicals to *Vibrio fischeri*. Environ Toxicol Chem 19:2348–2356.

Backhaus T, Arrhenius A, Blanck H. 2004. Toxicity of a mixture of dissimilarly acting substances to natural algal communities: predictive power and limitations of independent action and concentration addition. Environ Sci Technol 38:6363–6370.

Backhaus T, Scholze M, Grimme LH. 2000b. The single substance and mixture toxicity of quinolones to the bioluminescent bacterium *Vibrio fischeri*. Aquat Toxicol 49:49–61.

Bakker MI, Vorenhout M, Sijm DTHM, Kolloffel C. 1999. Dry deposition of atmospheric polycyclic aromatic hydrocarbons in three *Plantago* species. Environ Toxicol Chem 18:2289–2294.

Balakin KV, Ekins S, Bugrim A, Ivanenkov YA, Korolev D, Nikolsky YV, Skorenko AV, Ivashchenko AA, Savchuk NP, Nikolskaya T. 2004. Kohonen maps for prediction of binding to human cytochrome P450 3A4. Drug Metab Dispos 32:1183–1189.

Banks KE, Wood SH, Matthews C, Thuesen KA. 2003. Joint acute toxicity of diazinon and copper to *Ceriodaphnia dubia*. Environ Toxicol Chem 22:1562–1567.

Barahona LM, Loyo L, Guerrero M, Ramírez S, Romero I, Jarquin CV, Albores A. 2005. Ecotoxicological evaluation of diesel-contaminated soil before and after a bioremediation process. Environ Toxicol 20:100–109.

Barata C, Markich SJ, Baird DJ, Taylor G, Soares AMVM. 2002. Genetic variability in sub-lethal tolerance to mixtures of cadmium and zinc in clones of *Daphnia magna* Straus. Aquat Toxicol 60:85–99.

Barber MC. 2003. A review and comparison of models for predicting dynamic chemical bio-concentration in fish. Environ Toxicol Chem 22:1963–1992.

Bargagli R, Monaci F, Borghini F, Bravi F, Agnorelli C. 2002. Mosses and lichens as biomoni-tors of trace metals: a comparison study on *Hypneum cupressiforme* and *Parmelia caperata* in a former mining district in Italy. Environ Pollut 116:279–287.

Barton CN. 1993. Nonlinear statistical models for the joint action of toxins. Biometrics 49:95–105.

Barton HA, Creech JR, Godin CS, Randall GM, Seckel CS. 1995. Chloroethylene mixtures: pharmacokinetic modeling and *in vitro* metabolism of vinyl chloride, trichloroethylene, and trans-1,2-dichloroethylene in rat. Toxicol Appl Pharmacol 130:237–247.

Baumard P, Budzinski H, Garrigues P, Sorbe JC, Burgeot T, Bellocq J. 1998. Concentrations of PAHs (polycyclic aromatic hydrocarbons) in various marine organisms in relation to those in sediments and to trophic level. Mar Pollut Bull 36:951–960.

Bedaux JJM, Kooijman SALM. 1994. Statistical analysis of bioassays based on hazard model-ling. Environ Ecol Stat 1:303–314.

Belden JB, Gilliom RJ, Lydy MJ. 2007. How well can we predict the toxicity of pesticide mixtures to aquatic life. Integrated Environ Assess Manage 3:362–372.

Belden JB, Lydy MJ. 2006. Joint toxicity of chlorpyrifos and esfenvalerate to fathead minnows and midge larvae. Environ Toxicol Chem 25:623–629.

Belfroid A, Sikkenk M, Seinen W, Van Gestel K, Hermens J. 1994. The toxicokinetic behavior of chlorobenzenes in earthworm (*Eisenia andrei*) experiments in soil. Environ Toxicol Chem 13:93–99.

Beliveau M, Krishnan K. 2005. A spreadsheet program for modeling quantitative structure-pharmacokinetic relationships for inhaled volatile organics in humans. SAR QSAR Environ Res 16:63–77.

Beliveau M, Lipscomb J, Tardif R, Krishnan K. 2005. Quantitative structure-property rela-tionships for interspecies extrapolation of the inhalation pharmacokinetics of organic chemicals. Chem Res Toxicol 18:475–485.

Beliveau M, Tardif R, Krishnan K. 2003. Quantitative structure-property relationships for physiologically based pharmacokinetic modeling of volatile organic chemicals in rats. Toxicol Appl Pharmacol 189:221–232.

Belz RG, Cedergreen N, Sørensen H. 2008. Hormesis in mixtures—can it be predicted? Sci Total Environ 404:77–87.

Berenbaum MC. 1985. The expected effect of a combination of agents: the general solution. J Theor Biol 114:413–431.

Berenbaum MC. 1989. What is synergy? Pharmacol Rev 1989:93–141.

Berger U, Herzke D, Sandanger TM. 2004. Two trace analytical methods for determination of hydroxylated PCBs and other halogenated phenolic compounds in eggs from Norwegian birds of prey. Anal Chem 76:441–452.

Berkowitz GS, Obel J, Deych E, Lapinski R, Godbold J, Liu Z, Landrigan PJ, Wolff MS. 2003. Exposure to indoor pesticides during pregnancy in a multiethnic, urban cohort. Environ Health Perspect 111:79–84.

Bernillon P, Bois FY. 2000 Statistical issues in toxicokinetic modeling: a bayesian perspective. Environ Health Perspect 108(Suppl 5):883–893.

Bervoets L, Meregalli G, De Cooman W, Goddeeris B, Blust R. 2004. Caged midge larvae (*Chironomus riparius*) for the assessment of metal bioaccumulation from sediments *in situ*. Environ Toxicol Chem 23:443–454.

BFR (German Federal Institute for Risk Assessment). 2005. Documents of the BfR Forum on mul-tiple residues of pesticide residues in food. Available from: http://www.bfr.bund.de/cd/7078

Binder S, Sokal D, Maughn D. 1986. The use of tracer elements in estimating the amount of soil ingested by young children. Arch Environ Health 41:341–345.

Binderup ML, Dalgaard M, Dragsted LO, Hossaini A, Ladefoged O, Lam HR, Larsen JC, Madsen C, Meyer O, Rasmussen ES, Reffstrup TK, Soborg I, Vinggaard AM, Ostergard G. 2003. Combined actions and interactions of chemicals in mixtures: the toxicological effects to exposures of mixtures of industrial and environmental chemicals. Report 2003:12. Soborg (DK): Danish Veterinary and Food Administration.

Bisinoti MC, Jardim WF. 2003. Production of organic mercury from Hg^0: experiments using microcosms. J Brazil Chem Soc 14:244–248.

Bjorkman S. 2005. Prediction of drug disposition in infants and children by means of physiologically based pharmacokinetic (PBPK) modelling: theophylline and midazolam as model drugs. Br J Clin Pharmacol 59:691–704.

Blanck H. 2002. A critical review of procedures and approaches used for assessing pollution-induced community tolerance (PICT) in biotic communities. Human Ecol Risk Assess 8:1003–1034.

Bliss CI. 1939. The toxicity of poisons applied jointly. Ann J Appl Biol 26:585–615.

Bobbink R, Heil GW, Raessen M. 1992. Atmospheric deposition and canopy exchange processes in heathland ecosystems. Environ Pollut 75:29–37.

Bocquene G, Bellanger C, Cadiou Y, Galgani F. 1995. Joint action of combinations of pollutants on the aceetylcholinesterase activity of several marine species. Ecotoxicology 4:266–279.

Boedeker W, Altenburger R, Faust M, Grimme LH. 1990. Methods for the assessment of mixtures of plant protection substances (pesticides): mathematical analysis of combination effects in phytopharmacology and ecotoxicology. Nachrichtenblatt Deutschen Pflanzensch 42:70–78.

Boedeker W, Altenburger R, Faust M, Grimme LH. 1992. Synopsis of concepts and models for the quantitative analysis of combination effects: from biometrics to ecotoxicology. ACES 4:45–53.

Bond JA, Csanady GA, Gargas ML, Guengerich FP, Leavens T, Medinsky MA, Recio L. 1994. 1,3-Butadiene: linking metabolism, dosimetry, and mutation induction. Environ Health Perspect 102(Suppl 9):87–94.

Boom SP, Meyer I, Wouterse AC, Russel FG. 1998. A physiologically based kidney model for the renal clearance of ranitidine and the interaction with cimetidine and probenecid in the dog. Biopharm Drug Dispos 19:199–208.

Borgert CJ. 2007. Predicting interactions from mechanistic information: can omic data validate theories? Toxicol Appl Pharmacol 223:114–120.

Boxall ABA, Maltby L. 1997. The effects of motorway runoff on freshwater ecosystems. 3. Toxicant confirmation. Arch Environ Contam Toxicol 33:9–16.

Boyd EM. 1969. Dietary protein and pesticide toxicity in male weaning rats. Bull WHO 40:801–805.

Brack W, Atenburger R, Ensenbach U, Moder M, Segner H, Schüürmann G. 1999. Bioassay-directed identification of organic toxicants in river sediment in the industrial region of Bitterfeld (Germany)—a contribution to hazard assessment. Arch Environ Contam Toxicol 37:164–174.

Brian JV, Harris CA, Scholze M, Backhaus T, Booy P, Lamoree M, Pojana G, Jonkers N, Runnalls T, Bonfa A, Marcomini A, Sumpter JP. 2005. Accurate prediction of the response of freshwater fish to a mixture of estrogenic chemicals. Environ Health Perspect 113:721–728.

Brocklebank JR, Namdari R, Law FC. 1997. An oxytetracycline residue depletion study to assess the physiologically based pharmokinetic (PBPK) model in farmed Atlantic salmon. Can Vet J 38:645–646.

Broderius S, Kahl M. 1985. Acute toxicity of organic chemical mixtures to the fathead minnow. Aquat Toxicol 6:307–322.

Broughton RK, Osborn D, Shore RF, Wienburg CL, Wadsworth RA. 2003. Identifying pollution hot spots from polychlorinated biphenyl residues in birds of prey. Environ Toxicol Chem 22:2519–2524.

Brown RP, Delp MD, Lindstedt SL, Rhomberg LR, Beliles RP. 1997. Physiological parameter values for physiologically based pharmacokinetic models. Toxicol Ind Health 13:407–484.

Bulletti C, Flamigni C, Giacomucci E. 1996. Reproductive failure due to spontaneous abortion and recurrent miscarriage. Hum Reprod Update 2:118–136.

Bundy JG, Sidhu JK, Rana F, Spurgeon DJ, Svendsen C, Wren JF, Stürzenbaum SR, Morgan AJ, Kille P. 2008. "Systems toxicology" approach identifies coordinated metabolic responses to copper in a terrestrial non-model invertebrate, the earthworm *Lumbricus rubellus*. BMC Biol 6:25 (doi:10.1186/1741-7007-6-25).

Burkhard LP, Durhan EJ. 1991. Identification of nonpolar toxicants in effluents using toxicity-based fractionation with gas chromatography/mass spectrometry. Anal Chem 63:277–283.

Burton KW, Morgan E, Roig A. 1986. Interactive effects of cadmium, copper and nickel on the growth of Sitka Spruce and studies of metal uptake from nutrient solutions. New Phytol 103:549–557.

Cahill TM, Cousins I, Mackay D. 2003. General fugacity-based model to predict the environmental fate of multiple chemical species. Environ Toxicol Chem 22:483–493.

Calabrese EJ. 1991. Multiple chemical interactions. Part 4: Drugs; Part 5: The drug-pollutant interface. Chelsea (MI): Lewis Publishers, p 389–578.

Calabrese EJ. 2005. Paradigm lost, paradigm found: the re-emergence of hormesis as a fundamental dose response model in the toxicological sciences. Environ Pollut 138:8–411.

Callaghan A, Hirthe G, Fisher T, Crane M. 2001. Effect of short-term exposure to chlorpyrifos on developmental parameters and biochemical biomarkers in *Chironomus riparius* Meigen. Ecotoxicol Environ Safety 50:19–24.

Cape JN, Freersmith PH, Paterson IS, Parkinson JA, Wolfenden J. 1990. The nutritional status of *Picea abies* (L) Karst across Europe, and implications for forest decline. Trees Struct Funct 4:211–224.

Carlile DJ, Zomorodi K, Houston JB. 1997. Scaling factors to relate drug metabolic clearance in hepatic microsomes, isolated hepatocytes, and the intact liver: studies with induced livers involving diazepam. Drug Metab Dispos 25:903–911.

Casey M, Gennings C, Carter WH Jr, Moser VC, Simmons JE. 2005. Ds-optimal designs for studying combinations of chemicals using multiple fixed-ratio ray experiments. Environmetrics 16:129–147.

Cassee FR, Groten JP, Van Bladeren PJ, Feron VJ. 1998. Toxicological evaluation and risk assessment of chemical mixtures. Crit Rev Toxicol 28:73–101.

CDC. 2003. Second national report on human exposure to environmental chemicals. CDC/ NCEH Pub. 02-0716. Atlanta (GA): Department of Health and Human Services, Centers for Disease Control and Prevention.

CDC. 2005. Third national report on human exposure to environmental chemicals. CDC/ NCEH Pub. 05-0570. Atlanta (GA): Department of Health and Human Services, Centers for Disease Control and Prevention.

Cedergreen N, Kudsk P, Mathiassen SK, Sørensen H, Streibig JC. 2007. Reproducibility of binary-mixture toxicity studies. Environ Toxicol Chem 26:149–156.

Cedergreen N, Streibig JC. 2005. Can the choice of endpoint lead to contradictory results of mixture-toxicity experiments? Environ Toxicol Chem 24:1676–1683.

Cerklewski FL, Forbes RM. 1976. Influence of dietary zinc on lead toxicity in the rat. J Nutr 106:689–696.

Chapman PM. 1986. Sediment quality criteria from the sediment quality TRIAD—an example. Environ Toxicol Chem 5:957–964.

Chapman PM, McDonald BG, Lawrence GS. 2002. Weight-of-evidence issues and frameworks for sediment quality (and other) assessments. Human Ecol Risk Assess 8:1489–1515.

Chapman PM, Power EA, Burton GA Jr. 1992. Integrative assessments in aquatic ecosystems. In: Burton GA Jr, editor, Sediment toxicity assessment. Chelsea (MI): Lewis Publishers. p 313–340.

Chèvre N, Loepfe C, Singer H, Stamm C, Fenner K, Escher BI. 2006. Including mixtures in the determination of water quality criteria for herbicides in surface water. Environ Sci Technol 40:426–435.

Chien JY, Thumme KE, Slattery JT. 1997. Pharmacokinetic consequences of induction of CYP2E1 by ligand stabilization. Drug Metab Dispos 25:1165–1175.

Cho EA, Bailer J, Oris JT. 2003. Effect of methyl tert-butyl ether on the bioconcentration and photoinduced toxicity of fluoranthene in fathead minnow larvae (*Pimephales promelas*). Environ Sci Technol 37:1306–1310.

Chou TC, Talalay P. 1983. Analysis of combined drug effects—a new look at a very old problem. Trends Pharmacol Sci 4:450–454.

Ciucu A. 2002. Progress and perspectives in biosensors for environmental monitoring. Roum Biotechnol Lett 7:537–546.

Clark S, Bornschein RL, Pan W, Menrath W, Roda S, Grote J. 1996. The relationship between surface dust lead loadings on carpets and the blood lead of young children. Environ Geochem Health 18:143–146.

Clausing P, Brunekreef B, Van Wijnen JH. 1987. A method for estimating soil ingestion by children. Int Arch Occup Environ Health 59:73–82.

Clewell HJ III, Andersen ME. 1985. Risk assessment extrapolations and physiological modeling. Toxicol Ind Health 1:111–131.

Clewell RA, Merrill EA, Yu KO, Mahle DA, Sterner TR, Mattie DR, Robinson PJ, Fisher JW, Gearhart JM. 2003. Predicting fetal perchlorate dose and inhibition of iodide kinetics during gestation: a physiologically-based pharmacokinetic analysis of perchlorate and iodide kinetics in the rat. Toxicol Sci 73:235–255.

Clifford PA, Barchers DE, Ludwig DF, Sielken RL, Klingensmith JS, Graham RV, Banton MI. 1995. An approach to quantifying spatial components of exposure for ecological risk assessment. Environ Toxicol Chem 14:895–906.

Coffey T, Gennings C, Simmons JE, Herr WD. 2005. D-Optimal experimental designs to test for departure from additivity in a fixed-ratio mixture ray. Toxicol Sci 88:467–476.

Cogliano VJ. 1997. Plausible upper bounds: are their sums plausible? Risk Anal 17:77–84.

Cogliano VJ. 1998. Assessing the cancer risk from environmental PCBs. Environ Health Perspect 106:317–323.

Coombe VT, Moore KW, Hutchings MJ. 1999. TIE and TRE: an abbreviated guide to dealing with toxicity. Water Sci Technol 39:91–97.

COT. 2002. Risk assessment of mixtures of pesticides and similar substances. London: Committee on Toxicity of Chemicals in Food, Consumer Products and the Environment, UK Food Standards Agency. Available from: http://cot.food.gov.uk/pdfs/cotwig2000-2.pdf

Cotter MA, Policz DL, Pöch G, Dawson DA. 2000. Analysis of the combined osteolathyritic effects of beta-aminopropionitrile and diethyldithiocarbamate on *Xenopus* development. Toxicol Sci 58:144–152.

Cova D, Nebuloni C, Arnoldi A, Bassoli A, Trevisan M, DelRe AAM. 1996. N-nitrosation of triazines in human gastric juice. J Agric Food Chem 44:2852–2855.

Covington TR, Gentry PR, Van Landingham CB, Andersen ME, Kester JE, Clewell HJ. 2007. The use of Markov chain Monte Carlo uncertainty analysis to support a public health goal for perchloroethylene. Regul Toxicol Pharmacol 47:1–18.

Cowan CE, Versteeg DJ, Larson RJ, Kloeppersams PJ. 1995. Integrated approach for environmental assessment of new and existing substances. Regul Toxicol Pharmacol 21:3–31.

Crofton K, Craft ES, Hedge JM, Gennings C, Simmons JE, Carchman RA, Carter WH Jr, deVito JM. 2005. Thyroid-hormone-disrupting chemicals: evidence for dose-dependent additivity or synergism. Environ Health Perspect 113:1549–1554.

CSTEE. 2000. The available scientific approaches to assess the potential effects and risk of chemicals on terrestrial ecosystems. C2/JCD/csteeop/Ter91100/D(0). Brussels (BE): European Commission. p 178.

Cui Y, Zhu Y-G, Zhai R, Huang Y, Qiu Y, Liang J. 2005. Exposure to metal mixtures and human health impacts in a contaminated area in Nanning, China. Environ Intern 31:784–790.

Cuppen JGM, Crum SJH, Van den Heuvel HH, Smidt RA, Van den Brink PJ. 2002. The effects of a mixture of two insecticides on freshwater microcosms. I. Fate of insecticides and responses of macroinvertebrates. Ecotoxicology 11:19–34.

Damgaard IN, Skakkebaek NE, Toppari J, Virtanen HE, Shen H, Schramm KW, Petersen JH, Jensen TK, Main KM. 2006. Persistent pesticides in human breast milk and cryptorchidism. Environ Health Perspect 114:1133–1138.

Daskalakis KD. 1996. Variability of metal concentrations in oyster tissue and implications to biomonitoring. Mar Pollut Bull 32:794–801.

David RM, Clewell HJ, Gentry PR, Covington TR, Morgott DA, Marino DJ. 2006. Revised assessment of cancer risk to dichloromethane. II. Application of probabilistic methods to cancer risk determinations. Regul Toxicol Pharmacol 45:55–65.

Decaprio AP. 1997. Biomarkers: coming of age for environmental health and risk assessment. Environ Sci Technol 31:1837–1848.

De Graaf C, Vermeulen NP, Feenstra KA. 2005. Cytochrome P450 in silico: an integrative modeling approach. J Med Chem 48:2725–2755.

De Groot MH. 1986. Probability and statistics. 2nd ed. Reading (MA): Addison-Wesley Pub.

De Groot MJ, Ekins S. 2002. Pharmacophore modeling of cytochromes P450. Adv Drug Deliv Rev 54:367–383.

De Groot MJ, Kirton SB, Sutcliffe MJ. 2004. In silico methods for predicting ligand binding determinants of cytochromes P450. Curr Top Med Chem 4:1803–1824.

DeKoning EP, Karmaus W. 2000. PCB exposure in utero and via breast milk: a review. J Expo Anal Environ Epidemiol 10:285–293.

De Maagd PGJ, Van de Klundert ICM, Van Wezel AP, Opperhuizen A, Sijm DTHM. 1997. Lipid content and time-to-death-dependent lethal body burdens of naphthalene and 1,2,4-trichlorobenzene in fathead minnow (Pimephales promelas). Ecotoxicol Environ Safety 38:232–237.

De March BGE. 1987. Simple similar action and independent joint action—two similar models for the joint effects of toxicants applied as mixtures. Aquat Toxicol 9:291–304.

Deneer JW. 2000. Toxicity of mixtures of pesticides in aquatic systems. Pest Manage Sci 56:516–520.

Dennison JE, Andersen ME, Clewell HJ, Yang RSH. 2004a. Development of a physiologically based pharmacokinetic model for volatile fractions of gasoline using chemical lumping analyses. Environ Sci Technol 38:5674–5681.

Dennison JE, Andersen ME, Dobrev ID, Mumtaz MM, Yang RSH. 2004b. PBPK modeling of complex hydrocarbon mixtures: gasoline. Environ Toxicol Pharmacol 16:107–119.

Dennison JE, Andersen ME, Yang RSH. 2003. Characterization of the pharmacokinetics of gasoline using PBPK modeling with a complex mixture chemical lumping approach. Inhalation Toxicol 15:961–968.

De Rosa CT, El-Masri HE, Pohl H, Cibulas W, Mumtaz MM. 2004. Implications of chemical mixtures for public health practice. J Toxicol Environ Health 7:339–350.

De Rosa CT, Hansen H, Wilbur S, Pohl HR, El-Masri HA, Mumtaz MM. 2001. Interactions. In: Bingham E, Cohrssen B, Powell C, editors, Patty's toxicology. Vol. 1. New York: John Wiley & Sons, p 233–284.

De Zwart D. 2005. Ecological effects of pesticide use in the Netherlands: modeled and observed effects in the field ditch. IEAM 1:123–134.

De Zwart D, Dyer SD, Posthuma L, Hawkins CP. 2006. Use of predictive models to attribute potential effects of mixture toxicity and habitat alteration on the biological condition of fish assemblages. Ecol Appl 16:1295–1310.

De Zwart D, Posthuma L. 2005. Complex mixture toxicity for single and multiple species: proposed methodologies. Environ Toxicol Chem 24:2665–2676.

De Zwart D, Rutgers M, Notenboom J. 1998. Assessment of site-specific ecological risks of soil contamination: a design of an assessment methodology. Report nr 711701011. Bilthoven (NL): National Institute for Public Health and the Environment (RIVM).

De Zwart D, Sterkenburg A. 2002. Toxicity-based assessment of water quality. In: Posthuma L, Suter GW II, Traas TP, editors, Species sensitivity distributions in ecotoxicology. Boca Raton (FL): Lewis Publishers, p 383–402.

Dobrev I, Andersen ME, Yang RSH. 2001. Assessing interaction thresholds for trichloroethylene, tetrachloroethylene, and 1,1,1-trichloroethane using gas uptake studies and PBPK modeling. Arch Toxicol 75:134–144.

Dobrev I, Andersen ME, Yang RSH. 2002. In silico toxicology: simulating interaction thresholds for human exposure to mixtures of trichloroethylene, tetrachloroethylene, and 1,1,1-trichloroethane. Environ Health Perspect 110:1031–1039.

Dobson PD, Kell DB. 2008. Carrier-mediated cellular uptake of pharmaceutical drugs: an exception or the rule? Nature Rev Drug Discovery 7:205–220.

Dorne JLCM. 2007. Human variability in hepatic and renal elimination: implications for risk assessment. J Appl Toxicol 27:411–420.

Dorne JLCM, Papadopoulos A. 2008. Do uncertainty factors take into account toxicokinetic interactions? Conclusions and recommendations from the sixth framework project NOMIRACLE. Toxicol Lett 180:S90–S90.

Dorne JLCM, Ragas AMJ, Frampton GK, Spurgeon DS, Lewis DF. 2007b. Trends in human risk assessment of pharmaceuticals. Anal Bioanal Chem 387:1167–1172.

Dorne JLCM, Ragas AMJ, Lokke H. 2006. Harmonisation of uncertainty factors in human and ecological risk assessment. Toxicology 226:77–78.

Dorne JLCM, Skinner L, Frampton GK, Spurgeon DJ, Ragas AMJ. 2007a. Human and environmental risk assessment of pharmaceuticals: differences, similarities, lessons from toxicology. Anal Bioanal Chem 387:1259–1268.

Dorne JLCM, Walton K, Renwick AG. 2003. Human variability in CYP3A4 metabolism and CYP3A4-related uncertainty, factors for risk assessment. Food Chem Toxicol 41:201–224.

Dorne JLCM, Walton K, Renwick AG. 2005. Human variability in xenobiotic metabolism and pathway-related uncertainty factors for chemical risk assessment: a review. Food Chem Toxicol 43:203–216.

Drescher K, Bödeker W. 1995. Assessment of the combined effects of substances—the relationship between concentration addition and independent action. Biometrics 51:716–730.

D'Souza RW, Francis WR, Andersen ME. 1988. Physiological model for tissue glutathione depletion and increased resynthesis after ethylene dichloride exposure. J Pharmacol Exp Ther 245:563–568.

Durhan EJ, Norberg-King TJ, Burkhard LP. 1993. Methods for aquatic toxicity identification evaluations. Phase II toxicity identification evaluation procedures for samples exhibiting acute and chronic toxicity. EPA/600/R-92/080. Duluth (MN): Environmental Research Laboratory, Office of Research and Development, US Environmental Protection Agency.

EC. 2001. White paper: strategy for a future Chemicals Policy Commission of the European Communities. COM(2001) 88 final. Brussels (BE).

EC. 2003. Technical guidance document on risk assessment. Ispra (IT): European Chemicals Bureau (ECB), Institute for Health and Consumer Protection, European Commission, Joint Research Centre.

ECB. 2003a. Technical guidance document on risk assessment in support of Commission Directive 93/67/EEC on risk assessment for new notified substances; Commission Regulation (EC) No. 1488/94 on risk assessment for existing substances; Directive 98/8/EC of the European Parliament and of the Council concerning the placing of biocidal products on the market. Part II. Environmental risk assessment. Ispra (IT): European Commission–Joint Research Centre, Institute for Health and Consumer Protection, European Chemicals Bureau (ECB), Chap 3, Appendix 1.

ECB. 2003b. Technical guidance document on risk assessment in support of Commission Directive 93/67/EEC on risk assessment for new notified substances; Commission Regulation (EC) No. 1488/94 on risk assessment for existing substances; Directive 98/8/EC of the European Parliament and of the Council concerning the placing of biocidal products on the market. Part IV. Emission scenario documents. Ispra (IT): European Commission–Joint Research Centre, Institute for Health and Consumer Protection, European Chemicals Bureau (ECB), Chap 7.

Egeler P, Römbke J, Meller M, Knacker T, Nagel R. 1999. Bioaccumulation test with tubificid sludgeworms in artificial media—development of a standardisable method. Hydrobiologia 406:271–280.

Egnell AC, Houston JB, Boyer CS. 2005. Predictive models of CYP3A4 heteroactivation: *in vitro-in vivo* scaling and pharmacophore modeling. J Pharmacol Exp Ther 312:926–937.

Eide I, Neverdal G, Thorvaldsen B, Grung B, Kvalheim OM. 2002. Toxicological evaluation of complex mixtures by pattern recognition: correlating chemical fingerprints to mutagenicity. Environ Health Perspect 110(Suppl 6):985–988.

Ekins S. 2003. *In silico* approaches to predicting drug metabolism, toxicology and beyond. Biochem Soc Trans 31(Pt 3):611–614.

Ekins S, Andreyev S, Ryabov A, Kirillov E, Rakhmatulin EA, Sorokina S, Bugrim A, Nikolskaya T. 2006. A combined approach to drug metabolism and toxicity assessment. Drug Metab Dispos 34:495–503.

El-Masri HA, Constan AA, Ramsdell HS, Yang RSH. 1996b. Physiologically based pharmacodynamic modeling of an interaction threshold between trichloroethylene and 1,1-dichloroethylene in Fischer 344 rats. Toxicol Appl Pharmacol 141:124–132.

El-Masri HA, Tessari JD, Yang RSH. 1996c. Exploration of an interaction threshold for the joint toxicity of trichloroethylene and 1,1-dichloroethylene: utilization of a PBPK model. Arch Toxicol 70:527–539.

El-Masri HA, Thomas RS, Sabados GR, Phillips JK, Constan AA, Benjamin SA, Andersen, ME, Mehendale HM, Yang RSH. 1996a. Physiologically based pharmacokinetic/pharmacodynamic modeling of the toxicologic interaction between carbon tetrachloride and kepone. Arch Toxicol 70:704–713.

Emond C, Charbonneau M, Krishnan K. 2005. Physiologically based modeling of the accumulation in plasma and tissue lipids of a mixture of PCB congeners in female Sprague-Dawley rats. J Toxicol Environ Health A 68:1393–1412.

Escher BI, Hunziker RW, Schwarzenbach RP. 2001. Interaction of phenolic uncouplers in binary mixtures: concentration-additive and synergistic effects. Environ Sci Technol 35: 3905–3914.

Escher BI, Sigg L. 2004. Chemical speciation of organics and of metals at biological interfaces. In: Van Leeuwen HP, Köster W, editors, Physicochemical kinetics and transport at biointerfaces. Vol. 9. Chichester (UK): John Wiley. p 205–271.

Evans JS, Gray GM, Sielken RLJ, Smith AE, Valdez FC, Graham JD, 1994. Use of probabilistic expert judgment in uncertainty analysis of carcinogenic potency. Reg Toxicol Pharmacol 20:15–36.

Faessel HM, Slocum HK, Rustum YM, Greco WR. 1999. Folic acid-enhanced synergy for the combination of trimetrexate plus the glycinamide ribonucleotide formyltransferase inhibitor 4-[2-(2-amino-4-oxo-4,6,7,8-tetrahydro-3H-pyrimidino[5,4,6][1,4]thiazin-6-yl)-(S)-ethyl]-2,5-thienoyl amino-L-glutamic acid (AG2034)—comparison across sensitive and resistant human tumor cell lines. Biochem Pharmacol 57:567–577.

Fairman R, Mead CD, Williams WP. 1998. Environmental risk assessment: approaches, experiences and information sources. Environmental Issue Report 4. Copenhagen (DK): European Environmental Agency.

Faust M, Altenburger R, Backhaus T, Blanck H, Boedeker W, Gramatica P, Hamer V, Scholze M, Vighi M, Grimme LH. 2003. Joint algal toxicity of 16 dissimilarly acting chemicals is predictable by the concept of independent action. Aquat Toxicol 63:43–63.

Faust M, Altenburger R, Backhaus T, Boedeker W, Gramatica P, Hamer V, Scholze M, Vighi M, Grimme LH. 2001. Predicting the joint algal toxicity of multi-component s-triazine mixtures at low-effect concentrations of individual toxicants. Aquat Toxicol 56:13–32.

Fay M. 2005. Exposure to contaminant mixtures at US hazardous waste sites. In: Aral MM, Brebbia CA, Maslia ML, Sinks T, editors, Environmental exposure and health. WIT transactions on ecology and the environment. Vol. 85. Southhampton (UK): WIT Press. p 227–232.

Fay RM, Mumtaz MM 1996. Development of a priority list of chemical mixtures, occurring at 1188 hazardous waste sites, using HazDat database. Food Chem Toxicol 34:1163–1165.

Fenner K, Scheringer M, MacLeod M, Matthies M, McKone T, Stroebe M, Beyer A, Bonnell M, Le Gall AC, Klasmeier J, Mackay D, Van De Meent D, Pennington D, Scharenberg B, Suzuki N, Wania F. 2005. Comparing estimates of persistence and long-range transport potential among multimedia models. Environ Sci Technol 39:1932–1942.

Fernandes AR, Rose M, White S, Mortimer DN, Gem M. 2006. Dioxins and polychlorinated biphenyls (PCBs) in fish oil dietary supplements: occurrence and human exposure in the UK. Food Add Contam 23:939–947.

Fernandez MF, Araque P, Kiviranta H, Molina-Molina JM, Rantakokko P, Laine O, Vartiainen T, Olea N. 2007a. PBDEs and PBBs in the adipose tissue of women from Spain. Chemosphere 66:377–383.

Fernandez MF, Olmos B, Granada A, López-Espinosa MJ, Molina-Molina JM, Fernandez JM, Cruz M, Olea-Serrano F, Olea N. 2007b. Human exposure to endocrine disrupting chemicals and prenatal risk factors for cryptorchidism and hypospadias: a nested case-control study. Environ Health Perspect 115(Suppl 1):8–14.

Feron VJ, Groten JP. 2002. Toxicological evaluation of chemical mixtures. Food Chem Toxicol 40:825–839.

Feron VJ, Groten JP, van Zorge JA, Cassee FR, Jonker D, van Bladeren PJ. 1995. Toxicity studies in rats of simple mixtures of chemicals with the same or different target organs. Toxicol Lett 82/83:505–512.

Ferrario J, Byrne C, Lorber M, Saunders P, Leese W, Dupuy A, Winters D, Cleverly D, Schaum J, Pinsky P, Deyrup C, Ellis R, Walcott J. 1997. A statistical survey of dioxin-like compounds in the United States poultry fat. Organohalogen Compounds 32:245–251.

Ferrario J, Byrne C, McDaniel D, Dupuy A, Harless R. 1996. Determination of 2,3,7,8-chlorine substituted dibenzo-p-dioxins and furans at the part per trillion level in United States beef fat using high resolution gas chromatography/high resolution mass spectrometry. Anal Chem 68:647–652.

Ferreira KL, Burton JD, Coble HD. 1995. Physiological basis for antagonism of fluazifop-p by DPX-PE350. Weed Sci 43:184–191.

Fialkowski W, Klonowska-Olejnik M, Smith BD, Rainbow PS. 2003. Mayfly larvae (*Baetis rhodani* and *B. vernus*) as biomonitors of trace metal pollution in streams of a catchment draining a zinc and lead mining area of Upper Silesia, Poland. Environ Pollut 121:253–267.

Filser JG, Johanson G, Kessler W, Kreuzer PE, Stei P, Baur C, Csanady GA. 1993. A pharmacokinetic model to describe toxicokinetic interactions between 1,3-butadiene and styrene in rats: predictions for human exposure. IARC Sci Publ 127:65–78.

Filzek PDB, Spurgeon DJ, Broll G, Svendsen C, Hankard PK, Kammenga JE, Donker MH, Weeks JM. 2004. Pedological characterisation of sites along a transect from a primary cadmium/lead/zinc smelting works. Ecotoxicology 13:725–737.

Finney DJ. 1942. The analysis of toxicity tests on mixtures of poisons. Ann Appl Biol 29:82–94.

Fiore BJ, Anderson HA, Hanrahan MS, Olson LJ, Sonzogni WC. 1989. Sport fish consumption and body burden levels of chlorinated hydrocarbons: a study of Wisconsin anglers. Arch Environ Health 44:82–88.

Forbes VE, Palmqvist A, Bach L. 2006. The use and misuse of biomarkers in ecotoxicology. Environ Toxicol Chem 25:272–280.

Foster KL, MacKay D, Parkerton TF, Webster E, Milford L. 2005. Five-stage environmental exposure assessment strategy for mixtures: gasoline as a case study. Environ Sci Technol 39:2711–2718.

Fouchecourt MO, Beliveau M, Krishnan K. 2001. Quantitative structure-pharmacokinetic relationship modelling. Sci Total Environ 274:125–135.

Franchi M, Carrer P, Kotzias D, Rameckers E, Seppanen O, van Bronswijk JEMH, Viegi G, Gilder JA, Valovirta E. 2006. Working towards healthy air in dwellings in Europe. Allergy 61:864–868.

Fraysse B, Baudin J-P, Garnier-Laplace J, Adam C, Boudou A. 2002. Effects of Cd and Zn waterborne exposure on the uptake and depuration of 57Co, 110Ag and 134Cs by the Asiatic clam (*Corbicula fluminea*) and the zebra mussel (*Dreissena polymorpha*) — whole organism study. Environ Pollut 118:297–306.

Frederick CB, Potter DW, Chang-Mateu MI, Andersen ME. 1992. A physiologically based pharmacokinetic and pharmacodynamic model to describe the oral dosing of rats with ethyl acrylate and its implications for risk assessment. Toxicol Appl Pharmacol 114:246–260.

Fries GF, Marrow, GS. 1992. Influence of soil properties on the uptake of hexachlorobiphenyls by rats. Chemosphere 24:109–113.

Fryer M, Collins CD, Ferrier H, Colvile RN, Nieuwenhuijsen MJ. 2006. Human exposure modelling for chemical risk assessment: a review of current approaches and research and policy implications. Environ Sci Policy 9:261–274.

Fulton MH, Key PB. 2001. Acetylcholinesterase inhibition in estuarine fish and invertebrates as an indicator of organophosphorus insecticide exposure and effects. Environ Toxicol Chem 20:37–45.

Gabrielsson J, Weiner D. 2000. Pharmacokinetic and pharmacodynamic data analysis, concepts and applications. 3rd ed. Stockholm (SE): Swedish Pharmaceutical Press.

Gagne F, Blaise C. 2004. Review of biomarkers and new techniques for *in-situ* aquatic studies with bivalves. In: Thompson KC, Wadhia, K, Loibner A, editors, Environmental toxicity testing. Sheffield Analytical Chemistry Series. Oxford (UK): Blackwell Publishing, Chap 7.

Garcia-Ortega S, Holliman PJ, Jones DL. 2006. Toxicology and fate of Pestanal® and commercial propetamphos formulations in river and estuarine sediment. Sci Total Environ 366:826–836.

Gaumont Y, Kisliuk RL, Parsons JC, Greco WR. 1992. Quantitation of folic-acid enhancement of antifolate synergism. Cancer Res 52:2228–2235.

Gay JR, Korre A. 2006. A spatially-evaluated methodology for assessing risk to a population from contaminated land. Environ Pollut 142:227–234.

Gelman A, Bois FY, Jiang J. 1996. Physiological pharmacokinetic analysis using population modeling and informative prior distributions. J Am Stat Assoc 91:1400–1412.

Gennings C. 1995. An efficient experimental design for detecting departure from additivity in mixtures of many chemicals. Toxicology 105:189–197.

Gennings C. 1996. Economical designs for detecting and characterizing departure from additivity in mixtures of many chemicals. Food Chem Toxicol 34:1053–1058.

Gennings C, Carter WH. 1995. Utilizing concentration–response data from individual components to detect statistically significant departures from additivity in chemical mixtures. Biometrics 51:1264–1277.

Gennings C, Carter WH, Campain JA, Bae DS, Yang RSH. 2002. Statistical analysis of interactive cytotoxicity in human epidermal keratinocytes following exposure to a mixture of four metals. J Agric Biol Environ Stat 7:58–73.

Gennings C, Carter WH Jr, Carchman RA, Teuschler LK, Simmons JE, Carney EW. 2005. A unifying concept for assessing toxicological interactions: changes in slope. Toxicol Sci 88:287–297.

Gennings C, Carter WH, Caseya M, Moser V, Carchman R, Simmons JE. 2004. Analysis of functional effects of a mixture of five pesticides using a ray design. Environ Toxicol Pharmacol 18:115–125.

Gentry PR, Covington TR, Clewell HJ. 2003. Evaluation of the potential impact of pharmacokinetic differences on tissue dosimetry in offspring during pregnancy and lactation. Regul Toxicol Pharmacol 38:1–16.

Gerhardt A, Janssens de Bisthoven L, Guhr K, Soares AMVM, Pereira MJ. 2008. Phytotoxicity assessment of acid mine drainage: *Lemna gibba* bioassay and diatom community structure. Ecotoxicology 17:47–58.

Gerhardt A, Janssens de Bisthoven L, Soares AMVM. 2004. Macroinvertebrate response to acid mine drainage: community metrics and on-line behavioural toxicity bioassay. Environ Pollut 130:263–274.

Gerhardt A, Janssens de Bisthoven L, Soares AMVM. 2005. Effects of acid mine drainage and acidity on the activity of *Choroterpes picteti* (Ephemeroptera). Arch Environ Contam Toxicol 48:450–459.

Ginsberg G, Hattis D, Sonawane B. 2004. Incorporating pharmacokinetic differences between children and adults in assessing children's risks to environmental toxicants. Toxicol Appl Pharmacol 198:164–183.

Gobas FAPC, McCorquodale JR, Haffner GD. 1993. Intestinal absorption and biomagnification of organochlorines. Environ Toxicol Chem 12:567–576.

Goktepe I, Plhak LC. 2002. Comparative toxicity of two azadirachtin-based neem pesticides to *Daphnia pulex*. Environ Toxicol Chem 21:31–36.

Gough M. 1991. Human exposures from dioxin in soil—a meeting report. J Toxicol Environ Health 32:205–245.

Gouin T, Mackay D, Jones KC, Harner T, Meijer SN. 2004. Evidence for the "grasshopper" effect and fractionation during long-range atmospheric transport of organic contaminants. Environ Pollut 128:139–148.

Greco WR, Bravo G, Parsons JC. 1995. The search for synergy: a critical review from a response surface perspective. Pharmacol Rev 47:331–385.

Greco WR, Park HS, Rustum YM. 1990. An application of a new approach for the quantitation of drug synergism to the combination of cis-diamminedichloroplatinum and 1-b-D-arabinosefuranosylcytosine. Cancer Res 50:5318–5327.

Greco, WR, Unkelbach HD, Pöch G, Sühnel J, Kundi M, Boedeker W. 1992. Consensus on concepts and terminology for interaction assessment: the Saarselskä agreement. Arch Complex Environ Stud 4:65–69.

Grimme LH, Altenburger R, Boedeker W, Faust M. 1994. Kombinationswirkungen von Schadstoffen—Toxizität binärer Kombinationen von Pestiziden und Tensiden im Algenbiotest. Forschungsbericht Nr. 94-10207205 im Auftrag des Umweltbundesamtes.

Grote M, Brack W, Walter HA, Altenburger R. 2005. Confirmation of cause-effect relationships using effect-directed analysis for complex environmental samples. Environ Toxicol Chem 24:1420–1427.

Groten JP, Feron VJ, Suhnel J. 2001. Toxicology of simple and complex mixtures. Trends Pharm Sci 22:316–322.

Groten JP, Schoen ED, Van Bladeren PJ, Kuper CF, Van Zorge JA, Feron VJ. 1997. Subacute toxicity of a mixture of nine chemicals in rats: detecting interactive effects with a fractionated two-level factorial design. Fund Appl Toxicol 36:15–29.

Guha S, Jaffe PR, Peters CA. 1998. Bioavailability of mixtures of PAHs partitioned into the micellar phase of a nonionic surfactant. Environ Sci Technol 32:2317–2324.

Guha S, Peters CA, Jaffe PR. 1999. Multisubstrate biodegradation kinetics of naphthalene, phenanthrene, and pyrene mixtures. Biotechn Bioeng 65:491–499.

Gust KA, Fleeger JW. 2005. Exposure-related effects on Cd bioaccumulation explain toxicity of Cd-phenanthrene mixtures in *Hyalella azteca*. Environ Toxicol Chem 24:2918–2926.

Haanstra L, Doelman P, Oude Voshaar JH. 1985. The use of sigmoidal dose response curves in soil ecotoxicological research. Plant Soil 84:293–297.

Haas CN, Cidambi K, Kersten S, Wright K. 1996. Quantitative description of mixture toxicity: effect of level of response on interactions. Environ Toxicol Chem 15:1429–1437.

Haas CN, Kersten SP, Wright K, Frank MJ, Cidambi K. 1997. Generalization of independent response model for toxic mixtures. Chemosphere 34:699–710.

Hack CE. 2006. Bayesian analysis of physiologically based toxicokinetic and toxicodynamic models. Toxicology 221:241–248.

Haddad S, Charest-Tardif G, Krishnan K. 2000c. Physiologically based modeling of the maximal effect of metabolic interactions on the kinetics of components of complex chemical mixtures. J Toxicol Environ Health A 61:209–223.

Haddad S, Charest-Tardif G, Tardif R, Krishnan K. 2000b. Validation of a physiological modeling framework for simulating the toxicokinetics of chemicals in mixtures. Toxicol Appl Pharmacol 167:199–209.

Haddad S, Krishnan K. 1998. Physiological modeling of toxicokinetic interactions: implications for mixture risk assessment. Environ Health Perspect 106(Suppl 6):1377–1384.

Haddad S, Poulin P, Krishnan K. 2000a. Relative lipid content as the sole mechanistic determinant of the adipose tissue:blood partition coefficients of highly lipophilic organic chemicals. Chemosphere 40:839–843.

Haddad S, Tardif R, Charest-Tardif G, Krishnan K. 1999. Physiological modeling of the toxicokinetic interactions in a quaternary mixture of aromatic hydrocarbons. Toxicol Appl Pharmacol 161:249–257.

Haddad S, Withey J, Laparé S, Law FCP, Tardif R, Krishnan K. 1998. Physiologically-based pharmacokinetic modeling of pyrene in the rat. Environ Toxicol Pharmacol 5:245–255.

Hakooz N, Ito K, Rawden H, Gill H, Lemmers L, Boobis AR, Edwards RJ, Carlile DJ, Lake BG, Houston JB. 2006. Determination of a human hepatic microsomal scaling factor for predicting *in vivo* drug clearance. Pharm Res 23:533–539.

Harbers JV, Huijbregts MAJ, Posthuma L, Van de Meent D. 2006. Estimating the impact of high-production-volume chemicals on remote ecosystems by toxic pressure calculation. Environ Sci Technol 40:1573–1580.

Hass U, Scholze M, Christiansen S, Dalgaard M, Vinggaard AM, Axelstad M, Metzdorff SB, Kortenkamp A. 2007. Combined exposure to anti-androgens exacerbates disruption of sexual differentiation in the rat. Environ Health Perspect 115(Suppl 1):122–128.

Hassanin A, Johnston AE, Thomas GO, Jones KC. 2005. Time trends of atmospheric PBDEs inferred from archived UK herbage. Environ Sci Technol 39:2436–2441.

Hatzinger PB, Alexander M. 1995. Effect of aging of chemicals in soil on their biodegradability and extractability. Environ Sci Technol 29:537–545.

Hauser R, Chen Z, Pothier L, Ryan L, Altshul L. 2003a. The relationship between human semen parameters and environmental exposure to polychlorinated biphenyls and p,p'-DDE. Environ Health Perspect 111:1505–1511.

Hauser R, Singh NP, Chen Z, Pothier L, Altshul L. 2003b. Lack of an association between environmental exposure to polychlorinated biphenyls and p,p'-DDE and DNA damage in human sperm measured using the neutral comet assay. Hum Reprod 18:2525–2533.

Hawkins CP, Norris RH, Hogue JN, Feminella JW. 2000. Development and evaluation of predictive models for measuring the biological integrity of streams. Ecol Appl 10:1456–1477.

Haws NW, Ball WP, Bouwer EJ. 2006. Modeling and interpreting bioavailability of organic contaminant mixtures in subsurface environments. J Contam Hydrol 82:255–292.

Hearl FJ. 2005. Occupational exposure to chemical mixtures. Presented at the First International Conference on Environmental Exposure and Health, Atlanta (GA).

Hela DG, Konstantinou IK, Sakellarides TM, Lambropoulou DA, Akriotis T, Albanis TA. 2006. Persistent organochlorine contaminants in liver and fat of birds of prey from Greece. Arch Environ Contam Toxicol 50:603–613.

Hemond HF, Solo-Gabriele HM. 2004. Children's exposure to arsenic from CCA-treated wooden decks and playground structures. Risk Anal 24:51–64.

Hendriks AJ, Heikens A. 2001. The power of size. 2. Rate constants and equilibrium ratios for accumulation of inorganic substances related to species weight. Environ Toxicol Chem 20:1421–1437.

Hendriks AJ, Van der Linde A, Cornelissen G, Sijm D. 2001. The power of size. 1. Rate constants and equilibrium ratios for accumulation of organic substances related to octanol-water partition ratio and species weight. Environ Toxicol Chem 20:1399–1420.

Henning-de Jong I, Van Zelm R, Huijbregts MAJ, De Zwart D, Van der Linden TMA, Wintersen A, Posthuma L, Van de Meent D. 2008. Ranking of agricultural pesticides in the Rhine-Meuse-Scheldt Basin based on toxic pressure in marine ecosystems. Environ Toxicol Chem 27:737–745.

Hermens J, Busser F, Leeuwangh P, Musch A. 1985a. Quantitative structure–activity relationships and mixture toxicity of organic chemicals in *Photobacterium phosphoreum*: the Microtox test. Ecotoxicol Environ Safety 9:17–25.

Hermens J, Busser F, Leeuwangh P, Musch A. 1985c. Quantitative correlation studies between acute lethal toxicity of 15 organic halides to the guppy (*Poecilia reticulata*) and chemical reactivity towards 4-nitrobenzylpyridine. Toxicol Environ Chem 9:219–223.

Hermens J, Canton H, Janssen P, De Jong R. 1984. Quantitative structure–activity relationships and toxicity studies of mixtures of chemicals with anaesthetic potency: acute lethal and sublethal toxicity to *Daphnia magna*. Aquat Toxicol 5:143–154.

Hermens J, Leeuwangh P, Musch A. 1985b. Joint toxicity of mixtures of groups of organic aquatic pollutants to the guppy (*Poecilia reticulata*). Ecotoxicol Environ Safety 9:321–326.

Hertwich EG, McKone TE, Pease WS. 1999. Parameter uncertainty and variability in evaluative fate and exposure models. Risk Anal 19:1193–1204.

Hertzberg RC, MacDonell MM. 2002. Synergy and other ineffective mixture risk definitions. Sci Total Environ 288:31–42.

Hertzberg RC, Teuschler LK. 2002. Evaluating quantitative formulas for dose–response assessment of chemical mixtures. Environ Health Perspect 110:965–970.

Heugens EHW. 2003. Predicting effects of multiple stressors on aquatic biota. PhD thesis, University of Amsterdam (NL).

Heugens EHW, Hendriks AJ, Dekker T, van Straalen NM, Admiraal W. 2001. A review of the effects of multiple stressors on aquatic organisms and analysis of uncertainty factors for use in risk assessment. Crit Rev Toxicol 31:247–284.

Hewlett PS, Plackett RL. 1959. A unified theory for quantal responses to mixtures of drugs: non-interactive action. Biometrics 15:591–610.

Hickie BE, Mackay D, De Koning J. 1999. Lifetime pharmacokinetic model for hydrophobic contaminants in marine mammals. Environ Toxicol Chem 18:2622–2633.

Hill AV. 1910. The possible effects of the aggregation of the molecules of haemoglobin on its dissociation curves. J Physiol 40:iv–vii.

Hodgson E, Rose RL. 2005. Human metabolism and metabolic interactions of deployment-related chemicals. Drug Metab Rev 37:1–39.

Holford NH, Sheiner LB. 1981. Pharmacokinetic and pharmacodynamic modeling in vivo. Crit Rev Bioeng 5:273–322.

Hope BK. 2001. A case study comparing static and spatially explicit ecological exposure analysis methods. Risk Anal 21:1001–1010.

Hope BK. 2005. Performing spatially and temporally explicit ecological exposure assessments involving multiple stressors. Human Ecol Risk Assess 11:539–565.

Hopkin SP, Hardisty GN, Martin MH. 1986. The woodlouse Porcellio scaber as a "biological indicator" of zinc, cadmium, lead and copper pollution. Environ Pollut 11B:271–290.

Hopkin SP, Jones DT, Dietrich D. 1993. The terrestrial isopod Porcellio scaber as a monitor of the bioavailability of metals: towards a global "woodlouse watch" scheme. Sci Total Environ Suppl:357–365.

Houba VGJ, Lexmond TM, Novozamsky I, Van der Lee JJ. 1996. State of the art and future developments in soil analysis for bioavailability assessment. Sci Total Environ 178:21–28.

Houston JB, Galetin A. 2003. Progress towards prediction of human pharmacokinetic parameters from in vitro technologies. Drug Metab Rev 35:393–415.

Houston JB, Kenworthy KE. 2000. In vitro-in vivo scaling of CYP kinetic data not consistent with the classical Michaelis-Menten model. Drug Metab Dispos 28:246–254.

Houtman CJ, Cenijn PH, Hamers T, Lamoree MH, Legler J, Murk AJ, Brouwer A. 2004. Toxicological profiling of sediments using in vitro bioassays, with emphasis on endocrine disruption. Environ Toxicol Chem 23:32–40.

Huang XH, Qiu FR, Xie HT, Li J. 2005. Pharmacokinetic and pharmacodynamic interaction between irbesartan and hydrochlorothiazide in renal hypertensive dogs. J Cardiovasc Pharmacol 46:863–869.

Humphrey HEB. 1983. Evaluation of humans exposed to waterborne chemicals in the Great Lakes. Final report to the Environmental Protection Agency. Lansing (MI): Department of Public Health.

Hunter BA, Johnson MS, Thompson DJ. 1987. Ecotoxicology of copper and cadmium in a contaminated grassland ecosystem. 3. Small mammals. J Appl Ecol 24:601–614.

Hunter BA, Johnson MS, Thompson DJ. 1989. Ecotoxicology of copper and cadmium in a contaminated grassland ecosystem. 4. Tissue distribution and age accumulation in small mammals. J Appl Ecol 29:89–99.

Hutcheson MS, Pedersen D, Anastasa ND, Fitzgerald J, Silverman D. 1996. Beyond TPH: health-based evaluation of petroleum hydrocarbon exposures. Reg Toxicol Pharmacol 24:85–101.

Ibarluzea JJ, Fernandez MF, Santa-Marina L, Olea-Serrano MF, Rivas AM, Aurrekoetxea JJ, Exposito J, Lorenzo M, Torne P, Villalobos M, Pedraza V, Sasco AJ, Olea N. 2004. Breast cancer risk and the combined effect of environmental oestrogens. Cancer Causes Control 15:591–600.

IJC. 1983. An inventory of chemical substances identified in the Great Lakes ecosystem. Vols. 1–6. Windsor, Ontario (CA): International Joint Commission.

Irving EC, Baird DJ, Culp JM. 2003. Ecotoxicological responses of the mayfly *Baetis tricaudatus* to dietary and waterborne cadmium: implications for toxicity testing. Environ Toxicol Chem 22:1058–1064.

Isaacs KK, Evans MV, Harris TR. 2004. Visualization-based analysis for a mixed-inhibition binary PBPK model: determination of inhibition mechanism. J Pharmacokinet Pharmacodyn 31:215–242.

Ishigam M, Uchiyama M, Kondo T, Iwabuchi H, Inoue S, Takasaki W, Ikeda T, Komai T, Ito K, Sugiyama Y. 2001. Inhibition of *in vitro* metabolism of simvastatin by itraconazole in humans and prediction of *in vivo* drug-drug interactions. Pharm Res 18:622–631.

Ito K, Houston JB. 2004. Comparison of the use of liver models for predicting drug clearance using *in vitro* kinetic data from hepatic microsomes and isolated hepatocytes. Pharm Res 21:785–792.

Ito K, Houston JB. 2005. Prediction of human drug clearance from *in vitro* and preclinical data using physiologically based and empirical approaches. Pharm Res 22:103–112.

Ito N, Imaida K, Hasegawa R, Tsuda H. 1989a. Rapid bioassay methods for carcinogens and modifiers of hepatocarcinogenesis. Crit Rev Toxicol 19:385–415.

Ito N, Tatematsu M, Hasegawa R, Tsuda H. 1989b. Medium-term bioassay system for detection of carcinogens and modifiers of hepatocarcinogenesis utilizing the GST-P positive liver cell focus as an endpoint marker. Toxicol Pathol 17:630–641.

IUPAC 1997. IUPAC compendium of chemical terminology. 2nd ed. Triangle Park (NC): International Union of Pure and Applied Chemistry.

Jacobson JL, Jacobson SW. 1996. Intellectual impairment in children exposed to polychlorinated biphenyls *in utero*. N Engl J Med 335:783–789.

Jager T, Crommentuijn T, Van Gestel CAM, Kooijman SALM. 2004. Simultaneous modeling of multiple endpoints in life-cycle toxicity tests. Environ Sci Technol 38:2894–2900.

Jager T, Crommentuijn T, Van Gestel CAM, Kooijman SALM. 2007. Chronic exposure to chlorpyrifos reveals two modes of action in the springtail *Folsomia candida*. Environ Pollut 145:452–458.

Jager T, Fleuren RHLJ, Hogendoorn EA, De Korte G. 2003. Elucidating the routes of exposure for organic chemicals in the earthworm, *Eisenia andrei* (Oligochaeta). Environ Sci Technol 37:3399–3404.

Jager T, Heugens EHW, Kooijman SALM. 2006. Making sense of ecotoxicological test results: towards application of process-based models. Ecotoxicology 15:305–314.

Jager T, Kooijman SALM. 2005. Modeling receptor kinetics in the analysis of survival data for organophosphorus pesticides. Environ Sci Technol 39:8307–8314.

Jager T, Kooijman SALM. 2009. A biology-based approach for quantitative structure–activity relationships (QSARs) in ecotoxicity. Ecotoxicology 18:187–196.

Jager T, Posthuma L, De Zwart D, Van de Meent D. 2007. Novel view on predicting acute toxicity: decomposing toxicity data in species vulnerability and chemical potency. Ecotoxicol Environ Safety 67:311–322.

Jager T, Vandenbrouck T, Baas J, De Coen WM, Kooijman SALM. 2010. A biology-based approach for mixture toxicity of multiple endpoints over the life cycle. Ecotoxicology (doi:10.1007/s10646-009-0417-z).

Jager T, Van der Wal L, Fleuren RHLJ, Barendregt A, Hermens JLM. 2005. Bioaccumulation of organic chemicals in contaminated soils: evaluation of bioassays with earthworms. Environ Sci Technol 39:293–298.

Jang JY, Droz PO, Kim S. 2001. Biological monitoring of workers exposed to ethylbenzene and co-exposed to xylene. Int Arch Occup Environ Health 74:31–37.

Janssen MPM, Bruins A, De Vries TH, Van Straalen NM. 1991. Comparison of cadmium kinetics in four soil arthropod species. Arch Environ Contam Toxicol 20:305–312.

Janssen RPT, Posthuma L, Baerselman R, Den Hollander HA, Van Veen RPM, Peijnenburg WJGM. 1997. Equilibrium partitioning of heavy metals in Dutch field soils. II. Prediction of metal accumulation in earthworms. Environ Toxicol Chem 16:2479–2488.

Janssens de Bisthoven L, Gerhardt A, Soares AMVM. 2004. Effects of acid mine drainage on *Chironomus* spp. (Diptera) in laboratory and *in situ* bioassays with the multispecies freshwater biomonitor. Environ Toxicol Chem 23:1123–1128.

Janssens de Bisthoven L, Gerhardt A, Soares AMVM. 2005. Chironomidae as bioindicators of acid mine drainage stress. Hydrobiologia 532:181–191.

Janssens de Bisthoven L, Gerhardt A, Soares AMVM. 2006. Behavioural changes and acute toxicity of the freshwater shrimp *Atyaephyra desmaresti* Millet (Decapoda: Natantia) from exposure to acid mine drainage. Ecotoxicology 15:215–227.

Jaspers V, Covaci A, Maervoet J, Dauwe T, Voorspoels S, Schepens P, Eens M. 2005. Brominated flame retardants and organochlorine pollutants in eggs of little owls (*Athene noctua*) from Belgium. Environ Pollut 136:81–88.

Jensen J, Mesman M. (Eds). 2006. Ecological risk assessment of contaminated land. Decision support for site specific investigations. Report 711701047. Bilthoven (NL): National Institute for Public Health and the Environment (RIVM).

Johnston G, Walker CH, Dawson A. 1994. Interactive effects of prochloraz and malathion in pigeon, starling and hybrid red-legged partridge. Environ Toxicol Chem 13:115–120.

Jones DT, Hopkin SP. 1991. Biological monitoring of metal pollution in terrestrial ecosystems. In: Ravera O, editor, Terrestrial and aquatic ecosystems: perturbation and recovery. Chichester (UK): Ellis Horwood. p 148–152.

Jonker D, Woutersen RA, Feron VJ. 1996. Toxicity of mixtures of nephrotoxicants with similar or dissimilar mode of action. Food Chem Toxicol 34:1075–1082.

Jonker D, Woutersen RA, van Bladeren PJ, Til HP, Feron VJ. 1990. 4-Week oral toxicity study of a combination of eight chemicals in rats: comparison with the toxicity of the individual compounds. Food Chem Toxicol 28:623–631.

Jonker D, Woutersen RA, van Bladeren PJ, Til HP, Feron VJ. 1993. Subacute (4-wk) oral toxicity of a combination of four nephrotoxins in rats: comparison with the toxicity of the individual compounds. Food Chem Toxicol 31:125–136.

Jonker DM, Vermeij DA, Edelbroek PM, Voskuyl RA, Piotrovsky VK, Danhof M. 2003. Pharmacodynamic analysis of the interaction between tiagabine and midazolam with an allosteric model that incorporates signal transduction. Epilepsia 44:329–338.

Jonker MJ. 2003. Joint toxic effects on *Caenorhabditis elegans*: on the analysis and interpretation of mixture toxicity data. PhD thesis, Wageningen University, Wageningen (NL).

Jonker MJ, Svendsen C, Bedaux JJM, Bongers M, Kammenga JE. 2005. Significance testing of synergistic/antagonistic, dose level-dependent, or dose ratio-dependent effects in mixture dose–response analysis. Environ Toxicol Chem 24:2701–2713.

Jonker MJ, Sweijen RAJC, Kammenga JE. 2004. Toxicity of simple mixtures to the nematode *Caenorhabditis elegans* in relation to soil sorption. Environ Toxicol Chem 23:480–488.

Jonkers RE, Koopmans RP, Portier EJ, van Boxtel CJ. 1991. Debrisoquine phenotype and the pharmacokinetics and beta-2 receptor pharmacodynamics of metoprolol and its enantiomers. J Pharmacol Exp Ther 256:959–966.

Jonsson F, Johanson G. 2003. The Bayesian population approach to physiological toxicokinetic-toxicodynamic models—an example using the MCSim software. Toxicol Lett 138:143–150.

Jouraeva VA, Johnson DL, Hassett JP, Nowak DJ. 2002. Differences in accumulation of PAHs and metals on the leaves of *Tilia* × *euchlora* and *Pyrus calleryana*. Environ Pollut 120:331–338.

Kaag NHBM, Scholten MCT, Van Straalen NM. 1998. Factors affecting PAH residues in the lugworm *Arenicola marina*, a sediment feeding polychaete. J Sea Res 40:251–261.

Kammenga J, Dallinger R, Donker MH, Köhler HR, Simonsen V, Triebskorn R, Weeks JM. 2000. Biomarkers in terrestrial invertebrates: potential and limitations for ecotoxicological soil risk assessment. Rev Environ Contam Toxicol 164:93–147.

Kanamitsu S, Ito K, Green CE, Tyson CA, Shimada N, Sugiyama Y. 2000a. Prediction of *in vivo* interaction between triazolam and erythromycin based on *in vitro* studies using human liver microsomes and recombinant human CYP3A4. Pharm Res 17:419–426.

Kanamitsu SI, Ito K, Okuda H, Ogura K, Watabe T, Muro K, Sugiyama Y. 2000b. Prediction of *in vivo* drug-drug interactions based on mechanism-based inhibition from *in vitro* data: inhibition of 5-fluorouracil metabolism by (E)-5-(2-bromovinyl)uracil. Drug Metab Dispos 28:467–474.

Kapo KE, Burton GA Jr. 2006. A geographic information systems-based, weights of evidence approach for diagnosing aquatic ecosystem impairment. Environ Toxicol Chem 25:2237–2249.

Kavlock RJ, Daston GP, De Rosa C, Fenner-Crisp P, Gray LE, Kaattari S, Lucier G, Luster M, Mae MJ, Maczka C, Miller R, Moore J, Rolland R, Scott G, Sheehan DM, Sinks T, Tilson HA. 1996. Research needs for the risk assessment of health and environmental effects of endocrine disruptors: a report of the USEPA-sponsored workshop. Environ Health Perspect 104(Suppl 4):715–740.

Kedderis GL, Mason AD, Niang LL, Wilkes CR. 2006. Exposures and internal doses of trihalomethanes in humans: multi-route contributions from drinking water [Final]. EPA/600/R-06/087. Washington (DC): US Environmental Protection Agency.

Kelsey JW, Alexander M. 1997. Declining bioavailability and inappropriate estimation of risk of persistent compounds. Environ Toxicol Chem 16:582–585.

Kenntner N, Krone O, Altenkamp R, Tataruch F. 2003a. Environmental contaminants in liver and kidney of free-ranging northern goshawks (*Accipiter gentilis*) from three regions of Germany. Arch Environ Contam Toxicol 45:128–135.

Kenntner N, Krone O, Oehme G, Heidecke D, Tataruch F. 2003b. Organochlorine contaminants in body tissue of free-ranging white-tailed eagles from northern regions of Germany. Environ Toxicol Chem 22:1457–1464.

Keys DA, Schultz IR, Mahle DA, Fisher JW. 2004. A quantitative description of suicide inhibition of dichloroacetic acid in rats and mice. Toxicol Sci 82:381–393.

Kim SK, Oh JR, Shim WJ, Lee DH, Yim UH, Hong SH, Shin YB, Lee DS. 2002. Geographical distribution and accumulation features of organochlorine residues in bivalves from coastal areas of South Korea. Mar Pollut Bull 45:268–279.

King DJ, Lyne RL, Girling A, Peterson DR, Stephenson R, Short D. 1996. Environmental risk assessment of petroleum substances: the hydrocarbon block method. Report 96/52. Brussels: Concawe, Petroleum Products Ecology Group.

King JK, Harmon SM, Fu TT, Gladden JB. 2002. Mercury removal, methylmercury formation, and sulfate-reducing bacteria profiles in wetland mesocosms. Chemosphere 46:859–870.

Klaassen CD. 1996. Casarett and Doull's toxicology: the basic science of poisons. New York: McGraw-Hill.

Klein MT, Hou G, Quann R, Wei W, Liao KH, Yang RSH, Campain JA, Mazurek M, Broadbelt LJ. 2002. BioMOL: a computer-assisted biological modeling tool for complex chemical mixtures and biological processes at the molecular level. Environ Health Perspect 110(Suppl 6):1025–1029.

Kodell RL, Chen JJ. 1994 Reducing conservatism in risk estimation for mixtures of carcinogens. Risk Anal 14:327–332.

Kodell RL, Pounds JG. 1991. Assessing the toxicity of mixtures of chemicals. In: Krewski D, Franklin C, editors, Statistics in toxicology. New York: Gordon and Breach, p 559–591.

Könemann H. 1980. Structure–activity relationships and additivity in fish toxicities of environmental pollutants. Ecotoxicol Environ Safety 4:415–421.

Könemann H. 1981. Fish toxicity tests with mixtures of more than two chemicals: a proposal for a quantitative approach and experimental results. Toxicology 19:229–238.

Kooijman SALM. 1981. Parametric analyses of mortality rates in bioassays. Water Res 15:107–119.

Kooijman SALM. 1996. An alternative for NOEC exists, but the standard model has to be abandoned first. Oikos 75:310–316.

Kooijman SALM. 2000. Dynamic energy and mass budgets in biological systems. Cambridge (UK): Cambridge University Press.

Kooijman SALM. 2001. Quantitative aspects of metabolic organization: a discussion of concepts. Phil Trans R Soc London B 356:331–349.

Kooijman SALM, Bedaux JJM. 1996. Analysis of toxicity tests on *Daphnia* survival and reproduction. Water Res 30:1711–1723.

Kooistra L, Huijbregts MAJ, Ragas AMJ, Wehrens R, Leuven RSEW. 2005. Spatial variability and uncertainty in ecological risk assessment: a case study on the potential risk of cadmium for the little owl in a Dutch river flood plain. Environ Sci Technol 39:2177–2187.

Koppe JG. 1995. Nutrition and breast-feeding. Eur J Obstet Gynecol Reprod 61:73–78.

Kortenkamp A. 2007. Ten years of mixing cocktails—a review of combination effects of endocrine disrupting chemicals. Environ Health Perspect 115(Suppl 1):98–105.

Kortenkamp A, Altenburger R. 1998. Synergisms with mixtures of xenoestrogens: a reevaluation using the method of isoboles. Sci Total Environ 221:59–73.

Kortenkamp A, Faust M, Scholze M, Backhaus T. 2007. Low-level exposure to multiple chemicals: reason for human health concerns? Environ Health Perspect 115(Suppl 1):106–114.

Kosian PA, Makynen EA, Monson PD, Mount DR, Spacie A, Mekenyan OG, Ankley GT. 1998. Application of toxicity-based fractionation techniques and structure–activity relationship models for the identification of phototoxic polycyclic aromatic hydrocarbons in sediment pore water. Environ Toxicol Chem 17:1021–1033.

Kramarz P. 1999a. The dynamics of accumulation and decontamination of cadmium and zinc in carnivorous invertebrates. 2. The centipede *Lithobius mutabilis* Koch. Bull Environ Contam Toxicol 63:538–545.

Kramarz P. 1999b. The dynamics of accumulation and decontamination of cadmium and zinc in carnivorous invertebrates. 1. The ground beetle, *Poecilus cupreus* L. Bull Environ Contam Toxicol 63:531–537.

Krishnan K, Andersen ME, Clewell HJ, Yang RSH. 1994. Physiologically based pharmacokinetic modeling of chemical mixtures. In: Yang RSH, editor, Toxicology of chemical mixtures: case studies, mechanisms and novel approaches. New York: Academic Press. p 399–437.

Krishnan K, Andersen ME, Hayes AW. 2001. Physiologically based pharmacokinetic modeling in toxicology. 4th ed. Philadelphia (PA): Taylor and Francis. p 193–241.

Krishnan K, Brodeur J. 1991. Toxicological consequences of combined exposure to environmental pollutants. Arch Complex Environ Studies 3:1–106.

Krishnan K, Haddad S, Beliveau M, Tardif R. 2002. Physiological modeling and extrapolation of pharmacokinetic interactions from binary to more complex chemical mixtures. Environ Health Perspect 110(Suppl 6):989–994.

Krishnan K, Pelekis M. 1995. Hematotoxic interactions: occurrence, mechanisms and predictability. Toxicology 105:355–364.

Krull IS, Mills K, Hoffman G, Fine DH. 1980. The analysis of N-nitrosoatrazine and N-nitrosocarbaryl in whole mice. J Anal Toxicol 4:260–262.

Küster E, Dorusch F, Vogt C, Weiss H, Altenburger R. 2004. On line biomonitors used as a tool for toxicity reduction evaluation of *in situ* groundwater remediation techniques. Biosensors Bioelectronics 19:1711–1722.

Kwon CS, Penner D. 1995. The interaction of insecticides with herbicide activity. Weed Technol 9:119–124.

Landrum PF, Steevens JA, Gossiaux DC, McElroy M, Robinson S, Begnoche L, Chernyak S, Hickey J. 2004. Time-dependent lethal body residues for the toxicity of pentachlorobenzene to *Hyalella azteca*. Environ Toxicol Chem 23:1335–1343.

Lau CE, Wang Y, Falk JL. 1997. Differential reinforcement of low rate performance, pharmacokinetics and pharmacokinetic-pharmacodynamic modeling: independent interaction of alprazolam and caffeine. J Pharmacol Exp Ther 281:1013–1029.

Law FC, Abedini S, Kennedy CJ. 1991. A biologically based toxicokinetic model for pyrene in rainbow trout. Toxicol Appl Pharmacol 110:390–402.

Leavens TL, Bond JA. 1996. Pharmacokinetic model describing the disposition of butadiene and styrene in mice. Toxicology 113:310–313.

Lee JH, Landrum PF. 2006a. Application of multi-component damage assessment model (MDAM) for the toxicity of metabolized PAH in *Hyalella azteca*. Environ Sci Technol 40:1350–1357.

Lee JH, Landrum PF. 2006b. Development of a multi-component damage assessment model (MDAM) for time-dependent mixture toxicity with toxicokinetic interactions. Environ Sci Technol 40:1341–1349.

Lee JH, Landrum PF, Koh CH. 2002a. Prediction of time-dependent PAH toxicity in *Hyalella azteca* using a damage assessment model. Environ Sci Technol 36:3131–3138.

Lee JH, Landrum PF, Koh CH. 2002b. Toxicokinetics and time-dependent PAH toxicity in the amphipod *Hyalella azteca*. Environ Sci Technol 36:3124–3130.

Lee JS, Lee JH. 2005. Influence of acid volatile sulfides and simultaneously extracted metals on the bioavailability and toxicity of a mixture of sediment-associated Cd, Ni, and Zn to polychaetes *Neanthes arenaceodentata*. Sci Total Environ 338:229–241.

Legierse KCHM, Verhaar HJM, Vaes WHJ, De Bruijn JHM, Hermens JLM. 1999. Analysis of the time-dependent acute aquatic toxicity of organophosphorus pesticides: the critical target occupation model. Environ Sci Technol 33:917–925.

Lepper, P. 2005. Manual on the methodological framework to derive environmental quality standards for priority substances in accordance with Article 16 of the Water Framework Directive (2000/60/EC). Schmallenberg (DE): Fraunhofer-Institute, Molecular Biology and Applied Ecology.

Leslie HA, Hermens JLM, Kraak MHS. 2004. Baseline toxicity of a chlorobenzene mixture and total body residues measured and estimated with solid-phase microextraction. Environ Toxicol Chem 23:2017–2021.

Levsen K, Preiss A, Spraul M. 2003. Structure elucidation of unknown pollutants of environmental samples by coupling HPLC to NMR and MS. In: Namiesnik J, Chrzanowski W, Zmijewska P, editors, New horizons and challenges in environmental analysis and monitoring, Workshop, Gdansk (PL), August 18–29. p 150–180.

Lewtas J. 1985. Development of a comparative potency method for cancer risk assessment of complex mixtures using short-term *in vivo* and *in vitro* bioassays. Toxicol Ind Health 1:193–203.

Lewtas J. 1988. Genotoxicity of complex mixtures: strategies for the identification and comparative assessment of airborne mutagens and carcinogens from combustion sources. Fund Appl Toxicol 10:571–589.

Liao KH. 2004. Development and validation of a hybrid reaction network/physiologically based pharmacokinetic model of benzo[a]pyrene and its metabolites. PhD dissertation, Department of Chemical and Biological Engineering, Colorado State University, Fort Collins (CO).

Liao KH, Dobrev I, Dennison JE, Andersen ME, Reisfeld B, Reardon KF, Campain JA, Wei W, Klein MT, Quann RJ, Yang RSH. 2002. Application of biologically based computer modeling to simple or complex mixtures. Environ Health Perspect 110(Suppl 6):957–963.

Lichtenstein EP, Liang TT, Anderegg BN. 1973. Synergism of insecticides by herbicides. Science 181:847–849.

Linkov I, Burmistrov D, Cura J, Bridges TS. 2002. Risk-based management of contaminated sediments: consideration of spatial and temporal patterns in exposure modeling. Environ Sci Technol 36:238–246.

Lock K, Janssen CR. 2001. Zinc and cadmium body burdens in terrestrial oligochaetes: use and significance in environmental risk assessment. Environ Toxicol Chem 20:2067–2072.

Lock K, Janssen CR. 2003. Influence of ageing on zinc bioavailability in soils. Environ Pollut 126:371–374.

Loewe S, Muischneck H. 1926. Effect of combinations: mathematical basis of problem. Arch Exp Pathol Pharmakol 114:313–326.

Lohitnavy M, Lu Y, Lohitnavy O, Chubb LS, Hirono S, Yang RSH. 2008. A possible role of multidrug resistance-associated protein 2 (Mrp2) in hepatic excretion of PCB126, an environmental contaminant: PBPK/PD modeling. Toxicol Sci 104:27–39.

Loonen H, Muir DCG, Parsons JR, Govers HAJ. 1997. Bioaccumulation of polychlorinated dibenzo-p-dioxins in sediments by oligochaetes: influence of exposure pathway and contact time. Environ Toxicol Chem 16:1518–1525.

Loos M, Schipper AM, Schlink U, Strebel K, Ragas AMJ. 2010. Receptor-oriented approaches in wildlife and human exposure modelling: a comparative study. Environ Model Software 25:369–382.

Lorber M, Cleverly D, Schaum J, Phillips L, Schweer G, Leighton T. 1994. Development and validation of an air-to-beef food chain model for dioxin-like compounds. Sci Total Environ 156:39–65.

Lorber M, Saunders P, Ferrario J, Leese W, Winters D, Cleverly D, Schaum J, Deyrup C, Ellis R, Walcott J, Dupuy A, Byrne C, McDaniel D. 1997. A statistical survey of dioxin-like compounds in United States pork fat. Organohalogen Compounds 32:80–86.

Lorber MN, Winters DL, Griggs J, Cook R, Baker S, Ferrario J, Byrne C, Dupuy A, Schaum J. 1998. A national survey of dioxin-like compounds in the United States milk supply. Organohalogen Compounds 38:125–129.

Loureiro S, Soares AMVM, Nogueira AJA. 2005. Terrestrial avoidance behaviour tests as screening tool to assess soil contamination. Environ Pollut 138:121–131.

Lu Y, Lohitnavy M, Reddy M, Lohitnavy O, Eickman E, Ashley A, Gerjevic L, Xu Y, Conolly RB, Yang RSH. 2008. Quantitative analysis of liver GST-P foci promoted by a chemical mixture of hexachlorobenzene and PCB 126: implication of size-dependent cellular growth kinetics. Arch Toxicol 82:103–116.

Lucier GW, Rumbaugh RC, McCoy Z, Hass R, Harvan D, Albro P. 1986. Ingestion of soil contaminated with 2,3,7,8-tetrachlorodibenzo-p-dioxin (TCDD) alters hepatic enzyme activities in rats. Fund Appl Toxicol 6:364–371.

Luecke RH, Wosilait WD. 1979. Drug elimination interactions: analysis using a mathematical model. J Pharmacokinet Biopharm 7:629–641.

Lundin F, Lloyd J, Smith E. 1969. Mortality of uranium miners in relation to radiation exposure, hard-rock mining and cigarette smoking—1950 through 1967. Health Phys 16:571–578.

Luoma SN, Rainbow PS. 2005. Why is metal bioaccumulation so variable? Biodynamics as a unifying concept. Environ Sci Technol 39:1921–1931.

Lutz WK, Lutz RW, Andersen ME. 2006. Dose-incidence relationships derived from superposition of distributions of individual susceptibility on mechanism-based dose responses for biological effects. Toxicol Sci 90:33–38.

Lydy MJ, Linck SL. 2003. Assessing the impact of triazine herbicides on organophosphate insecticide toxicity to the earthworm *Eisenia fetida*. Arch Environ Contam Toxicol 45:343–349.

Lyons M, Yang RSH, Mayeno AN, Reisfeld B. 2008. Computational toxicology of chloroform: reverse dosimetry using Bayesian inference, Markov chain Monte Carlo simulation, and human biomonitoring data. Environ Health Perspect 116:1040–1046.

Mackay D, Fraser A. 2000. Bioaccumulation of persistent organic chemicals: mechanisms and models. Environ Pollut 110:375–391.

Mackay D, Paterson S, Shiu WY. 1992a. Generic models for evaluating the regional fate of chemicals. Chemosphere 24:695–717.

Mackay D, Puig H, McCarty LS. 1992b. An equation describing the time course and variability in uptake and toxicity of narcotic chemicals to fish. Environ Toxicol Chem 11:941–951.

MacLeod M, Fraser AJ, Mackay D. 2002. Evaluating and expressing the propagation of uncertainty in chemical fate and bioaccumulation models. Environ Toxicol Chem 21:700–709.

MADEP. 2002. Characterizing risks posed by petroleum contaminated sites: implementation of the MADEP VPH/EPH approach. Boston (MA): Massachusetts Department of Environmental Protection. Available from: http://www.mass.gov/dep/cleanup/laws/policies.htm#02-411

MADEP. 2003. Updated petroleum hydrocarbon fraction toxicity values for the VPH/EPH/APH methodology. Boston (MA): Massachusetts Department of Environmental Protection. Available from: http://www.mass.gov/dep/water/drinking/standards/pethydro.htm

Mahmood I. 2002. Prediction of clearance in humans from *in vitro* human liver microsomes and allometric scaling: a comparative study of the two approaches. Drug Metabol Drug Interact 19:49–64.

Main KM, Kiviranta H, Virtanen HE, Sundquist E, Tuomisto JT, Tuomisto J, Vartiainen T, Skakkebaek NE, Toppari J. 2007. Flame retardants in placenta and breast milk and cryptorchidism in newborn boys. Environ Health Perspect 115:1519–1526.

Manceau A, Tamura N, Celestre RS, MacDowell AA, Geoffroy N, Sposito G, Padmore HA. 2003. Molecular-scale speciation of Zn and Ni in soil ferromanganese nodules from loess soils of the Mississippi Basin. Environ Sci Technol 37:75–80.

Mandema JW, Heijligers-Feijen CD, Tukker E, De Boer AG, Danhof M. 1992a. Modeling of the effect site equilibration kinetics and pharmacodynamics of racemic baclofen and its enantiomers using quantitative EEG effect measures. J Pharmacol Exp Ther 261:88–95.

Mandema JW, Kuck MT, Danhof M. 1992b. *In vivo* modeling of the pharmacodynamic interaction between benzodiazepines which differ in intrinsic efficacy. J Pharmacol Exp Ther 261:56–61.

Manno M, Rezzadore M, Grossi M, Sbrana C. 1996. Potentiation of occupational carbon tetrachloride toxicity by ethanol abuse. Hum Exp Toxicol 15:294–300.

Marigomez I, Kortabitarte M, Dussart GBJ. 1998. Tissue-level biomarkers in sentinel slugs as cost-effective tools to assess metal pollution in soils. Arch Environ Contam Toxicol 34:167–176.

Marino DJ, Clewell HJ, Gentry PR, Covington TR, Hack CE, David RM, Morgott DA. 2006. Revised assessment of cancer risk to dichloromethane. Part I. Bayesian PBPK and dose–response modeling in mice. Regul Toxicol Pharmacol 45:44–54.

Marinussen MPJC, Van der Zee SEATM. 1996. Conceptual approach to estimating the effects of home-range size on the exposure of organisms to spatially variable soil contamination. Ecol Model 87:83–89.

Martín-Díaz ML, Blasco J, de Canales MG, Sales D, DelValls TA. 2005a. Bioaccumulation and toxicity of dissolved heavy metals from the Guadalquivir Estuary after the Aznalcollar mining spill using *Ruditapes philippinarum*. Arch Environ Contam Toxicol 48:233–241.

Martín-Díaz ML, Villena-Lincoln A, Bamber S, Blasco J, DelValls TÁ. 2005b. An integrated approach using bioaccumulation and biomarker measurements in female shore crab, *Carcinus maenas*. Chemosphere 58:615–626.

Matscheko N, Lundstedt S, Svensson L, Harju M, Tysklind M. 2002. Accumulation and elimination of 16 polycylic aromatic compounds in the earthworm (*Eisenia fetida*). Environ Toxicol Chem 21:1724–1729.

Mattsson JL. 2007. Mixtures in the real world: the importance of plant self-defense toxicants, mycotoxins, and the human diet. Toxicol Appl Pharmacol 223:125–132.

Mayeno AN, Yang RSH, Reisfeld B. 2005. Biochemical reaction network modeling: a new tool for predicting metabolism of chemical mixtures. Environ Sci Techol 39:5363–5371.

Mayer P, Holmstrup M. 2008. Passive dosing of soil invertebrates with polycyclic aromatic hydrocarbons: limited chemical activity explains toxicity cutoff. Environ Sci Technol 42:7516–7521.

Mayer P, Tolls J, Hermens L, Mackay D. 2003. Equilibrium sampling devices. Environ Sci Technol 37:184A–191A.

McCarty LS, Borgert CJ. 2006. Review of the toxicity of chemical mixtures: theory, policy and regulatory practice. Reg Toxicol Pharmacol 45:119–143.

McCarty LS, Mackay D. 1993. Enhancing ecotoxicological modelling and assessment: body residues and modes of toxic action. Environl Sci Technol 27:1719–1728.

McCarty LS, Ozburn GW, Smith AD, Dixon DG 1992. Toxicokinetic modeling of mixtures of organic chemicals. Environ Toxicol Chem 11:1037–1047.

Mehendale HM. 1984. Potentiation of halomethane hepatotoxicity: chlordecone and carbon tetrachloride. Fund Appl Toxicol 4:295–308.

Mehendale HM. 1991. Role of hepatocellular regeneration and hepatolobular healing in the final outcome of liver injury: a two-stage model of toxicity. Biochem Pharmacol 42:1155–1162.

Mehendale HM. 1994. Mechanism of the interactive amplification of halomethane hepatotoxicity and lethality by other chemicals. In: Yang RSH, editor, Toxicology of chemical mixtures: case studies, mechanisms, and novel approaches. San Diego (CA): Academic Press. p 299–334.

Meili M, Bishop K, Bringmark L, Johansson K, Munthe J, Sverdrup H, De Vries W. 2003. Critical levels of atmospheric pollution: criteria and concepts for operational modelling of mercury in forest and lake ecosystems. Sci Total Environ 304:83–106.

Mendoza G, Gutierrez L, Pozo-Gallardo K, Fuentes-Rios D, Montory M, Urrutia R, Barra R. 2006. Polychlorinated biphenyls (PCBs) in mussels along the Chilean Coast. Environ Sci Pollut Res 13:67–74.

Mesman M, Rutgers M, Peijnenburg WJGM, Bogte JJ, Dirven-Van Breemen ME, De Zwart D, Posthuma L, Schouten AJ. 2003. Site-specific ecological risk assessment: the Triad approach in practice. In: Conference proceedings of CONSOIL: 8th International FKZ/TNO Conference on Contaminated Soil, Ghent (BE), May 12–16, 2003. p 649–656.

Metzdorff SB, Dalgaard M, Christiansen S, Axelstad M, Hass U, Kiersgaard MK, Scholze M, Kortenkamp A, Vinggaard AM. 2007. Dysgenesis and histological changes of genitals and perturbations of gene expression in male rats after *in utero* exposure to antiandrogens. Toxicol Sci 98:87–98.

Mileson BE, Chambers JE, Chen WL, Dettbarn W, Ehrich M, Eldefrawi AT, Gaylor DW, Hamernik K, Hodgson E, Karczmar AG, Padilla S, Pope CN, Richardson RJ, Saunders DR, Sheets LP, Sultatos LG, Wallace KB. 1998. Common mechanism of toxicity: a case study of organophosphorus pesticides. Toxicol Sci 41:8–20.

Miners JO, Knights KM, Houston JB, Mackenzie PI. 2006. *In vitro-in vivo* correlation for drugs and other compounds eliminated by glucuronidation in humans: pitfalls and promises. Biochem Pharmacol 71:1531–1539.

Minh TB, Kunisue T, Yen NTH, Watanabe M, Tanabe S, Hue ND, Qui V. 2002. Persistent organochlorine residues and their bioaccumulation profiles in resident and migratory birds from North Vietnam. Environ Toxicol Chem 21:2108–2118.

Monirith I, Ueno D, Takahashi S, Nakata H, Sudaryanto A, Subramanian A, Karuppiah S, Ismail A, Muchtar M, Zheng JS, Richardson BJ, Prudente M, Hue ND, Tana TS, Tkalin AV, Tanabe S. 2003. Asia-Pacific mussel watch: monitoring contamination of persistent organochlorine compounds in coastal waters of Asian countries. Mar Pollut Bull 46:281–300.

Monosson E. 2005. Chemical mixtures: considering the evolution of toxicology and chemical assessment. Environ Health Perspect 113:383–390.

Mood AM, Graybill FA, Boes DC. 1974. Introduction to the theory of statistics. 3rd ed. Auckland (NZ): McGraw-Hill Book Company.

Moolgavkar SH, Luebeck G. 1990. Two-event model for carcinogenesis: biological, mathematical, and statistical considerations. Risk Anal 10:323–341.

Moolgavkar SH, Venzon DJ. 2000. Two-event model for carcinogenesis. Math Biosci 47:55–77.

Morgan JE, Morgan AJ. 1993. Seasonal changes in the tissue-metal (Cd, Zn and Pb) concentrations in two ecophysiologically dissimilar earthworm species: pollution-monitoring implications. Environ Pollut 82:1–7.

Mould DR, DeFeo TM, Reele S, Milla G, Limjuco R, Crews T, Choma N, Patel IH. 1995. Simultaneous modeling of the pharmacokinetics and pharmacodynamics of midazolam and diazepam. Clin Pharmacol Ther 58:35–43.

Mount DI, Anderson-Carnahan DM. 1988. Methods for aquatic toxicity identification evaluations. Phase I. Toxicity characterization procedures. EPA/600/3-88/034. Duluth (MN): Environmental Research Laboratory, Office of Research and Development, US Environmental Protection Agency.

Mount DI, Anderson-Carnahan L. 1989. Methods for aquatic toxicity identification evaluations. Phase II toxicity identification procedures. EPA/600/3-88/035. Washington (DC): US Environmental Protection Agency.

Mount DI, Norberg-King TJ. 1993. Methods for aquatic toxicity identification evaluations. Phase II toxicity identification evaluation procedures for samples exhibiting acute and chronic toxicity. EPA/600/R-92/081. Duluth (MN): Environmental Research Laboratory, Office of Research and Development, US Environmental Protection Agency.

Mu X, LeBlanc GA. 2004. Synergistic interaction of endocrine-disrupting chemicals: model development using an ecdysone receptor antagonist and a hormone synthesis inhibitor. Environ Toxicol Chem 23:1085–1091.

Muenchow G. 1986. Ecological use of failure time analysis. Ecology 67:246–250.

Mulder C, Aldenberg T, De Zwart D, Van Wijnen HJ, Breure AM. 2005. Evaluating the impact of pollution on plant-Lepidoptera relationships. Environmetrics 16:357–373.

Mulder C, Breure A. 2006. Impact of heavy metal pollution on plants and leaf-miners. Environ Chem Lett 4:83–86.

Munns WRJ, Kroes R, Veith G, Suter GWI, Damstra T, Water MD. 2003a. Approaches for integrated risk assessment. Human Ecol Risk Assess 9:267–272.

Munns WRJ, Suter GWI, Damstra T, Kroes R, Reiter W, Marafante E. 2003b. Integrated risk assessment—results of an international workshop. Human Ecol Risk Assess 9:379–386.

Murk AJ, Leonards PEG, Bulder AS, Jonas AS, Rozemeijer MJC, Denison MS, Koeman JH, Brouwer A. 1997. The CALUX (chemical-activated luciferase expression) assay adapted and validated for measuring TCDD equivalents in blood plasma. Environ Toxicol Chem 16:1583–1589.

Nagel R, Loskill R, editors. 1991. Bioaccumulation in aquatic systems. Contributions to the assessment. Weinheim (DE): VCH.

Namdari R. 1998. A physiologically based toxicokinetic model of pyrene and its major metabolites in starry flounder, *Platichthys stellatus*. Thesis dissertation, Burnaby, British Columbia (CA): Simon Fraser University.

Narotsky MG, Weller EA, Chinchilli VM, Kavlock RJ. 1995. Nonadditive developmental toxicity in mixtures of trichloroethylene, di(2-ethylhexyl) phthalate, and heptachlor in a 5×5×5 design. Fund Appl Toxicol 27:203–216.

National Health and Environmental Effect Research. 2005. Wildlife reserach strategy. EPA 600/R-04/050. Research Triangle Park (NC): Office of Research and Development, US Environmental Protection Agency.

National Oceanic and Atmospheric Administration. 2002. Contaminant trends in US National Estuarine Research Reserves. Silver Springs (MD): NOAA.

Nesnow S. 1990. Mouse skin tumours and human lung cancer: relationships with complex environmental emissions. In: Vainio H, Sorsa M, McMichael AJ, editors, Complex mixtures and cancer risk. Lyon (FR): IARC Scientific Publication, p 44–54.

Nesnow S, Mass MJ, Ross JA, Galati AJ, Lambert GR, Gennings C, Carter WH, Stoner GD. 1998. Lung tumorigenic interactions in strain A/J mice of five environmental polycyclic aromatic hydrocarbons. Environ Health Perspect 106:1337–1346.

Neter N, Kutner HK, Nachtsheim CJ, Wasserman W. 1996. Applied linear statistical models. 4th ed. Boston: WCB/McGraw-Hill.

Newman MC, McCloskey JT. 1996. Time-to-event analyses of ecotoxicity data: ecotoxicology 5:187–196.

Newman MC, McCloskey JT. 2000. The individual tolerance concept is not the sole explanation for the probit dose-effect model. Environ Toxicol Chem 19:520–526.

Newton I, Wyllie I. 1992. Recovery of a Sparrowhawk population in relation to declining pesticide contamination. J Appl Ecol 29:476–484.

Newton I, Wyllie I, Asher A. 1991. Mortality causes in British barn owls *Tyto alba*, with a discussion of aldrin dieldrin poisoning. Ibis 133:162–169.

Newton I, Wyllie I, Asher A. 1993. Long-term trends in organochlorine and mercury residues in some predatory birds in Britain. Environ Pollut 79:143–151.

Nichols JW, Fitzsimmons PN, Whiteman FW. 2004a. A physiologically based toxicokinetic model for dietary uptake of hydrophobic organic compounds by fish. II. Simulation of chronic exposure scenarios. Toxicol Sci 77:219–229.

Nichols JW, Fitzsimmons PN, Whiteman FW, Dawson TD, Babeu L, Juenemann J. 2004b. A physiologically based toxicokinetic model for dietary uptake of hydrophobic organic compounds by fish. I. Feeding studies with 2,2',5,5'-tetrachlorobiphenyl. Toxicol Sci 77:206–218.

Nichols JW, McKim JM, Andersen ME, Gargas ML, Clewell HJ, Erickson RJ. 1990. A physiologically based toxicokinetic model for the uptake and disposition of waterborne organic chemicals in fish. Toxicol Appl Pharmacol 106:433–447.

Nicholson JK, Kendall MD, Osborn, D. 1983. Cadmium and nephrotoxicity. Nature 304:633–635.

[NIOSH] National Institute for Occupational Safety and Health. 1976. Criteria for a recommended standard for occupational exposure to methylene chloride. Cincinnati (OH): National Institute for Occupational Safety and Health.

Nisbet RM, Muller EB, Lika K, Kooijman SALM. 2000. From molecules to ecosystems through dynamic energy budget models. J Anim Ecol 69:913–926.

Nong A, McCarver DG, Hines RN, Krishnan K. 2006. Modeling interchild differences in pharmacokinetics on the basis of subject-specific data on physiology and hepatic CYP2E1 levels: a case study with toluene. Toxicol Appl Pharmacol 214:78–87.

Norberg-King TJ, Mount DI, Amato JR, Jensen DA, Thompson JA. 1992. Toxicity identification evaluation: characterization of chronically toxic effluents, phase I. USEPA/600/6-91/005F. Duluth (MN): Environmental Research Laboratory, Office of Research and Development, US Environmental Protection Agency.

Norum U, Lai VWM, Cullen WR. 2005. Trace element distribution during the reproductive cycle of female and male spiny and Pacific scallops, with implications for biomonitoring. Mar Pollut Bull 50:175–184.

[NRC] National Research Council. 1983. Risk assessment in the federal government: managing the process. Washington (DC): Committee on the Institutional Means for Assessment of Risks to Public Health, Commission on Life Sciences, National Research Council, National Academy Press.

[NRC] National Research Council. 1989. Mixtures. In: Drinking water and health. Vol. 9. Washington (DC): Safe Drinking Water Committee, National Research Council, National Academy of Sciences, National Academy Press.

Obach RS. 1997. Nonspecific binding to microsomes: impact on scale-up of *in vitro* intrinsic clearance to hepatic clearance as assessed through examination of warfarin, imipramine, and propranolol. Drug Metab Dispos 25:1359–1369.

Obach RS, Walsky RL, Venkatakrishnan K, Gaman EA, Houston JB, Tremaine LM. 2006. The utility of *in vitro* cytochrome P450 inhibition data in the prediction of drug-drug interactions. J Pharmacol Exp Ther 316:336–348.

[OECD] Organisation for Economic Co-operation and Development. 1999. Compendium of estimation methods to quantify releases to the environment for use in pollutant release and transfer registries. Paris: Organisation for Economic Cooperation and Development.

[OECD] Organisation for Economic Co-operation and Development. 2002a. Resource compendium of PRTR release estimation techniques. Part 1. Summary of point source techniques. ENV/JM/MONO(2002)20. Paris: Organisation for Economic Cooperation and Development, Environment Directorate, Joint Meeting of the Chemicals Committee and the Working Party on Chemicals, Pesticides and Biotechnology.

[OECD] Organisation for Economic Co-operation and Development. 2002b. Emission scenario document on textile finishing industry. Paris: Organisation for Economic Cooperation and Development, Environment Directorate.

[OECD] Organisation for Economic Co-operation and Development. 2002c. Emission scenario document on industrial surfactants [Draft]. Paris: Organisation for Economic Cooperation and Development, Environment Directorate.

[OECD] Organisation for Economic Co-operation and Development. 2004. Guidance document on the use of multimedia models for estimating overall environmental persistance and long-range transport. OECD Series on Testing and Assessment No. 45, ENV/JM/MONO(2004)5. Paris: Organisation for Economic Cooperation and Development, Environment Directorate, Joint Meeting of the Chemicals Committee and the Working Party on Chemicals, Pesticides and Biotechnology.

[OECD] Organisation for Economic Co-operation and Development. 2006. Comparison of emission estimation methods used in pollutant release and transfer registers and emission scenario documents: case study of pulp and paper and textile sectors. OECD Series on Testing and Assessment No. 52, ENV/JM/MONO(2006)6. Paris: Organisation for Economic Cooperation and Development, Environment Directorate, Joint Meeting of the Chemicals Committee and the Working Party on Chemicals, Pesticides and Biotechnology.

Office of Emergency and Remedial Response. 1991. Risk assessment guidance for Superfund: human health evaluation manual: risk evaluation of remedial alternatives. Vol. 1, Part C, Publication 9285.7-01C. Washington (DC): US Environmental Protection Agency.

O'Halloran K. 2006. Toxicological considerations of contaminants in the terrestrial environment for ecological risk assessment. Human Ecol Risk Assess 12:74–83.

Oomen AG, Sips A, Groten JP, Sijm D, Tolls J. 2000. Mobilization of PCBs and lindane from soil during *in vitro* digestion and their distribution among bile salt micelles and proteins of human digestive fluid and the soil. Environ Sci Technol 34:297–303.

Oomen AG, Tolls J, Kruidenier M, Bosgra SSD, Sips A, Groten JP. 2001. Availability of polychlorinated biphenyls (PCBs) and lindane for uptake by intestinal Caco-2 cells. Environ Health Perspect 109:731–737.

Oomen AG, Tolls J, Sips A, Groten JP. 2003. *In vitro* intestinal lead uptake and transport in relation to speciation. Archiv Environ Contam Toxicol 44:116–124.

[OSHA] Occupational Safety and Health Administration. 1993. Air contaminants; rule. 29 CFR 1910.1000. Occupational Safety and Health Administration. Federal Register 58(124):35338–35351.

[OSHA] Occupational Safety and Health Administration. 2001. OSHA regulations (standards—29 CFR): air contaminants (standard number: 1910.1000). Washington (DC): Occupational Safety and Health Administration, US Department of Labor.

Ou YC, Conolly RB, Thomas R, Gustafson DL, Long ME, Dovrev ID, Chubb LS, Xu Y, Lapidot S, Andersen ME, Yang RSH. 2003. Stochastic simulation of hepatic preneoplasic foci development for four chlorobenzene congeners in a medium-term bioassay. Toxicol Sci 73:301–314.

Ou YC, Conolly RB, Thomas RS, Xu Y, Andersen ME, Chubb LS, Pitot HC, Yang RSH. 2001. A clonal growth model: time-course simulations of liver foci growth following penta- or hexachlorobenzene treatment in a medium-term bioassay. Cancer Res 61:1879–1889.

Page DS, Boehm PD, Brown JS, Neff JM, Burns WA, Bence AE. 2005. Mussels document loss of bioavailable polycyclic aromatic hydrocarbons and the return to baseline conditions for oiled shorelines in Prince William Sound, Alaska. Mar Environ Res 60:422–436.

Paquin PR, Gorsuch JW, Apte S, Batley GE, Bowles KC, Campbell PGC, Delos CG, Di Toro DM, Dwyer RL, Galvez F, Gensemer RW, Goss GG, Hogstrand C, Janssen CR, McGeer JC, Naddy RB, Playle RC, Santore RC, Schneider U, Stubblefield WA, Wood CM, Wu KB. 2002a. The biotic ligand model, a historical overview. Comp Biochem Physiol C 133:3–35.

Paquin PR, Zoltay V, Winfield RP, Wu KB, Mathew R, Santore RC, Di Toro DM. 2002b. Extension of the biotic ligand model of acute toxicity to a physiologically-based model of the survival time of rainbow trout (*Oncorhynchus mykiss*) exposed to silver. Comp Biochem Physiol C 133:305–343.

Payne J, Scholze M, Kortenkamp A. 2001. Mixtures of four organochlorines enhance human breast cancer cell proliferation. Environ Health Perspect 109:391–397.

Peijnenburg WJGM, Jager T. 2003. Monitoring approaches to assess bioaccessibility and bioavailability of metals: matrix issues. Ecotoxicol Environ Safety 56:63–77.

Pelekis M, Krishnan K. 1997. Assessing the relevance of rodent data on chemical interactions for health risk assessment purposes: a case study with dichloromethane-toluene mixture. Regul Toxicol Pharmacol 25:79–86.

Pereira R, Ribeiro R, Goncalves F. 2004. Scalp hair analysis as a tool in assessing human exposure to heavy metals (S. Domingos mine, Portugal). Sci Total Environ 327:81–92.

Pierik FH, Burdorf A, Deddens JA, Juttmann RE, Weber RFA. 2004. Maternal and paternal risk factors for cryptorchidism and hypospadias: a case-control study in newborn boys. Environ Health Perspect 112:1570–1576.

Pieters BJ, Jager T, Kraak MHS, Admiraal W. 2006. Modeling responses of *Daphnia magna* to pesticide pulse exposure under varying food conditions: intrinsic versus apparent sensitivity. Ecotoxicology 15:601–608.

Pilling ED, Bromley-Challenor KA, Walker CH, Jepson PC. 1995. Mechanism of synergism between the pyrethroid insecticide lambda-cyhalothrin and the imidazole fungicide prochloraz, in the honeybee (*Apis mellifera* L.). Pest Biochem Physiol 51:1–11.

Plackett RL, Hewlett, PS. 1952. Quantal responses to mixtures of poisons. J Royal Stat Soc Ser B 14:141-163.

Plackett RL, Hewlett PS. 1963a. A unified theory for quantal responses to mixtures of drugs: the fitting to data of certain models for two non-interactive drugs with complete positive correlation of tolerances. Biometrics 19:517–531.

Plackett RL, Hewlett PS. 1963b. Quantal response to mixtures of poisons. J R Stat Soc B 14:141–163.

Playle RC. 2004. Using multiple metal-gill binding models and the toxic unit concept to help reconcile multiple-metal toxicity results. Aquat Toxicol 67:359–370.

Pöch G. 1993. Combined effects of drugs and toxic agents. New York: Springer-Verlag.

Poet TS, Kousba AA, Dennison SL, Timchalk C. 2004. Physiologically based pharmacokinetic/pharmacodynamic model for the organophosphorus pesticide diazinon. Neurotoxicology 25:1013–1030.

Pohl H, Hibbs B. 1996. Breast-feeding exposure of infants to environmental contaminants—a public health risk assessment viewpoint: chlorinated dibenzo-p-dioxins and chlorinated dibenzofurans. Toxicol Ind Health 12:593–611.

Pohl HR, McClure P, De Rosa CT. 2004. Persistent chemicals found in breast milk and their possible interactions. Environ Toxicol Pharmacol 18:259–266.

Pohl HR, Roney N, Wilbur S, Hansen H, De Rosa CT, 2003. Six interaction profiles for simple mixtures. Chemosphere 53:183–197.

Pohl HR, Tylenda CA. 2000. Breast-feeding exposure of infants to selected pesticides: a public health viewpoint. Toxicol Ind Health 16:65–77.

Pohl HR, van Engelen J, Wilson J, Sips A. 2005. Risk assessment of chemicals and pharmaceuticals in the pediatric population: a workshop report. Regul Toxicol Pharmacol 42:83–95.

Poiger H, Schlatter C. 1986. Pharmacokinetics of 2,3,7,8-TCDD in man. Chemosphere 15:1489–1494.

Poirier L, Berthet B, Amiard JC, Jeantet AY, Amiard-Triquet C. 2006. A suitable model for the biomonitoring of trace metal bioavailabilities in estuarine sediments: the annelid polychaete *Nereis diversicolor*. J Mar Biol Assoc UK 86:71–82.

Posthuma L, Baerselman R, Van Veen RPM, Dirven-van Breemen EM. 1997. Single and joint toxic effects of copper and zinc on reproduction of *Enchytraeus crypticus* in relation to sorption of metals in soils. Ecotoxicol Environ Safety 38:108–121.

Posthuma L, De Zwart D. 2006. Predicted effects of toxicant mixtures are confirmed by changes in fish species assemblages in Ohio, USA, rivers. Environ Toxicol Chem 25:1094–1105.

Posthuma L, De Zwart D, Wintersen A, Lijzen J, Swartjes F, Cuypers C, Van Noort P, Harmsen J, Groenenberg BJ. 2006. Beslissen over bagger op bodem. Deel 1. Systeembenadering, model en praktijkvoorbeelden. Report 711701044. Bilthoven (NL): National Institute for Public Health and the Environment (RIVM).

Posthuma L, Richards S, De Zwart D, Dyer SD, Sibley P, Hickey C, Altenburger R. 2008. Mixture extrapolation approaches. In: Solomon KR, Brock TCM, De Zwart D, Dyer SD, Posthuma L, Richards S, Sanderson H, Sibley R, Van den Brink PJ, editors, Extrapolation practice for ecotoxicological effect characterization of chemicals. Results of the EXPECT workshop, February 2005, St. Petersburg, FL, USA. Pensacola (FL): SETAC Press.

Posthuma L, Traas TP, Suter GW II, editors. 2002. Species sensitivity distributions in ecotoxicology. Boca Raton (FL): Lewis Publishers.

Posthuma L, Van Straalen NM. 1993. Heavy-metal adaptation in terrestrial invertebrates: a review of occurrence, genetics, physiology and ecological consequences. Comp Biochem Physiol C 106:11–38.

Posthumus R, Traas TP, Peijnenburg W, Hulzebos EM. 2005. External validation of EPIWIN biodegradation models. SAR QSAR Environ Res 16:135–148.

Poulin P, Schoenlein K, Theil FP. 2001. Prediction of adipose tissue: plasma partition coefficients for structurally unrelated drugs. J Pharm Sci 90:436–447.

Poulin P, Theil FP. 2002. Prediction of pharmacokinetics prior to in vivo studies. 1. Mechanism-based prediction of volume of distribution. J Pharm Sci 91:129–156.

Price K, Haddad S, Krishnan K. 2003a. Physiological modeling of age-specific changes in the pharmacokinetics of organic chemicals in children. J Toxicol Environ Health A 66:417–433.

Price K, Krishnan K. 2005. An integrated QSAR-PBPK model for simulating pharmacokinetics of chemicals in mixtures. 44th Annual Meeting of the Society of Toxicology, New Orleans (LA), March 6–10.

Price PS, Conolly RB, Chaisson CF, Gross EA, Young JS, Mathis ET, Tedder DR. 2003b. Modeling interindividual variation in physiological factors used in PBPK models of humans. Crit Rev Toxicol 33:469–503.

Psaty BM, Furberg CD, Ray WA, Weiss NS. 2004. Potential for conflict of interest in the evaluation of suspected adverse drug reactions. J Am Med Assoc 292:2622–2631.

Purcell KJ, Cason GH, Gargas ML, Andersen ME, Travis CC. 1990. In vivo metabolic interactions of benzene and toluene. Toxicol Lett 52:141–152.

Putzrath RM. 2000. Reducing uncertainty of risk estimates for mixtures of chemicals within regulatory constraints. Regul Toxicol Pharmacol 31:44–52.

Ra JS, Lee BC, Chang NI, Kim SD. 2006. Estimating the combined toxicity by two-step prediction model on the complicated chemical mixtures from wastewater treatment plant effluents. Environ Toxicol Chem 25:2107–2113.

Ragas AMJ, Etienne RS, Willemsen FH, Van de Meent D. 1999. Assessing model uncertainty for environmental decision making: a case study of the coherence of independently derived environmental quality objectives for air and water. Environ Toxicol Chem 18:1856–1867.

Rainbow PS. 2002. Trace metal concentrations in aquatic invertebrates: why and so what? Environ Pollut 120:497–507.

Rainbow PS, Fialkowski W, Sokolowski A, Smith BD, Wolowicz M. 2004. Geographical and seasonal variation of trace metal bioavailabilities in the Gulf of Gdansk, Baltic Sea using mussels (Mytilus trossulus) and barnacles (Balanus improvisus) as biomonitors. Mar Biol 144:271–286.

Rajapakse N, Silva E, Kortenkamp A. 2002. Combining xenoestrogens at levels below individual no-observed effect concentrations dramatically enhances steroid hormone action. Environ Health Perspect 110:917–921.

Raymond JW, Rogers TN, Shonnard DR, Kline AA. 2001. A review of structure-based biodegradation estimation methods. J Hazardous Mater 84:189–215.

Read HJ, Martin MH. 1993. The effects of heavy metals on populations of small mammals from woodlands in Avon (England); with particular emphasis on metal concentrations in Sorex araneus L. and Sorex minutus L. Chemosphere 27:2197–2211.

Redding LE, Sohn MD, McKone TE, Chen JW, Wang SL, Hsieh DPH, Yang RSH. 2008. Population physiologically-based pharmacokinetic modeling for the human lactational transfer of PCB 153 with consideration of worldwide human biomonitoring results. Environ Health Perspect 116:1629–1634.

Reffstrup TK. 2002. Combined actions of pesticides in food. Report 2002:19. Soborg (DK): Danish Veterinary and Food Administration.

Regnell O. 1994. The effect of pH and dissolved oxygen levels on methylation and partitioning of mercury in freshwater model systems. Environ Pollut 84:7–13.

Reichenberg F, Mayer P. 2006 Two complementary sides of bioavailability: accessibility and chemical activity of organic contaminants in sediments and soils. Environ Toxicol Chem 25:1239–1245.

Reisfeld B, Mayeno AN, Lyons MA, Yang RSH. 2007. Physiologically-based pharmacokinetic and pharmacodynamic modeling, in computational toxicology. In: Ekins S, editor, Risk assessment for pharmaceutical and environmental chemicals. Hoboken (NJ): John Wiley & Sons, p 33–69.

Reisfeld B, Yang RSH. 2004. A reaction network model for CYP2E1-mediated metabolism of toxicant mixtures. Environ Toxicol Pharmacol 18:173–179.

Renn O, Benighaus C. 2006. Framing the perception of cumulative stressors, especially chemical risks. Report on approaches to the characterization of knowledge of risks, uncertainties and ambiguity and their use and quality assurance in the IP domain. EU FP6 Project NOMIRACLE, Deliverable 4.3.2. Stuttgart (DE): Dialogik.

Renwick AG, Hayes AW. 2001. Toxicokinetics: pharmacokinetics in toxicology. 4th ed. Philadelphia (PA): Taylor and Francis.

Reynders H, Van Campenhout K, Bervoets L, De Coen WM, Blust R. 2006. Dynamics of cadmium accumulation and effects in common carp (*Cyprinus carpio*) during simultaneous exposure to water and food (*Tubifex tubifex*). Environ Toxicol Chem 25:1558–1567.

Rice GE, Teuschler LK, Bull RJ, Feder PI, Simmons JE. 2009. Evaluating the similarity of complex drinking water disinfection by-product mixtures: overview of the issues. J Toxicol Environ Health A 72:429–436.

Rider CV, Furr J, Wilson VS, Gray LE Jr. 2008. A mixture of seven antiandrogens induces reproductive malformations in rats. Int J Androl 31:249–262.

Riley RJ, McGinnity DF, Austin RP. 2005. A unified model for predicting human hepatic, metabolic clearance from *in vitro* intrinsic clearance data in hepatocytes and microsomes. Drug Metab Dispos 33:1304–1311.

Riviere JE, Brooks JD. 2007. Prediction of dermal absorption from complex chemical mixtures: incorporation of vehicle effects and interactions into a QSPR framework. SAR QSAR Environ Res 18:31–44.

Rodgers T, Leahy D, Rowland M. 2005. Physiologically based pharmacokinetic modelling. 1. Predicting the tissue distribution of moderate-to-strong bases. J Pharm Sci 94:1259–1276.

Rodgers T, Rowland M. 2006. Physiologically based pharmacokinetic modelling. 2. Predicting the tissue distribution of acids, very weak bases, neutrals and zwitterions. J Pharm Sci 95:1238–1257.

Roelofs D, Mariën J, Van Straalen NM. 2007. Differential gene expression profiles associated with heavy metal tolerance in the soil insect *Orchesella cincta*. Insect Biochem Mol Biol 37:287–295.

Rogan WJ, Gladen BC, McKinney JD, Carreras N, Hardy P, Thullen J, Tingelstad J, Tully M. 1986. Polychlorinated biphenyls (PCBs) and dichlorophenyl dichloroethene (DDE) in human milk: effects of maternal factors and previous lactation. Am J Public Health 76:172–177.

Romijn CAFM, Luttik R, Canton J. 1993a. Presentation of a general algorithm to include effect assessment on secondary poisoning in the derivation of environmental quality criteria. 2. Terrestrial food chains. Ecotoxicol Environ Safety 26:61–83.

Romijn CAFM, Luttik R, Van de Meent D, Sloof W, Canton J. 1993b. Presentation of a general algorithm to include effect assessment on secondary poisoning in the derivation of environmental quality criteria. 1. Aquatic food chains. Ecotoxicol Environ Safety 26:61–83.

Ross HLB. 1996. The interaction of chemical mixtures and their implications on water quality guidelines. Hon thesis, University of Technology, Sydney, NSW (AU).

Ross HLB, Warne MStJ. 1997. Most chemical mixtures have additive aquatic toxicity. In Proceedings of the Third Annual Conference of the Australasian Society for Ecotoxicology, Brisbane (AU), July 17–19, p 30.

Rowland M, Tozer TN. 1995. Clinical pharmacokinetics concepts and applications. 3rd ed. Media (PA): Williams & Williams.

Rozman KK, Doull J. 2000. Dose and time as variables of toxicity. Toxicology 144:169–178.

Russel FG, Wouterse AC, Van Ginneken CA. 1987. Physiologically based pharmacokinetic model for the renal clearance of salicyluric acid and the interaction with phenolsulfonphthalein in the dog. Drug Metab Dispos 15:695–701.

Russel FG, Wouterse AC, Van Ginneken CA. 1989. Physiologically based pharmacokinetic model for the renal clearance of iodopyracet and the interaction with probenecid in the dog. Biopharm Drug Dispos 10:137–152.

Sanchez-Dardon J, Voccia I, Hontela A, Chilmonczyk S, Dunier M, Boermans H, Blakley B, Fournier M. 1999. Immunomodulation by heavy metals tested individually or in mixtures in rainbow trout (*Oncorhynchus mykiss*) exposed *in vivo*. Environ Toxicol Chem 18:1492–1497.

Sarangapani R, Teeguarden J, Plotzke KP, McKim JM Jr, Andersen ME. 2002. Dose–response modeling of cytochrome p450 induction in rats by octamethylcyclotetrasiloxane. Toxicol Sci 67:159–172.

Schecter A, Gasiewicz TA. 1987a. Health hazard assessment of chlorinated dioxins and dibenzofurans contained in human milk. Chemosphere 16:2147–2154.

Schecter A, Gasiewicz TA. 1987b. Human breast milk levels of dioxins and dibenzofurans: significance with respect to current risk assessments. ACS Symp Ser 338:162–173.

Schecter A, Li L. 1997. Dioxins, dibenzofurans, dioxin-like PCBs, and DDE in US fast food, 1995. Chemosphere 34:1449–1457.

Schecter A, Startin J, Wright C, Kelly M, Papke O, Lis A, Ball M, Olson JR. 1994. Congener-specific levels of dioxins and dibenzofurans in US food and estimated daily dioxin toxic equivalent intake. Environ Health Perspect 102:962–966.

Schmider J, von Moltke LL, Shader RI, Harmatz JS, Greenblatt DJ. 1999. Extrapolating *in vitro* data on drug metabolism to *in vivo* pharmacokinetics: evaluation of the pharmacokinetic interaction between amitriptyline and fluoxetine. Drug Metab Rev 31:545–560.

Scholz NL, Truelove NK, Labenia JS, Baldwin DH, Collier TK. 2006. Dose-additive inhibition of chinook salmon acetylcholinesterase activity by mixtures of organophosphate and carbamate insecticides. Environ Toxicol Chem 25:1200–1207.

Scholze M, Boedeker W, Faust M, Backhaus T, Altenburger R, Grimme LH. 2001. A general best-fit method for concentration–response curves and the estimation of low-effect concentrations. Environ Toxicol Chem 20:448–457.

Schramm KW. 1990. Exams 2—Exposure analysis modeling system. Toxicol Environ Chem 26:73–82.

Schuler LJ, Landrum PF, Lydy MJ. 2006. Comparative toxicity of fluoranthene and pentachlorobenzene to three freshwater invertebrates. Environ Toxicol Chem 25:985–994.

Scott Fordsmand JJ, Krogh PH, Weeks JM. 2000. Responses of *Folsomia fimetaria* (Collembola: Isotomidae) to copper under different soil copper contamination histories in relation to risk assessment. Environ Toxicol Chem 19:1297–1303.

Segel IH. 1975. Enzyme kinetics: behavior and analysis of rapid equilibrium and steady-state enzyme systems. Toronto (CA): John Wiley & Sons.

Selikoff IJ, Seidman H, Hammond C. 1980. Mortality effects of cigarette smoking among site asbestos factory workers. J Natl Cancer Inst 65:507–513.

[SETAC] Society of Environmental Toxicology and Chemistry. 2004. Technical issue paper: whole effluent toxicity testing. Pensacola (FL): Society of Environmental Toxicology and Chemistry.

Shakman RA. 1974. Nutritional influences on the toxicity of environmental pollutants. Arch Environ Health 28:105–113.

Sharma-Shanti S, Schat H, Vooijs R, Van Heerwaarden LM. 1999. Combination toxicology of copper, zinc, and cadmium in binary mixtures: concentration-dependent antagonistic, nonadditive, and synergistic effects on root growth in *Silene vulgaris*. Environ Toxicol Chem 18:348–355.

Shin KH, Ahn Y, Kim KW. 2005. Toxic effect of biosurfactant addition on the biodegradation of phenanthrene. Environ Toxicol Chem 24:2768–2774.

Shiverick KT, Slikker W, Rogerson SJ, Miller RK. 2003. Drugs and the placenta—a workshop report. Placenta 24:S55–S59.

Siegrist M, Cvetkovich G. 2001. Better negative than positive? Evidence of a bias for negative information about possible health dangers. Risk Anal 21:199–206.

Sijm DTHM, Van der Linde A. 1995. Size-dependent bioconcentration kinetics of hydrophobic organic chemicals in fish based on diffusive mass transfer and allometric relationships. Environ Sci Technol 29:2769–2777.

Silva CAR, Rainbow PS, Smith BD. 2003. Biomonitoring of trace metal contamination in mangrove-lined Brazilian coastal systems using the oyster *Crassostrea rhizophorae*: comparative study of regions affected by oil, salt pond and shrimp farming activities. Hydrobiologia 501:199–206.

Silva E, Rajapakse N, Kortenkamp A. 2002. Something from "nothing"—eight weak estrogenic chemicals combined at concentrations below NOECs produce significant mixture effects. Environ Sci Technol 36:1751–1756.

Simmons JE, Richardson SD, Speth TF, Miltner RJ, Rice G, Schenck K, Hunter III ES, Teuschler LK. 2002. Development of a research strategy for integrated technology-based toxicological and chemical evaluation of complex mixtures of drinking water disinfection byproducts. Environ Health Perspect 110:1013–1024.

Sjodin A, Jones RS, Focant JF, Lapeza C, Wang RY, McGahee EE III, Zhang Y, Turner WE, Slazyk B, Needham LL, Patterson DG Jr. 2004. Retrospective time-trend study of polybrominated diphenyl ether and polybrominated and polychlorinated biphenyl levels in human serum from the United States. Environ Health Perspect 112:654–658.

Skaggs SM, Foti RS, Fisher MB. 2006. A streamlined method to predict hepatic clearance using human liver microsomes in the presence of human plasma. J Pharmacol Toxicol Methods 53:284–290.

Slaveykova VI, Wilkinson KJ. 2005. Predicting the bioavailability of metals and metal complexes: critical review of the biotic ligand model. Environ Chem 2:9–24.

Slooff W, De Zwart D. 1991. The pT-value as environmental policy indicator for the exposure to toxic substances. Report nr. 719102 003. Bilthoven (NL): National Institute for Public Health and the Environment (RIVM).

Smit CE, Van Gestel CAM. 1998. Effects of soil type, prepercolation, and ageing on bioaccumulation and toxicity of zinc for the springtail *Folsomia candida*. Environ Toxicol Chem 17:1132–1141.

Sokal RR, Rohlf FJ. 1995. Biometry, the principles and practice of statistics in biological research. 3rd ed. San Francisco (CA): Freeman.

Sole M, Porte C, Barcelo D, Albaiges J. 2000. Bivalves residue analysis for the assessment of coastal pollution in the Ebro Delta (NW Mediterranean). Mar Pollut Bull 40:746–753.

Solomon KR, Brock TCM, De Zwart D, Dyer SD, Posthuma L, Richards S, Sanderson H, Sibley P, Van den Brink PJ. 2008. Extrapolation practice for ecotoxicological effect characterization of chemicals. Pensacola (FL): SETAC Press.

Sonzogni W, Maack L, Degenhardt D, Anderson H, Fiore B. 1991. Polychlorinated biphenyl congeners in blood of Wisconsin sport fish consumers. Arch Environ Contam Toxicol 20:56–60.

Sørensen PB, Vorkamp K, Thomsen M. 2004. Persistent organic pollutants (POPs) in the Greenland environment—long-term temporal changes and effects on eggs of a bird of prey. NERI Technical Report 509. Silkeborg (DK): National Environment Research Institute.

Speijers GJA, Speijers MHM. 2004. Combined toxic effects of mycotoxins. Toxicol Lett 153:91–98.

Sprague JB. 1970. Measurement of pollutant toxicity to fish. II. Utilizing and applying bioassay results. Water Res 4:3–32.

Spurgeon DJ, Hopkin SP. 1996. Risk assessment of the threat of secondary poisoning by metals of predators of earthworms in the vicinity of a primary smelting works. Sci Total Environ 187:167–183.

Squillace PJ, Scott JC, Moran MJ, Nolan T, Koplin DW. 2002. VOCs, pesticides, nitrate, and their mixtures in groundwater used for drinking water in the United States. Environ Sci Technol 36:1923–1930.

Stark JD. 2006. Toxicity endpoints used in risk assessment: what do they really mean? SETAC Globe 7(2):29–30.

Staunton S. 2004. Sensitivity analysis of the distribution coefficient, Kd, of nickel with changing soil chemical properties. Geoderma 122:281–290.

Steen Redeker E, Bervoets L, Blust R. 2004. Dynamic model for the accumulation of cadmium and zinc from water and sediment by the aquatic oligochaete, *Tubifex tubifex*. Environ Sci Technol 38:6193–6200.

Steen Redeker E, Blust R. 2004. Accumulation and toxicity of cadmium in the aquatic oligochaete *Tubifex tubifex*: a kinetic modeling approach. Environ Sci Technol 38:537–543.

Steevens JA, Benson WH. 1999. Toxicological interactions of chlorpyrifos and methyl mercury in the amphipod, *Hyalella azteca*. Toxicol Sci 52:168–177.

Stefanelli P, Ausili A, Di Muccio A, Fossi C, Di Muccio S, Rossi S, Colasanti A. 2004. Organochlorine compounds in tissues of swordfish (*Xiphias gladius*) from Mediterranean Sea and Azores islands. Mar Pollut Bull 49:938–950.

Sterner TR, Robinson PJ, Mattie DR, Burton GA. 2005. The toxicology of chemical mixtures risk assessment for human and ecological receptors. AFRL-HE-WP-TR-2005-0173. Wright-Patterson AFB (OH): Air Force Research Laboratory, Human Effectiveness Directorate, Biosciences and Protection Division, Applied Biotechnology Branch.

Stork LG, Gennings C, Carter WH Jr, Teuschler LK, Carney EW. 2008. Empirical evaluation of sufficient similarity in dose–response for environmental risk assessment of chemical mixtures. J Agric Biol Environ Stat 13:313–333.

Strenkoski-Nix LC, Forrest A, Schentag JJ, Nix DE. 1998. Pharmacodynamic interactions of ciprofloxacin, piperacillin, and piperacillin/tazobactam in healthy volunteers. J Clin Pharmacol 38:1063–1071.

Struijs J. 2003. Evaluatie van pT. De bepaling van toxische stress in Rijkswateren. Report nr 860703001. Bilthoven (NL): National Institute for Public Health and the Environment (RIVM).

Sudaryanto A, Takahashi S, Monirith I, Ismail A, Muchtar M, Zheng J, Richardson BJ, Subramanian A, Prudente M, Hue ND, Tanabe S. 2002. Asia-Pacific mussel watch: monitoring of butyltin contamination in coastal waters of Asian developing countries. Environ Toxicol Chem 21:2119–2130.

Sugita O, Sawada Y, Sugiyama Y, Iga T, Hanano M. 1982. Physiologically based pharmacokinetics of drug-drug interaction: a study of tolbutamide-sulfonamide interaction in rats. J Pharmacokinet Biopharm 10:297–316.

Sühnel J. 1992. Assessment of interaction of biologically active agents by means of the isobole approach: fundamental assumptions and recent developments. ACES 4:35–44.

Suter II GW, Munns WR Jr, Sekizawa W. 2003. Types of integration in risk assessment and management, and why they are needed. Human Ecol Risk Assess 9:273–279.

Suzuki H, Iwatsubo T, Sugiyama Y. 1995. Applications and prospects for physiologically based pharmacokinetic (PB-PK) models involving pharmaceutical agents. Toxicol Lett 82/83:349–355.

Svenson A, Sanden B, Dalhammar G, Remberger M, Kaj L. 2000. Toxicity identification and evaluation of nitrification inhibitors in wastewaters. Environ Toxicol 15:527–532.

Swain S, Wren J, Stürzenbaum SR, Kille P, Morgan AJ, Jager T, Jonker MJ, Hankard PK, Svendsen C, Owen J, Hedley BA, Blaxter M, Spurgeon DJ. 2010. Linking toxicants mechanism of action and physiological mode of action in *Caenorhabditis elegans*. BMC Biol.

Swan SH, Kruse RL, Liu F, Barr DB, Drobnis EZ, Redmon JB, Wang C, Brazil C, Overstreet JW, the Study for Future Families Research Group. 2003. Semen quality in relation to biomarkers of pesticide exposure. Environ Health Perspect 111:1478–1484.

Swan SH, Main KM, Liu F, Stewart SL, Kruse RL, Calafat AM, Mao CS, Redmon JB, Ternand CL, Sullivan S, Teague JL, the Study for Future Families Research Group. 2005. Decrease in anogenital distance among male infants with prenatal phthalate exposure. Environ Health Perspect 113:1056–1061.

Tanabe S. 2000. Asia-Pacific mussel watch progress report. Mar Pollut Bull 40:651–651.

Tardif R, Charest-Tardif G. 1999. The importance of measured end-points in demonstrating the occurrence of interactions: a case study with methylchloroform and m-xylene. Toxicol Sci 49:312–317.

Tardif R, Charest-Tardif G, Brodeur J, Krishnan K. 1997. Physiologically based pharmacokinetic modeling of a ternary mixture of alkyl benzenes in rats and humans. Toxicol Appl Pharmacol 144:120–134.

Tardif R, Lapare S, Charest-Tardif G, Brodeur J, Krishnan K. 1995. Physiologically-based pharmacokinetic modeling of a mixture of toluene and xylene in humans. Risk Anal 15:335–342.

Tardif R, Lapare S, Krishnan K, Brodeur J. 1993. Physiologically based modeling of the toxicokinetic interaction between toluene and m-xylene in the rat. Toxicol Appl Pharmacol 120:266–273.

Ter Laak TL, Agbo SO, Barendregt A, Hermens JLM. 2006. Freely dissolved concentrations of PAHs in soil pore water: measurements via solid-phase extraction and consequences for soil tests. Environ Sci Technol 40:1307–1313.

Teuschler LK. 2007. Deciding which chemical mixtures risk assessment methods work best for what mixtures. Toxicol Appl Pharmacol 223:139–147.

Teuschler LK, Gennings C, Stiteler WM, Hertzberg RC, Colman JT, Thiyagarajah A, Lipscomb JC, Hartley WR, Simmons JE. 2000. A multiple-purpose design approach to the evaluation of risks from mixtures of disinfection by-products. Drug Chem Toxicol 23:307–321.

Teuschler LK, Klaunig J, Carney E, Chambers J, Conolly R, Gennings C, Giesy J, Hertzberg R, Klaassen C, Kodell R, Paustenbach D, Yang R. 2002. Support of science-based decisions concerning the evaluation of the toxicology of mixtures: a new beginning. Reg Toxicol Pharmacol 36:34–39.

Teuschler LK, Rice GE, Wilkes CR, Lipscomb JC, Power FW. 2004. A feasibility study of cumulative risk assessment methods for drinking water disinfection by-product mixtures. J Toxicol Environ Health A 67:755–777.

Theil FP, Guentert TW, Haddad S, Poulin P. 2003. Utility of physiologically based pharmacokinetic models to drug development and rational drug discovery candidate selection. Toxicol Lett 138:29–49.

Thomas GO, Wilkinson M, Hodson S, Jones KC. 2006. Organohalogen chemicals in human blood from the United Kingdom. Environ Pollut 141:30–41.

Thomas RS, Conolly RB, Gustafson DL, Long ME, Benjamin SA, Yang RSH. 2000. A physiologically based pharmacodynamic analysis of hepatic foci within a medium-term liver bioassay using pentachlorobenzene as a promoter and diethylnitrosamine as an initiator. Toxicol Appl Pharmacol 166:128–137.

Thomsen M, Sørensen PB, Fauser P, Ragas A, Peirano F. 2006. Prioritised listing of VOCs/semi-VOCs including test scenarios. Report D.1.2.2 from EC FP6-IP NoMiracle, restricted.

Thorpe KL, Gross-Sorokin M, Johnson I, Brighty G, Tyler C. 2006. An assessment of the model of concentration addition for predicting the estrogenic activity of chemical mixtures in wastewater treatment works effluents. Environ Health Perspect 114(Suppl 1):90–97.

Thrall KD, Poet TS. 2000. Determination of biokinetic interactions in chemical mixtures using real-time breath analysis and physiologically based pharmacokinetic modeling. J Toxicol Environ Health A 59:653–670.

Timchalk C, Poet TS. 2008. Development of a physiologically based pharmacokinetic and pharmacodynamic model to determine dosimetry and cholinesterase inhibition for a binary mixture of chlorpyrifos and diazinon in the rat. Neurotoxicology 29:428–443.

Tinwell H, Ashby J. 2004. Sensitivity of the immature rat uterotrophic assay to mixtures of estrogens. Environ Health Perspect 112:575–582.

Toose L, Woodfine DG, MacLeod M, Mackay D, Gouin J. 2004. BETR-World: a geographically explicit model of chemical fate: application to transport of alpha-HCH to the Arctic. Environ Pollut 128:223–240.

Tozer TN, Rowland M. 2006. Introduction to pharmacokinetics and pharmacodynamics: the quantitative basis of drug therapy. Baltimore (MD): Lippincott Williams & Wilkins.

Trapp S, Matthies M. 1995. Generic one-compartment model for uptake of organic chemicals by foliar vegetation. Environ Sci Technol 29:2333–2338.

Trapp S, McFarlane JC. 1995. Plant contamination: modeling and simulation of organic chemical processes. Boca Raton (FL): Lewis Publishers.

Tuk B, van Gool T, Danhof M. 2002. Mechanism-based pharmacodynamic modeling of the interaction of midazolam, bretazenil, and zolpidem with ethanol. J Pharmacokinet Pharmacodyn 29:235–250.

Umbreit TH, Hesse EJ, Gallo MA. 1986a. Differential bioavailability of TCDD from contaminated soils. Abstracts Am Chem Soc 191:47.

Umbreit TH, Hesse EJ, Gallo MA. 1986b. Comparative toxicity of TCDD contaminated soil from Times Beach, Missouri, and Newark, New Jersey. Chemosphere 15:2121–2124.

[USEPA] US Environmental Protection Agency. 1986. Guidelines for health risk assessment of chemical mixtures. US Environmental Protection Agency. Federal Register 51(185):34014–34025.

[USEPA]. US Environmental Protection Agency. 1989a. Exposure factors handbook. USEPA/600/8-89/043. Washington (DC): Office of Health and Environmental Assessment: US Environmental Protection Agency.

[USEPA] US Environmental Protection Agency. 1989b. Risk assessment guidance for superfund: human health evaluation manual. Vol. 1, Part A, EPA/540/1-89/002. Washington (DC): Office of Emergency and Remedial Response: US Environmental Protection Agency.

[USEPA] US Environmental Protection Agency. 1991. Methods for aquatic toxicity identification evaluations. Phase I. Toxicity characterization procedures. 2nd ed., EPA/600/6-91/003. Washington (DC): Office of Research and Development: US Environmental Protection Agency.

[USEPA] US Environmental Protection Agency. 1994. EPA's national water quality inventory: 1992. Report to Congress. Report 841-R-94-001. Washington (DC): US Environmental Protection Agency.

[USEPA] US Environmental Protection Agency. 1995. Whole effluent toxicity: guidelines establishing test procedures for the analysis of pollutants. Office of Science and Technology, US Environmental Protection Agency. Federal Register 60(199):53529–53544.

[USEPA] US Environmental Protection Agency. 1996. PCBs: cancer dose–response assessment and application to environmental mixtures. EPA/600/P-96/001F. Washington (DC): US Environmental Protection Agency, National Center for Environmental Assessment.

[USEPA] US Environmental Protection Agency. 1997. Ecological risk assessment guidance for Superfund, process for designing and conducting ecological risk assessments. EPA 540-R-97-006. Washington (DC): Office of Solid Waste and Emergency Response: US Environmental Protection Agency.

[USEPA] US Environmental Protection Agency. 1998. Guidelines for ecological risk assessment. EPA/630/R-95/002F. US Environmental Protection Agency, Risk Assessment Forum. Federal Register 63(93):26846–26924.

[USEPA] US Environmental Protection Agency. 1999. Residual risk report to Congress. EPA-453/R-99-00. Triangle Park (NC): Office of Air Quality, Planning and Standards: US Environmental Protection Agency.

[USEPA] US Environmental Protection Agency. 2000a. Exposure and human health reassessment of 2,3,7,8-tetrachlorodibenzo-p-dioxin (TCDD) and related compounds [draft final]. Part II, EPA/600P-00/001(September). Washington (DC).

[USEPA] US Environmental Protection Agency. 2000b. Supplementary guidance for conducting health risk assessment of chemical mixtures. EPA/630/R-00/002, ORD/NCEA. Cincinnati (OH): US Environmental Protection Agency.

[USEPA] US Environmental Protection Agency. 2002a. Child-specific exposure factors handbook. EPA-600-P-00-002B, NTIS PB2003-101678. Washington (DC): National Center for Environmental Assessment, Office of Research and Development.

[USEPA] US Environmental Protection Agency. 2002b. Guidance on cumulative risk assessment of pesticide chemicals that have a common mechanism of toxicity. Washington (DC): Office of Pesticide Programs. Available from: http://www.epa.gov/oppfead1/trac/science/cumulative_guidance.pdf

[USEPA] US Environmental Protection Agency. 2002c. Organophosphate pesticides: revised cumulative risk assessment. Available from: http://www.epa.gov/pesticides/cumulative/rra-op/

[USEPA] US Environmental Protection Agency. 2002d. Methods for measuring the acute toxicity of effluents and receiving waters to freshwater and marine organisms. 5th ed., EPA-821-R-02-012. Washington (DC): US Environmental Protection Agency.

[USEPA] US Environmental Protection Agency. 2002e. Short-term methods for estimating the chronic toxicity of effluents and receiving waters to freshwater organisms 4th ed., EPA-821-R-02-013. Washington (DC): Office of Water: US Environmental Protection Agency.

[USEPA] US Environmental Protection Agency. 2002f. Short-term methods for estimating the chronic toxicity of effluents and receiving waters to marine and estuarine organisms. 3rd ed., EPA-821-R-02-014. Washington (DC): Office of Water: US Environmental Protection Agency.

[USEPA] US Environmental Protection Agency. 2003a. The feasibility of performing cumulative risk assessments for mixtures of disinfection by-products in drinking water. EPA/600/R-03/051, ORD/NCEA. Cincinnati (OH): US Environmental Protection Agency.

[USEPA] US Environmental Protection Agency. 2003b. Developing relative potency factors for pesticide mixtures: biostatistical analyses of joint dose–response. EPA/600/R-03/052, ORD/NCEA. Cincinnati (OH): US Environmental Protection Agency.

[USEPA] US Environmental Protection Agency. 2004. National whole effluent toxicity (WET) implementation guidance under the NPDES program [Draft]. EPA 832-B-04-003. Washington (DC): US Environmental Protection Agency. Office of Wastewater Management.

[USEPA] US Environmental Protection Agency. 2005. ECOTOX database. US Environmental Protection Agency. Available from: http://cfpub.epa.gov/ecotox/

[USEPA] US Environmental Protection Agency. 2006a. Exposures and internal doses of trihalomethanes in humans: multi-route contributions from drinking water. EPA/600/R-06/087, ORD/NCEA. Cincinnati (OH): US Environmental Protection Agency.

[USEPA] US Environmental Protection Agency. 2006b. Available from: http://www.epa.gov/iaq/voc.html; http://www.epa.gov/iaq/pubs/insidest.html

[USEPA] US Environmental Protection Agency. 2008. Integrated Risk Information System (IRIS). Available from: http://cfpub.epa.gov/ncea/iris/index.cfm

[US PCCRARM] Presidential/Congressional Commission on Risk Assessment and Risk Management. 1997. Framework for environmental health risk management. Washington (DC): US Presidential/Congressional Commission on Risk Assessment and Risk Management.

Van Brummelen TC, Van Gestel CAM, Verweij RA. 1996a. Long-term toxicity of five polycyclic aromatic hydrocarbons for the terrestrial isopods *Oniscus asellus* and *Porcellio scaber*. Environ Toxicol Chem 15:1199–1210.

Van Brummelen TC, Verweij RA, Wedzinga SA, Van Gestel CAM. 1996b. Polycyclic aromatic hydrocarbons in earthworms and isopods from contaminated forest soils. Chemosphere 32:315–341.

Van den Berg M, Birnbaum L, Bosveld ATC, Brunström B, Cook P, Feeley M, Giesy JP, Hanberg A, Hasegawa R, Kennedy SW, Kubiak T, Larsen JC, van Leeuwen FXR, Liem AKD, Nolt C, Peterson RE, Poellinger L, Safe S, Schrenk D, Tillitt D, Tysklind M, Younes M, Waern F, Zacharewski T. 1998. Toxic equivalency factors (TEFs) for PCBs, PCDDs, PCDFs for humans and wildlife. Environ Health Perspect 106:775–792.

Van den Berg M, Birnbaum LS, Denison M, DeVito M, Farland W, Feeley M, Fiedler H, Hakansson H, Hanberg A, Haws L, Rose M, Safe S, Schrenk D, Tohyama C, Tritscher A, Tuomisto J, Tysklind M, Walker N, Peterson RE. 2006. Review: the 2005 World Health Organization reevaluation of human and mammalian toxic equivalency factors for dioxins and dioxin-like compounds. Toxicol Sci 93:223–241.

Van den Brink PJ, Roelsma J, Van Nes EH, Scheffer M, Brock TCM. 2002. PERPEST model, a case-based reasoning approach to predict ecological risks of pesticides. Environ Toxicol Chem 21:2500–2506.

Van der Geest HG, Greve GD, Boivin ME, Kraak MHS, Van Gestel CAM. 2000. Mixture toxicity of copper and diazinon to larvae of the mayfly (*Ephoron virgo*) judging additivity at different effect levels. Environ Toxicol Chem 19:2900–2905.

Van der Oost R, Beyer J, Vermeulen NPE. 2003. Fish bioaccumulation and biomarkers in environmental risk assessment: a review. Environ Toxicol Pharmacol 13:57–149.

Van Ewijk PH, Hoekstra JA. 1993. Calculation of the EC50 and its confidence interval when subtoxic stimulus is present. Ecotoxicol Environ Safety 25:25–32.

van Gestel CAM, Hensbergen PJ. 1997. Interaction of Cd and Zn toxicity for *Folsomia candida* Willem (Collembola: Isotomidae) in relation to bioavailability in soil. Environ Toxicol Chem 16:1177–1186.

Van Leeuwen CJ, Hermens JLM, editors. 1995. Risk assessment of chemicals: an introduction. 2nd ed. Dordrecht (NL): Kluwer Academic Publishers.

Van Leeuwen CJ, Vermeire T, editors. 2007. Risk assessment of chemicals: an introduction. Dordrecht (NL): Kluwer Academic Publishers.

Van Leeuwen IMM, Zonneveld C, Kooijman SALM. 2003. The embedded tumour: host physiology is important for the evaluation of tumour growth. Br J Cancer 89:2254–2263.

Van Meeuwen JA, van den Berg M, Sanderson JT, Verhoef A, Piersma AH. 2007. Estrogenic effects of mixtures of phyto- and synthetic chemicals on uterine growth of prepubertal rats. Toxicol Lett 170:165–176.

Van Vlaardingen PL, Traas TP, Wintersen AM, Aldenberg T. 2004. ETX 2.0. A program to calculate risk limits and fraction affected, based on normal species sensitivity distributions. Report 601501028/2004. Bilthoven (NL): National Institute for Public Health and the Environment (RIVM).

Van Wezel AP, De Vries DAM, Sijm DTHM, Opperhuizen A. 1996. Use of the lethal body burden in the evaluation of mixture toxicity. Ecotoxicol Environ Safety 35:236–241.

Van Wijk RJ, Postma JF, Van Houwelingen H. 1994. Joint toxicity of ethyleneamines to algae, daphnids and fish. Environ Toxicol Chem 13:167–171.

Venkatakrishnan K, Von Moltke LL, Greenblatt DJ. 2000. Effects of the antifungal agents on oxidative drug metabolism: clinical relevance. Clin Pharmacokinet 38:111–180.

Venkatakrishnan K, Von Moltke LL, Greenblatt DJ. 2001. Human drug metabolism and the cytochromes P450: application and relevance of *in vitro* models. J Clin Pharmacol 41:1149–1179.

Verhaar HJM, Van Leeuwen CJ, Hermens J. 1992. Classifying environmental pollutants. 1. Structure–activity relationships for prediction of aquatic toxicity. Chemosphere 25:471–491.

Vijver MG, Van Gestel CAM, Lanno RP, Van Straalen NM, Peijnenburg WJGM. 2004. Internal metal sequestration and its ecotoxicological relevance: a review. Environ Sci Technol 38:4705–4712.

Vijver MG, Vink JPM, Jager T, Wolterbeek HT, Van Straalen NM, Van Gestel CAM. 2005. Biphasic elimination and uptake kinetics of Zn and Cd in the earthworm *Lumbricus rubellus* exposed to contaminated floodplain soil. Soil Biol Biochem 37:1843–1851.

Vink K, Dewi L, Bedaux J, Tompot A, Hermans M, Van Straalen NM. 1995. The importance of exposure route when testing the toxicity of pesticides to saprotrophic isopods. Environ Toxicol Chem 14:1225–1232.

Voet D, Voet JG. 2004. Biochemistry. 3rd ed. Toronto (CA): John Wiley & Sons.

VROM. 1989. Premises for risk management. Risk limits in the context of environmental policy. Parliament session 1988–1989, 21137, no 5. The Hague (NL): Ministry of Housing, Spatial Planning and the Environment (VROM).

Wade MG, Foster WG, Younglai EV, McMahon A, Leingartner K, Yagminas A, Blakey D, Fournier M, Desaulniers D, Hughes CL. 2002. Effects of subchronic exposure to a complex mixture of persistent contaminants in male rats: systemic, immune, and reproductive effects. Toxicol Sci 67:131–143.

Walker CH, Hopkin SP, Sibly RM, Peakall DB. 2001. Principles of ecotoxicology. London: Taylor & Francis.

Walker CH, Johnston GO. 1993. Potentiation of pesticide toxicity in birds: role of cytochrome P-450. Biochem Soc Trans 21:1066–1068.

Walter H, Consolaro F, Gramatica P, Scholze M, Altenburger R. 2002. Mixture toxicity of priority pollutants at no observed effect concentrations (NOECs). Ecotoxicology 11:299–310.

Wania F, Mackay D. 1996. Tracking the distribution of persistent organic pollutants. Environ Sci Technol 30:A390–A396.

Warne MStJ. 2003. A review of the ecotoxicity of mixtures, approaches to, and recommendations for, their management. In: Langley A, Gilbey M, Kennedy B, editors, Proceedings
of the Fifth National Workshop on the Assessment of Site Contamination. Adelaide
(AU): National Environment Protection Council Service Corporation, p 253–276.

Wassenberg DM, Di Guilio RT. 2004. Synergystic embryotoxicity of polycyclic aromatic
hydrocarbons aryl hydrocarbon receptor agonists with cytochrome P4501A inhibitors
in *Fundulus heteroclitus*. Environ Health Perspect 112:1658–1664.

Watts AW, Ballestero, TP, Gardener KH. 2006. Uptake of polycyclic aromatic hydrocarbons
(PAHs) in salt march plants *Spartina alterniflora* grown in contaminated sediments.
Chemosphere 62:1253–1260.

Weis BK, Balshawl D, Barr JR, Brown D, Ellisman M, Liov P, Omenn G, Potter JD, Smith MT,
Sohn L, Suk WA, Sumner S, Swenberg J, Walt DR, Watkins S, Thompson C, Wilson SH.
2005. Personalized exposure assessment: promising approaches for human environmental health research. Environ Health Perspect 113:840–848.

White JC, Pignatello JJ. 1999. Influence of bisolute competition on the desorption kinetics of
polycyclic aromatic hydrocarbons in soil. Environ Sci Technol 33:4292–4298.

Whitfield J. 2001. Vital signs. Nature 411:989–990.

WHO. 2001. Integrated risk assessment. Report prepared for the WHO/UNEP/ILO International
Programme on Chemical Safety. WHO/IPCS/IRA/01/12. Geneva (CH): World Health
Organization, International Programme on Chemical Safety.

Whyatt RM, Camann DE, Kinney PL, Reyes A, Dietrich J, Diaz D, Holmes D, Perera FP.
2002. Residential pesticide use during pregnancy among urban minority women.
Environ Health Perspect 110:507–514.

Willett KL, Wassenberg D, Lienesch L, Reichert W, Di Guilio RT. 2001. *In vivo* and *in vitro*
inhibition of CYP1A-dependent activity in *Fundulus heteroclitus* by the polynuclear
aromatic hydrocarbon fluoranthene. Toxicol Appl Pharmacol 177:264–271.

Winters D, Cleverly D, Meier K, Dupuy A, Byrne C, Deyrup C, Ellis R, Ferrario J, Harless
R, Lesse W, Lorber M, McDaniel D, Schaum J, Walcott J. 1996. A statistical survey of
dioxin-like compounds in the United States beef. Chemosphere 32:469–478.

Wintersen A, Posthuma L, De Zwart D. 2004. The RIVM e-toxBase. A database for storage,
retrieval and export of ecotoxicity data. Bilthoven (NL): National Institute for Public
Health and the Environment (RIVM).

Witherow LE, Houston JB. 1999. Sigmoidal kinetics of CYP3A substrates: an approach for
scaling dextromethorphan metabolism in hepatic microsomes and isolated hepatocytes
to predict *in vivo* clearance in rat. J Pharmacol Exp Ther 290:58–65.

Woo YT, Lai D, McLain JL, Manibusan MK, Dellarco V. 2002. Use of mechanism-based
structure–activity relationships analysis in carcinogenic potential ranking for drinking
water disinfection by-products. Environ Health Perspect 110:75–87.

Wright JF, Sutcliffe DW, Furse MT, editors. 2000. Assessing the biological quality of fresh waters
RIVPACS and other techniques. Ambleside (UK): The Freshwater Biological Association.

Yang RSH, editor. 1994. Toxicology of chemical mixtures: case studies, mechanisms, and
novel approaches. San Diego (CA): Academic Press.

Yang RSH. 1997. Toxicologic interactions of chemical mixtures. In: Bond JA, editor,
Comprehensive toxicology. Vol. 1. Oxford (UK): Elsevier Science Ltd. p 189–203.

Yang RSH, Andersen ME. 2005. Physiologically based pharmacokinetic modeling of
chemical mixtures. In: Reddy MB, Yang RSH, Clewell HJ III, Andersen ME, editors,
Physiologically based pharmacokinetics: science and applications. New York: John
Wiley & Sons, p 349–373.

Yang RSH, El-Masri HA, Thomas RS, Dobrev ID, Dennison JE, Bae DS, Campain JA,
Liao KH, Reisfeld B, Andersen ME, Mumtaz M. 2004. Chemical mixture toxicology:
from descriptive to mechanistic, and going on to *in silico* toxicology. Environ Toxicol
Pharmacol 18:65–81.

Yang RSH, Mayeno AN, Liao KH, Reardon KF, Reisfeld B. 2006. Integration of PBPK and reaction network modeling: predictive xenobiotic metabolomics. ALTEX 23(Special Issue 2):373–379.

Yang RSH, Mayeno AN, Lyons M, Reisfeld B. 2010. The application of physiologically-based pharmacokinetics (PBPK), Bayesian population PBPK modeling, and biochemical reaction network (BRN) modeling to chemical mixture toxicology. In: Mumtaz M, editor, Principles and practices of mixture toxicology. Hoboken (NJ): John Wiley & Sons.

Young JF, Wosilait WD, Luecke RH. 2001. Analysis of methylmercury disposition in humans utilizing a PBPK model and animal pharmacokinetic data. J Toxicol Environ Health A 63:19–52.

Yu X, Johanson G, Ichihara G, Shibata E, Kamijima M, Ono Y, Takeuchi Y. 1998. Physiologically based pharmacokinetic modeling of metabolic interactions between n-hexane and toluene in humans. J Occup Health 40:293–301.

Zhang H. 2004. In-situ speciation of Ni and Zn in freshwaters: comparison between DGT measurements and speciation models. Environ Sci Technol 38:1421–1427.

Zhang H, Zhao FJ, Sun B, Davison W, McGrath SP, 2001. A new method to measure effective soil solution concentration predicts copper availability to plants. Environ Sci Technol 35:2602–2607.

Zhao Y, Newman MC. 2004. Shortcomings of the laboratory-derived median lethal concentration for predicting mortality in field populations: exposure duration and latent mortality. Environ Toxicol Chem 23:2147–2153.

Zhao Y, Newman MC. 2007. The theory underlying dose–response models influences predictions for intermittent exposure. Environ Toxicol Chem 26:543–547.

Zhou S, Kestell P, Paxton JW. 2002. Predicting pharmacokinetics and drug interactions in patients from in vitro and in vivo models: the experience with 5,6-dimethylxanthenone-4-acetic acid (DMXAA), an anti-cancer drug eliminated mainly by conjugation. Drug Metab Rev 34:751–790.

Zwick M, Renn O. 1998. Wahrnehmung und Bewertung von Technik in Baden-Württemberg. Eine Präsentation der Akademie für Technikfolgenabschätzung in Baden-Württemberg. Stuttgart (DE): Akademie für Technikfolgenabschätzung in Baden-Württemberg. Available from: http://elib.uni-stuttgart.de/opus/volltexte/2004/1765/

Index

Other Titles from the Society of Environmental Toxicology and Chemistry (SETAC)

Derivation and Use of Environmental Quality and Human Health Standards for Chemical Substances in Water and Soil
Crane, Matthiessen, Maycock, Merrington, Whitehouse, editors
2009

Aquatic Macrophyte Risk Assessment for Pesticides
Maltby, Arnold, Arts, Davies, Heimbach, Pickl, Poulsen, editors
2009

Copper: Environmental Fate, Effects, Transport and Models: Papers from Environmental Toxicology and Chemistry, 1982 to 2008 and Integrated Environmental Assessment and Management, 2005 to 2008
Gorsuch, Arnold, Santore, Smith, Reiley, editors
2009

Veterinary Medicines in the Environment
Crane, Boxall, Barrett
2008

Relevance of Ambient Water Quality Criteria for Ephemeral and Effluent-dependent Watercourses of the Arid Western United States
Gensemer, Meyerhof, Ramage, Curley
2008

Extrapolation Practice for Ecotoxicological Effect Characterization of Chemicals
Solomon, Brock, De Zwart, Dyer, Posthuma, Richards, Sanderson, Sibley, van den Brink, editors
2008

Environmental Life Cycle Costing
Hunkeler, Lichtenvort, Rebitzer, editors
2008

Valuation of Ecological Resources: Integration of Ecology and Socioeconomics in Environmental Decision Making
Stahl, Kapustka, Munns, Bruins, editors
2007

Genomics in Regulatory Ecotoxicology: Applications and Challenges
Ankley, Miracle, Perkins, Daston, editors
2007

Population-Level Ecological Risk Assessment
Barnthouse, Munns, Sorensen, editors
2007

SETAC

A Professional Society for Environmental Scientists and Engineers and Related Disciplines Concerned with Environmental Quality

The Society of Environmental Toxicology and Chemistry (SETAC), with offices currently in North America and Europe, is a nonprofit, professional society established to provide a forum for individuals and institutions engaged in the study of environmental problems, management and regulation of natural resources, education, research and development, and manufacturing and distribution.

Specific goals of the society are

- Promote research, education, and training in the environmental sciences.
- Promote the systematic application of all relevant scientific disciplines to the evaluation of chemical hazards.
- Participate in the scientific interpretation of issues concerned with hazard assessment and risk analysis.
- Support the development of ecologically acceptable practices and principles.
- Provide a forum (meetings and publications) for communication among professionals in government, business, academia, and other segments of society involved in the use, protection, and management of our environment.

These goals are pursued through the conduct of numerous activities, which include:

- Hold annual meetings with study and workshop sessions, platform and poster papers, and achievement and merit awards.
- Sponsor a monthly scientific journal, a newsletter, and special technical publications.
- Provide funds for education and training through the SETAC Scholarship/Fellowship Program.
- Organize and sponsor chapters to provide a forum for the presentation of scientific data and for the interchange and study of information about local concerns.
- Provide advice and counsel to technical and nontechnical persons through a number of standing and ad hoc committees.

SETAC membership currently is composed of more than 5000 individuals from government, academia, business, and public-interest groups with technical backgrounds in chemistry, toxicology, biology, ecology, atmospheric sciences, health sciences, earth sciences, and engineering.

If you have training in these or related disciplines and are engaged in the study, use, or management of environmental resources, SETAC can fulfill your professional affiliation needs.

All members receive a newsletter highlighting environmental topics and SETAC activities and reduced fees for the Annual Meeting and SETAC special publications.

All members except Students and Senior Active Members receive monthly issues of Environmental Toxicology and Chemistry (ET&C) and Integrated Environmental Assessment and Management (IEAM), peer-reviewed journals of the Society. Student and Senior Active Members may subscribe to the journal. Members may hold office and, with the Emeritus Members, constitute the voting membership.

If you desire further information, contact the appropriate SETAC Office.

1010 North 12th Avenue	Avenue de la Toison d'Or 67
Pensacola, Florida 32501-3367 USA	B-1060 Brussels, Belgium
T 850 469 1500 F 850 469 9778	T 32 2 772 72 81 F 32 2 770 53 86
E setac@setac.org	E setac@setaceu.org

www.setac.org
Environmental Quality Through Science®

T - #0110 - 101024 - C0 - 234/156/17 [19] - CB - 9781439830086 - Gloss Lamination